Edited by
Ulrich Bröckel, Willi Meier,
and Gerhard Wagner

Product Design and Engineering

Related Titles

Bröckel, U., Meier, W., Wagner, G. (eds.)

Product Design and Engineering

Best Practices

2007
ISBN: 978-3-527-31529-1

Rähse, W.

Industrial Product Design of Solids and Liquids

A Practical Guide

2014
ISBN: 978-3-527-33335-6

Jameel, F., Hershenson, S. (eds.)

Formulation and Process Development Strategies for Manufacturing Biopharmaceuticals

2010
ISBN: 978-1-118-12473-4

Norton, I. (ed.)

Practical Food Rheology - An Interpretive Approach

2011
ISBN: 978-1-405-19978-0

Tadros, T.F.

Rheology of Dispersions

Principles and Applications

2010
ISBN: 978-3-527-32003-5

Edited by Ulrich Bröckel, Willi Meier, and Gerhard Wagner

Product Design and Engineering

Formulation of Gels and Pastes

WILEY-VCH

Verlag GmbH & Co. KGaA

The Editors

Prof. Dr.-Ing Ulrich Bröckel
HS Trier
Umwelt-Campus Birkenfeld
Campus Allee 9916
55761 Birkenfeld
Germany

Dr. Willi Meier
DECHEMA e.V.
Theodor-Heuss-Allee 25
60486 Frankfurt
Germany

Dr.-Ing Gerhard Wagner
Global R&D Director DSM
Biotechnology Center
Alexander Fleminglaan 1
2613 AX Delft
The Netherlands

and

Dianastrasse 12
4310 Rheinfelden
Switzerland

Library of Congress Card No.: applied for

British Library Cataloguing-in-Publication Data
A catalogue record for this book is available from the British Library.

Bibliographic information published by the Deutsche Nationalbibliothek
The Deutsche Nationalbibliothek lists this publication in the Deutsche Nationalbibliografie; detailed bibliographic data are available on the Internet at <http://dnb.d-nb.de>.

© 2013 Wiley-VCH Verlag GmbH & Co. KGaA, Boschstr.12, 69469 Weinheim, Germany

Print ISBN: 978-3-527-33220-5
ePDF ISBN: 978-3-527-65477-2
ePub ISBN: 978-3-527-65476-5
Mobi ISBN: 978-3-527-65475-8
oBook ISBN: 978-3-527-65474-1

Cover Design Formgeber, Mannheim
Typesetting Laserwords Private Limited, Chennai, India
Printing and Binding Markono Print Media Pte Ltd, Singapore

Printed in Singapore
Printed on acid-free paper

1007 10 646 X

Contents

Navam Hettiarachchy, Arvind Kannan, Christian Schäfer, and Gerhard Wagner

List of Contributors

Muhammad Mohsin Azim
University of Leipzig
Institute of Chemical Technology
Linnéstraße 3-4
04103 Leipzig
Germany

Arjen Bot
Unilever R&D Vlaardingen
Olivier van Noortlaan 120
3133 AT Vlaardingen
The Netherlands

Ulrich Bröckel
Institute for
Micro-Process-Engineering and
Particle Technology (IMiP)
Umwelt-Campus Birkenfeld
Campus Allee 24
55761 Birkenfeld
Germany

Rüdiger Brummer
Beiersdorf AG
R&D cosmid
Unnastrasse 48
20245 Hamburg
Germany

Elisa Conte
Separation Processes Process
Technology
AkzoNobel Research,
Development and Innovation
Zuthphenseweg 10
7418 AJ Deventer
The Netherlands

Eckhard Flöter
Technical University of Berlin
Institute of Food Process
Engineering
Königin-Luise-Straße 22
14195 Berlin
Germany

Rafiqul Gani
Technical University of Denmark
Department of Chemical and
Biochemical Engineering
Søltofts Plads
Building 229
2800 Lyngby
Denmark

Kristina Georgieva
Karlsruher Institut für
Technologie (KIT)
Institut für Mechanische
Verfahrenstechnik und Mechanik
Geb. 30.70
Straße am Forum 8
76131 Karlsruhe
Germany

Navam Hettiarachchy
University of Arkansas
Food Science Department
2650 N Young Avenue
Fayetteville, AR 72704
USA

Arvind Kannan
University of Arkansas
Food Science Department
2650 N Young Avenue
Fayetteville, AR 72704
USA

Heike P. Karbstein-Schuchmann
Karlsruhe Institute of Technology
(KIT)
Institute of Food Process
Engineering
Kaiserstrasse 12
76131 Karlsruhe
Germany

Georgios M. Kontogeorgis
Technical University of Denmark
Department of Chemical and
Biochemical Engineering
Søltofts Plads
Building 229
2800 Lyngby
Denmark

Michele Mattei
Technical University of Denmark
Department of Chemical and
Biochemical Engineering
Building 229
2800 Lyngby
Denmark

Willi Meier
DECHEMA e.V.
Theodor-Heuss-Allee 25
60486 Frankfurt
Germany

Wilfried Rähse
ATS License GmbH
R&D Cosmeceuticals
Bahlenstr. 168
40589 Düsseldorf
Germany

Muhammad Ramzan
University of Leipzig
Institute of Chemical Technology
Linnéstraße 3-4
04103 Leipzig
Germany

Henelyta Santos Ribeiro
Unilever R&D Vlaardingen
Olivier van Noortlaan 120
3133 AT Vlaardingen
The Netherlands

Richard Sass
DECHEMA e.V.
Theodor-Heuss-Allee 25
60486 Frankfurt
Germany

Christian Schäfer
DSM Nutritional Products Ltd.
R&D Centre Formulation and
Application
4002 Basel
Switzerland

Anne Schmidt
University of Leipzig
Institute of Chemical Technology
Linnéstraße 3-4
04103 Leipzig
Germany

Annegret Stark
University of Leipzig
Institute of Chemical Technology
Linnéstraße 3-4
04103 Leipzig
Germany

Tharwat F. Tadros
89 Nash Grove Lane
Wokingham
Berkshire RG40 4HE
UK

Gerhard Wagner
Global R&D Director DSM
Biotechnology Center
Alexander Fleminglaan 1
2613 AX Delft
The Netherlands

and

Dianastrasse 12
4310 Rheinfelden
Switzerland

Martin Wild
University of Leipzig
Institute of Chemical Technology
Linnéstraße 3-4
04103 Leipzig
Germany

Norbert Willenbacher
Karlsruher Institut für
Technologie (KIT)
Institut für Mechanische
Verfahrenstechnik und Mechanik
Geb. 30.70
Straße am Forum 8
76131 Karlsruhe
Germany

Introduction

Gerhard Wagner, Willi Meier, and Ulrich Bröckel

What Is Product Design and Engineering?

Product design is, in principle, a term that describes a very broad variety of designs, ranging from industry design of goods, for example cars, measurement instruments and furniture, to scientific and technological product design. Scientific and technological product design can be subdivided into four main clusters:

- *Biological product design*: This includes, for example, product design for proteins, probiotics [1], or enzymes for various specific purposes such as biomass utilization [2] or food processing [3], animal feed utilization, agricultural, brewery, fuel and textile finishing applications.
- *Chemical product design*: [4] according to Cussler and Moggridge this aims for new speciality chemicals or pharmaceuticals. Other examples of chemical product design are modified starches, polyols [5] or polymers with tailor-made properties.
- *Food design* [6] or *food engineering* [7]: this covers a wide range of technologies and raw materials. Numerous books [8] and journals are focusing on this subject.
- *Physical product design*: this avoids a modification of the chemical structure of the active components (e.g., vitamins, pharmaceuticals, pigments, etc.) but modifies instead the physical properties of the components, for example particle size, shape, surface morphology or crystal habit. Physical product design also includes the wide field of formulation and Galenic formulation in which additives and excipients change, for example, the appearance, stability, bioavailability or even the purpose of the products over a very wide range.

All four clusters of product design have in common a need for a multidisciplinary approach based on biology, chemistry and physics combined with engineering skills and sciences; furthermore, disciplines like nutrition, pharmacy and ergonomics and form design are crucial to ensure a suitable product. Product design is more of an iterative than a linear process; the understanding of the different disciplines is therefore important at the different steps of product design depending of the scale of the product design. Figure 1 shows how the disciplines impact on the physical properties relevant to product qualities within the different size ranges.

Product Design and Engineering: Formulation of Gels and Pastes, First Edition.
Edited by Ulrich Bröckel, Willi Meier, and Gerhard Wagner.

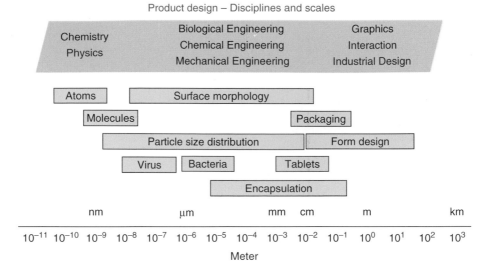

Figure 1 Product design – disciplines and scales.

In the first two volumes of this book series we focused mainly on basic technologies and solids. The very important aspect of liquid-like or gel-type applications have been excluded so far. In this third volume of our book series we will, therefore, elaborate these applications in more detail.

Product design has become a growing field of interest during recent years. The reasons for this are manifold. Looking at the markets of the chemical engineering field today, we observe well-established and quite saturated markets. Breakthrough innovation with brand new products is difficult to accomplish. Product design is, therefore, crucial to improve the following properties of a product:

- handling
- applicability
- appearance
- stability
- performance, activity, or bioavailability.

With product design, existing products can be optimized, improved and positioned in the market, prolonging their life cycle or differentiating and making them applicable for new markets.

Product design has been seen as a paradigm shift in process design away from the unit operation concept to a new interdisciplinary thinking [9], speeding up the development process. But, obviously, there is still a long way to go before enough knowledge and basic understanding of the complex interactions of multicomponent systems is gathered in order to calculate the composition and the processing conditions in order to predict the performance of a new formulation using a computer.

Why This Book?

As mentioned in the introduction of the first two volumes of this series, product design is much more than a buzz word and is of the highest importance for related industries. For example, Aspirin® as a highly developed pharmaceutical bulk chemical is available in new formulations every few years [10]. The same is true for formulated and encapsulated carotenoid products, which also show a continuous development in their product form and composition. In addition, beverages, like soft drinks, energy drinks and coffee-based beverages or washing powder, laundry products or detergents are subject to on-going development, owing to customers expecting an improved taste or a better performance of the new product. These kinds of improvements ensure the lifecycle management of products.

Several company strategies clearly state the importance of product design and engineering for the future of process industries. Customer demands have to be recognized and turned into products with the help of well-established processes and technologies. Fulfilling customer needs will automatically lead the business to new applications and new markets were real growth is created.

For such examples we elaborated in Volumes I and II different technologies, raw materials and additives. One area we were not able to investigate in more detail, even though we touched on emulsion technologies, is liquid and gel-type applications, which are very important for the broad field of, among other industries, the life science industry.

Volume I describes the basics and fundamentals of chemical engineering that are essential for product design and engineering. This enabled us to give an overview of the basic knowledge and related activities. The second volume describes recent applications that turn the technologies described in Volume I into customer oriented products. Volume II shows some examples of these new products with an introductory chapter on product design fundamentals. The superior behavior of, for example, coffee, aspirin and carotenoid products is crucial. The taste of coffee, the bioavailability of aspirin and carotenoid products and the UV absorption of polymers can be adjusted. Product design offers opportunities to change, to adopt, and to improve products.

The intention of this, the third volume of this series is to discuss in more detail the basics of rheology and how product design is carried out in liquid and gel-type applications. Differentiation and product design is essential for raw materials and additives. The behavior of, for example, starches and gelatins is designed and changed by product design to give a specific texture and/or performance of the final product.

The structure of the third volume of this series is illustrated in Figure 2. The fundamentals are the basics in rheology, which are important for describing and quantifying the properties of gels or pastes.

This volume starts with the essential chapter entitled "Rheology of Disperse Systems" (Chapter 1) while Chapter 3 gives an insight view into "Rheology Modifiers, Thickeners, and Gels." The stability of dispersions is crucial for the shelf life of customer related products. This problem is discussed in the chapter

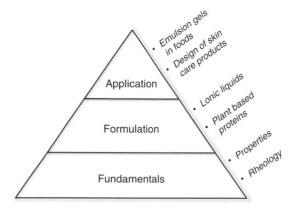

Figure 2 Structure and content of the third volume.

on the "Use of Rheological Techniques for Assessment and Prediction of Stability of Dispersions (Suspensions and Emulsions)" (Chapter 4). The specific chapter "Rheology of Cosmetic Emulsions" (Chapter 2) focuses on products closer to cosmetic daily life application. A theoretical approach can be found in the chapters on the "Prediction of Thermophysical Properties of Liquid Products" (Chapter 5) and "Sources of Thermophysical Properties for the Efficient Use in Product Design" (Chapter 6). A more forward-looking contribution is "Trends in Ionic Liquid Research" (Chapter 7). The possibilities in terms of modifying the viscosity of liquid formulations are given in "Gelling of Plant Based Proteins" (Chapter 8) and "Enzymatic Texturized Plant Proteins for the Food Industry" (Chapter 9). Examples of some important applications are given in the "Design of Skin Care Products" (Chapter 10) and "Emulsion Gels in Foods" (Chapter 11).

A complete prediction of product formulation based on scientific knowledge is not possible given the current state of the art. Very specific and often still empirical knowledge and specific trials in the laboratories are still needed to ensure the design of a product. Product design and engineering it is a very multifaceted area. Nevertheless, this book aims to give a good overview of the different fields of technologies and successful instances for liquid and gel-type applications.

As product design will become increasingly important in the near future the teaching of students in this field should be intensified given that:

- profound physical and chemical knowledge is needed;
- product design is the interaction of multiple disciplines.

It is our intention to contribute with this book series to the on-going improvements in technologies, fundamentals and discussions in the community about teaching product design. Finally, without the highly qualified contributions of the persons most important for this book, our authors, this volume would still only be a nice idea.

References

1. Sutton, A. (2008) Product development of probiotics as biological drugs. *Clin. Infect. Dis.*, **46**, 128–132.
2. Merino, S.T. and Cherry, J. (2007) Progress and challenges in enzyme development for biomass utilization. *Adv. Biochem. Eng. Biotechnol.*, **108**, 95–120.
3. Rastall, R.A. (2007) *Novel Enzyme Technology for Food Applications*, Woodhead Press, Cambridge. ISBN: 9781420043969.
4. Cussler, E.L. and Moggridge, G.D. (2011) *Chemical Product Design*, 2nd edn, Cambridge University Press, Cambridge. ISBN: 978-0-521-16822-9.
5. Calorie Control Council (2013) Polyols Information Source *http://www.polyol.org/* (accessed 8 March 2013).
6. Stummerer, S. and Hablesreiter, M. (2010) *Food Design XL*, Springer, New York. ISBN: 978-3-211-99230-2.
7. Ortega-Rivas, E. *et al.* (2005) *Food Powders*, Springer, New York. ISBN: 978-0-306-47806-2.
8. Norton, I.T. *et al.* (2013) *Formulation Engineering of Foods*, Wiley-Blackwell. ISBN: 10: 0470672900.
9. Costa, R. *et al.* (2006) Chemical product engineering: an emerging paradigm within chemical engineering. *AIChE J.*, **52** (6), 1976–1986.
10. Bayer HealthCare LLC (2009) Bayer Group Aspirin *http://www.aspirin. com/scripts/pages/en/aspirin_history/ index.php* (accessed 8 March 2013).

1
Rheology of Disperse Systems

Norbert Willenbacher and Kristina Georgieva

1.1
Introduction

The rheology of disperse systems is an important processing parameter. Being able to characterize and manipulate the flow behavior of dispersions one can ensure their optimal performance. Waterborne automotive coatings, for example, should exhibit a distinct low-shear viscosity necessary to provide good leveling but to avoid sagging at the same time. Then, a strong degree of shear thinning is needed to guarantee good pump- and sprayability. The rheological properties of dispersions, especially at high solids content, are complex and strongly dependent on the applied forces and flow kinematics. Adding particles does not simply increase the viscosity of the liquid as a result of the hydrodynamic disturbance of the flow; it also can be a reason for deviation from Newtonian behavior, including shear rate dependent viscosity, elasticity, and time-dependent rheological behavior or even the occurrence of an apparent yield stress. In colloidal systems particle interactions play a crucial role. Depending on whether attractive or repulsive interactions dominate, the particles can form different structures that determine the rheological behavior of the material. In the case of attractive particle interactions loose flocs with fractal structure can be formed, immobilizing part of the continuous phase. This leads to a larger effective particle volume fraction and, correspondingly, to an increase in viscosity. Above a critical volume fraction a sample-spanning network forms, which results in a highly elastic, gel-like behavior, and an apparent yield stress. Shear-induced breakup and recovery of floc structure leads to thixotropic behavior. Electrostatic or steric repulsion between particles defines an excluded volume that is not accessible by other particles. This corresponds to an increase in effective volume fraction and accordingly to an increase in viscosity. Crystalline or gel-like states occur at particle concentrations lower than the maximum packing fraction.

Characterization of the microstructure and flow properties of dispersions is essential for understanding and controlling their rheological behavior. In this chapter we first introduce methods and techniques for standard rheological tests and then characterize the rheology of hard sphere, repulsive, and attractive particles. The effect of particle size distribution on the rheology of highly concentrated

Product Design and Engineering: Formulation of Gels and Pastes, First Edition.
Edited by Ulrich Bröckel, Willi Meier, and Gerhard Wagner.
© 2013 Wiley-VCH Verlag GmbH & Co. KGaA. Published 2013 by Wiley-VCH Verlag GmbH & Co. KGaA.

dispersions and the shear thickening phenomenon will be discussed with respect to the influence of colloidal interactions on these phenomena. Finally, typical features of emulsion rheology will be discussed with special emphasis on the distinct differences between dispersion and emulsion rheology.

1.2
Basics of Rheology

According to its definition, rheology is the science of the deformation and flow of matter. The rheological behavior of materials can be regarded as being between two extremes: Newtonian viscous fluids, typically liquids consisting of small molecules, and Hookean elastic solids, like, for example, rubber. However, most real materials exhibit mechanical behavior with both viscous and elastic characteristics. Such materials are termed *viscoelastic*. Before considering the more complex viscoelastic behavior, let us first elucidate the flow properties of ideally viscous and ideally elastic materials.

Isaac Newton first introduced the notion of viscosity as a constant of proportionality between the force per unit area (shear stress) required to produce a steady simple shear flow and the resulting velocity gradient in the direction perpendicular to the flow direction (shear rate):

$$\sigma = \eta \dot{\gamma} \tag{1.1}$$

where $\sigma = F/A$ is the shear stress, η the viscosity, and the $\dot{\gamma} = v/h$ is the shear rate. Here A is the surface area of the sheared fluid volume on which the shear force F is acting and h is the height of the volume element over which the fluid layer velocity v varies from its minimum to its maximum value. A fluid that obeys this linear relation is called *Newtonian*, which means that its viscosity is independent of shear rate for the shear rates applied. Glycerin, water, and mineral oils are typical examples of Newtonian liquids. Newtonian behavior is also characterized by constant viscosity with respect to the time of shearing and an immediate relaxation of the shear stress after cessation of flow. Furthermore, the viscosities measured in different flow kinematics are always proportional to one another.

Materials such as dispersions, emulsions, and polymer solutions often exhibit flow properties distinctly different from Newtonian behavior and the viscosity decreases or increases with increasing shear rate, which is referred to a shear thinning and shear thickening, respectively. Figure 1.1a,b shows the general shape of the curves representing the variation of viscosity as a function of shear rate and the corresponding graphs of shear stress as a function of shear rate.

Materials with a yield stress behave as solids at rest and start to flow only when the applied external forces overcome the internal structural forces. Soft matter, such as, for example, dispersions or emulsions, does not exhibit a yield stress in this strict sense. Instead, these materials often show a drastic change of viscosity by orders of magnitude within a narrow shear stress range and this is usually termed an *"apparent" yield stress* (Figure 1.2a,b). Dispersions with attractive interactions,

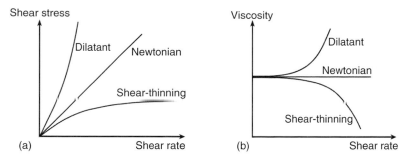

Figure 1.1 Typical flow curves for Newtonian, shear thinning and shear thickening (dilatant) fluids: (a) shear stress as a function of shear rate; (b) viscosity as a function of shear rate.

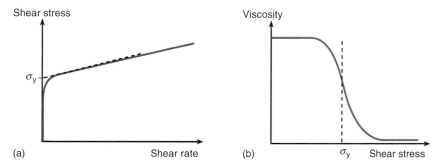

Figure 1.2 Flow curve of a material with an apparent yield stress σ_y: (a) shear stress as a function of shear rate; (b) viscosity as a function of shear stress.

such as emulsions and foams, clay suspensions, and ketchup, are typical examples of materials with an apparent yield stress. Note that there are various methods for yield stress determination and the measured value may differ depending on the method and instrument used.

The flow history of a material should also be taken into account when making predictions of the flow behavior. Two important phenomena related to the time-dependent flow behavior are thixotropy and rheopexy. For materials showing thixotropic behavior the viscosity gradually decreases with time under constant shear rate or shear stress followed by a gradual structural recovery when the stress is removed. The thixotropic behavior can be identified by measuring the shear stress as a function of increasing and decreasing shear rate. Figure 1.3 shows a hysteresis typical for a thixotropic fluid. Examples of thixotropic materials include coating formulations, ketchup, and concentrated dispersions in the two-phase region (Section 1.4.1.1). The term *rheopexy* is defined as shear-thickening followed by a gradual structural recovery when the shearing is stopped. Tadros pointed out that rheopexy should not be confused with anti-thixotropy, which is the time dependent shear thickening [1]. However, rheopectic materials are not very common and will not be discussed here.

Shear stress

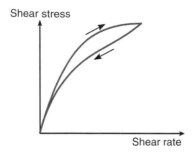

Shear rate

Figure 1.3 Flow curve of a thixotropic material.

So far we have considered the flow behavior of viscous fluids in terms of Newton's law and a nonlinear change of viscosity with applied stress that can occur either instantaneously or over a long period of time. At the other extreme is the ideal elastic behavior of solids, which can be described by Hooke's law of elasticity:

$$\sigma = G\gamma \tag{1.2}$$

where γ is the shear deformation (also termed *strain*) and G is the shear modulus characterizing the rigidity of a material. The shear modulus of an ideal elastic solid is independent of the shear stress and duration of the shear load. As soon as a deformation is applied a constant corresponding stress occurs instantaneously. In viscoelastic materials stress relaxes gradually over time at constant deformation and eventually vanishes for viscoelastic liquids. When the stress relaxation is proportional to the strain we are talking about the so-called linear viscoelastic regime. Above a critical strain the apparent shear modulus becomes strain dependent. This is the so-called nonlinear viscoelastic regime. The linear viscoelastic material properties are in general very sensitive to microstructural changes and interactions in complex fluids.

A dynamic test or small amplitude oscillatory shear (SAOS) test is the most widely used rheological measurement to investigate the linear viscoelastic behavior of a fluid, since it has a superior accuracy compared to step strain or step stress experiments. When a sinusoidal oscillatory shear strain is applied with amplitude γ_0 and angular frequency ω the deformation $\gamma(t)$ can be written as:

$$\gamma(t) = \gamma_0 \sin(\omega t) \tag{1.3}$$

where t denotes the time. The shear rate is the time derivative of the shear strain and then reads as follows:

$$\dot{\gamma}(t) = \frac{d\gamma(t)}{dt} = \gamma_0 \omega \cos(\omega t) \tag{1.4}$$

A linear viscoelastic fluid responds with a sinusoidal course of shear stress $\sigma(t)$ with amplitude σ_0 and angular frequency ω, but is phase shifted by an angle δ compared to the imposed strain:

$$\sigma(t) = \sigma_0 \sin(\omega t + \delta) \tag{1.5}$$

Depending on material behavior, the phase shift angle δ occurs between $0°$ and $90°$. For ideal elastic materials the phase shift disappears, that is, $\delta = 0$, while for ideal viscous liquids $\delta = 90°$. The shear modulus can be written in complex form:

$$G^*(\omega) = G'(\omega) + iG''(\omega) \tag{1.6}$$

with the storage modulus G' and loss modulus G''. G' is a measure of the energy stored by the material during a cycle of deformation and represents the elastic behavior of the material, while G'' is a measure of the energy dissipated or lost as heat during the shear cycle and represents the viscous behavior of the material. The terms G' and G'' can be expressed as sine and cosine function of the phase shift angle δ:

$$G'(\omega) = \frac{\sigma_0}{\gamma_0} \cos \delta \tag{1.7}$$

$$G''(\omega) = \frac{\sigma_0}{\gamma_0} \sin \delta \tag{1.8}$$

Hence the tangent of the phase shift δ represents the ratio of loss and storage modulus:

$$\tan \delta = \frac{G''(\omega)}{G'(\omega)} \tag{1.9}$$

Analogous to the complex shear modulus we can define a complex viscosity η^*:

$$\eta^*(\omega) = \frac{\sigma(t)}{\dot{\gamma}(t)} = \eta'(\omega) + i\eta''(\omega) \tag{1.10}$$

with:

$$\eta'(\omega) = \frac{G''(\omega)}{\omega} \quad \text{and} \quad \eta''(\omega) = \frac{G'(\omega)}{\omega} \tag{1.11}$$

The viscoelastic properties of a fluid can be characterized by oscillatory measurements, performing amplitude- and frequency-sweep. The oscillatory test of an unknown sample should begin with an amplitude sweep, that is, variation of the amplitude at constant frequency. Up to a limiting strain γ_c the structure of the tested fluid remains stable and G' as well as G'' is independent of the strain amplitude. The linear viscoelastic range may depend on the angular frequency ω; often, γ_c decreases weakly with increasing frequency.

Frequency sweeps are used to examine the time-dependent material response. For this purpose the frequency is varied using constant amplitude within the linear viscoelastic range. At an appropriately high angular frequency ω, that is, short-term behavior, the samples show an increased rigidity and hence $G' > G''$. At lower frequencies (long-term behavior) stress can relax via long-range reorganization of the microstructure and the viscous behavior dominates and, correspondingly, $G'' > G'$.

1.3
Experimental Methods of Rheology

Rheometers can be categorized according to the flow type in which material properties are investigated: simple shear and extensional flow. Shear rheometers can be divided into rotational rheometers, in which the shear is generated between fixed and moving solid surfaces, and pressure driven like the capillary rheometer, in which the shear is generated by a pressure difference along the channel through which the material flows. Extensional rheometers are far less developed than shear rheometers because of the difficulties in generating homogeneous extensional flows, especially for liquids with low viscosity. Many different experimental techniques have been developed to characterize the elongational properties of fluids and predict their processing and application behavior, including converging channel flow [2], opposed jets [3], filament stretching [4], and capillary breakup [5, 6] techniques. However, knowledge about the extensional rheology of complex fluids like dispersions and emulsions is still very limited.

1.3.1
Rotational Rheometry

Rotational instruments are used to characterize materials in steady or oscillatory shear flow. Basically there are two different modes of flow: controlled shear rate and controlled shear stress. Three types of measuring systems are commonly used in modern rotational rheometry, namely, concentric cylinder, parallel plate, and cone-and-plate. Typical shear rates that can be measured with rotational rheometers are in the range 10^{-3} to $10^3\,\mathrm{s}^{-1}$.

1.3.1.1 Concentric Cylinder Measuring System
As shown in Figure 1.4a, a cylinder measuring system consists of an outer cylinder (cup) and an inner cylinder (bob). There are two modes of operation depending on whether the cup or the bob is rotating. The Searle method corresponds to a

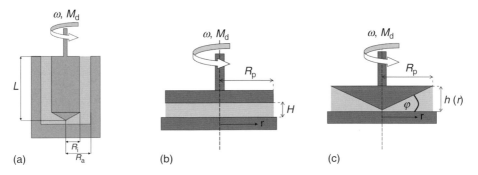

Figure 1.4 Schematic representation of (a) concentric cylinder, (b) parallel-plate, and (c) cone-and-plate measuring system.

rotating bob and stationary cup, while in the Couette mode the cup is set in motion and the bob is fixed. The gap between the two concentric cylinders should be small enough so that the sample confined in the gap experiences a constant shear rate. This requirement is fulfilled and the gap is classified as "narrow" when the ratio of the inner to the outer cylinder radius is greater than 0.97.

When the bob is rotating at an angular velocity ω the shear rate is given by:

$$\dot{\gamma} = 2\omega \frac{R_\mathrm{a}^2}{R_\mathrm{a}^2 - R_\mathrm{i}^2} \tag{1.12}$$

where R_i and R_a are the radii of the bob and the cup, respectively. If the torque measured on the bob is M_d, the shear stress σ in the sample is given by:

$$\sigma = \frac{M_\mathrm{d}}{2\pi R_\mathrm{i}^2 L} \tag{1.13}$$

where L is the effective immersed length of the bob.

Having the shear rate $\dot{\gamma}$ and shear stress σ, the sample viscosity η can be calculated according to Equation 1.1 For these calculations we ignore any end effects, which are actually likely to occur as a result of the different shearing conditions in the liquid covering the ends of the cylinders. To minimize the end effect the ratio of the length L to the gap between cylinders is maintained at greater than 100 and the shape of the bottom of the bob is designed as a cone with an angle α, which is chosen so that the shear rate in the bottom area matches that in the narrow gap between the concentric cylinders.

The concentric cylinder measuring system is especially suitable for low-viscous liquids, since it can be designed to offer a large shear area and at high shear rates the sample is not expelled from the gap. Other advantages of this geometry are that sample evaporation is of minor relevance since the surface area is small compared to the sample volume, the temperature can be easily controlled due to the large contact area, and even if suspensions exhibit sedimentation and particle concentration varies along the vertical direction the measured viscosity is a good approximation of the true value.

1.3.1.2 Parallel-Plate Measuring System

The parallel plate geometry is shown in Figure 1.4b. The sample, confined within the gap of height H between the two parallel plates, is sheared by the rotation of one of the plates at angular velocity ω. Thereby, the circumferential velocity v depends on the distance from the plate at rest h and the distance r from the rotational axis:

$$v(r, h) = r\omega \frac{h}{H} \tag{1.14}$$

and thus:

$$\dot{\gamma}(r) = \frac{v}{h} = \frac{r\omega}{H} \tag{1.15}$$

The shear rate $\dot{\gamma}$ at constant ω is not constant within the gap. Typically, the calculations and analysis of rheological results in parallel-plate measuring systems are related to the maximum shear rate value at the rim of the plate ($r = R_\mathrm{p}$). The shear rate can be varied over a wide range by changing the gap height H and the angular velocity ω.

The shear stress σ is a function of the shear rate $\dot{\gamma}$, which is not constant within the gap. Thus, to relate the shear stress to the total torque an expression for the $\sigma(\dot{\gamma})$ dependence is necessary. For Newtonian liquids the shear stress depends linearly on the shear rate and can be expressed as follows:

$$\sigma(R) = \frac{2M_d}{\pi R_p^3} \tag{1.16}$$

This expression is called the *apparent shear stress*. For non-Newtonian fluids Giesekus and Langer [7] developed a simple approximate single point method to correct the shear rate data, based on the idea that the true and apparent shear stress must be equal at some position near the wall. It was found that this occurs at the position where $r/R_p = 0.76$ and this holds for a wide range of liquids.

The parallel-plate measuring system allows for measurements of suspensions with large particles by using large gap heights. On the other hand, by operating at small gaps the viscosity can be obtained at relatively high shear rates. Small gaps also allow for a reduction of errors due to edge effects and secondary flows. Wall slip effects can be corrected by performing measurements at different gap heights. Rough plates are often used to minimize wall slip effects. Note that for sedimenting suspensions the viscosity is systematically underestimated since the upper rotating plate moves on a fluid layer with reduced particle loading.

1.3.1.3 Cone-and-Plate Measuring System

A cone-and-plate geometry is shown schematically in Figure 1.4c. The sample is contained between a rotating flat cone and a stationary plate. Note that the apex of the cone is cut off to avoid friction between the rotating cone and the lower plate. The gap angle φ is usually between 0.3° and 6° and the cone radius R_p is between 10 and 30 mm. The gap h increases linearly with the distance r from the rotation axis:

$$h(r) = r \tan \varphi \tag{1.17}$$

The circumferential velocity v also increases with increasing distance r:

$$v(r) = r\omega \tag{1.18}$$

Hence the shear rate is constant within the entire gap and does not depend on the radius r:

$$\dot{\gamma} = \frac{dv(r)}{dh(r)} = \frac{\omega}{\tan \varphi} \approx \frac{\omega}{\varphi} \tag{1.19}$$

The shear stress is related to the torque M_d on the cone:

$$\sigma = \frac{3M_d}{2\pi R_p^3} \tag{1.20}$$

A great advantage of the cone-and-plate geometry is that the shear rate remains constant und thus provides homogenous shear conditions in the entire shear gap.

The limited maximum particle size of the investigated sample, difficulties with avoiding solvent evaporation, and temperature gradients in the sample as well as concentration gradients due to sedimentation are typical disadvantages of the cone-and-plate measuring system.

1.3.2
Capillary Rheometer

Figure 1.5 shows a schematic diagram of a piston driven capillary rheometer. A piston drives the sample to flow at constant flow rate from a reservoir through a straight capillary tube of length L. Generally, capillaries with circular (radius R) or rectangular (width B and height H) cross-sections are used. The measured pressure drop Δp along the capillary and the flow rate Q are used to evaluate the shear stress, shear rate, and, correspondingly, viscosity of the sample.

Pressure driven flows through a capillary have a maximum velocity at the center and maximum shear rate at the wall of the capillary, that is, the deformation is essentially inhomogeneous. Assuming Newtonian behavior and fully developed, incompressible, laminar, steady flow, the apparent wall shear stress σ_a in a circular capillary with radius R is related to the pressure drop Δp by:

$$\sigma_a = \frac{\Delta p R}{2L} \tag{1.21}$$

and the apparent or Newtonian shear rate at the wall can be calculated on the basis of measured flow rate according to:

$$\dot{\gamma}_a = \frac{4Q}{\pi R^3} \tag{1.22}$$

Therefore, we can evaluate the viscosity in terms of an apparent viscosity based on Newton's postulate (Equation 1.1).

To obtain the true shear rate in the case of non-Newtonian fluids the Weissenberg–Rabinowitch correction [8] for non-parabolic velocity profiles should

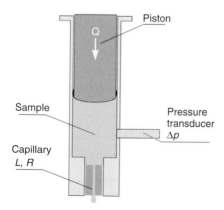

Figure 1.5 Schematic representation of a controlled flow rate capillary rheometer.

be taken into account. A simpler method to determine the true shear rate has been developed by Giesekus and Langer [7] as well as Schümmer and Worthoff [9]. Their single point method is based on the idea that the true and apparent shear rate must be equivalent at a certain radial position near the wall and thus the true shear rate $\dot{\gamma}$ is given simply by:

$$\dot{\gamma} = 0.83\dot{\gamma}_a \tag{1.23}$$

Note that this approximation does no differ significantly from the Weissenberg–Rabinowitch correction for weakly shear thinning fluids.

Other possible sources of error in capillary flow experiments are entrance effects, slippage at the capillary wall, and viscous heating effects. Furthermore, the pressure drop Δp is difficult to measure directly in the capillary. Therefore, it is usually detected by a transducer mounted above the entrance of the capillary. Hence, the measured pressure includes not only the pressure loss due to the laminar flow in the die but also the entrance pressure loss due to rearrangement of the streamlines at the entrance and the exit of the capillary. Bagley [10] proposed a correction that accounts for these additional pressure losses but for practical purposes it is sufficient to use a single capillary die with sufficiently large L/R ratio, typically $L/R \geq 60$ [8].

For highly concentrated suspensions wall slip effects, due to shear induced particle migration (only for very large particles), and specific particle–wall interactions have to be considered. If the slip velocity is directly proportional to the applied stress it is possible to correct the apparent wall shear rate according to the procedure developed by Mooney [11], which compares the flow curves determined with dies of different radii but similar L/R.

The major advantage of the capillary rheometer is that the flow properties of fluids can be characterized under high shear conditions (up to $\dot{\gamma} = 10^6 \, \text{s}^{-1}$) and process-relevant temperatures (up to 400 °C). Another advantage is that the capillary flow is closed and has no free surface so that edge effects, solvent evaporation, and other problems that trouble rotational rheometry can be avoided.

1.4
Rheology of Colloidal Suspensions

The flow behavior of colloidal (often also termed *Brownian*) dispersions is controlled by the balance between hydrodynamic and thermodynamic interactions as well as Brownian particle motion. Thermodynamic interactions mainly include electrostatic and steric repulsion and van der Waals attraction. The relative importance of individual forces can be assessed on the basis of dimensionless groups, which can be used to scale rheological data. In this section we first consider dispersions of Brownian hard sphere particles and elucidate the effect of particle volume fraction, size, and shape of particles on dispersion rheology. Then, we take into account the effect of repulsive and attractive interactions on the microstructure of suspensions and its corresponding rheological response. Special attention will be paid to the rheological behavior of concentrated dispersions.

1.4.1
Hard Spheres

Hard-sphere dispersions are idealized model systems where no thermodynamic or colloidal particle–particle interactions are present unless these particles come into contact. In that sense, they represent the first step from ideal gases towards real fluids. Even such simple systems can show complex rheological behavior. The parameters controlling dispersion rheology will be discussed below.

1.4.1.1 Viscosity of Suspensions of Spheres in Newtonian Media

Hard-sphere dispersions exist in the liquid, crystalline, or glassy state depending on the particle volume fraction similar to the temperature-dependent phase transition of atomic or molecular systems. Figure 1.6 demonstrates schematically the hard-sphere phase diagram in terms of particle volume fraction ϕ, constructed by means of light diffraction measurements [12]. At a low volume fraction ϕ particles can diffuse freely and there is no long-range ordering in particle position, that is, the dispersion is in the fluid state, while with increasing concentration above $\phi = 0.50$ crystalline and liquid phases coexist in equilibrium and the fraction of crystalline phase increases until the sample is fully crystalline at $\phi = 0.55$. With further increasing particle volume fraction, crystallization becomes slower due to reduced particle mobility. At a critical volume fraction $\phi = 0.58$ particle mobility is so strongly reduced that no ordered structure can be formed and the dispersion remains in the disordered glassy (immobile) state. Crystalline ordering only occurs if all particles are of equal size, otherwise disordered gel-like structures form at $\phi > 0.5$.

The phase states of hard sphere dispersions are reflected in their characteristic flow curves. Figure 1.7 demonstrates the general features of the shear rate dependence of viscosity at various particle concentrations. At volume fractions up to $\phi = 0.50$ the dispersion is in the liquid state and a low-shear Newtonian plateau is observed for the viscosity. The low-shear viscosity, as well as the shear thinning, increases with increasing particle volume fraction ϕ. In the two-phase region colloidal hard-sphere dispersions may show thixotropic behavior (see the

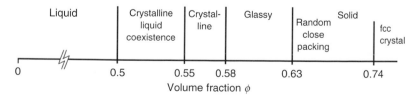

Figure 1.6 Hard-sphere phase-diagram constructed from light diffraction measurements [12].

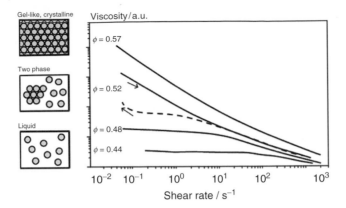

Gel-like, crystalline

Two phase

Liquid

Figure 1.7 Viscosity versus shear rate for hard sphere dispersions at various volume fractions ϕ. The downward and upward arrows indicate the viscosity measurement with increasing and consequent decreasing shear rate, respectively.

curves in Figure 1.7 at $\phi = 0.52$), due to the shear induced destruction and subsequent recovery of sample structure, associated with coexisting liquid and crystalline domains. The degree of thixotropy, if any, depends on the measuring conditions. For a particle volume fraction of $\phi \geq 0.55$ dispersions are in the crystalline or gel-like state and show shear thinning behavior in the whole shear rate range investigated. On the other hand, thixotropy vanishes since no long-range particle rearrangements are possible due to the dense particle packing.

Viscosity in the low shear Newtonian plateau, referred to as *zero-shear viscosity* η_0, depends only on the total volume occupied by the particles and is independent of particle size. The solvent viscosity η_s always acts as a constant pre-factor, and in the following we will focus on the relative viscosity $\eta_r = \eta/\eta_s$. Various models describing the volume fraction dependence of the zero-shear viscosity have been proposed. The classical model of Einstein [13, 14] for infinitely dilute, non-interacting hard spheres showed that single particles increase the viscosity of the dispersion medium as a linear function of the volume fraction ϕ according to the equation:

$$\eta_r = (1 + 2.5\phi) \tag{1.24}$$

The Einstein equation applies to $\phi < 0.01$, assuring that the flow around a particle does not influence the velocity field of any other particle. At higher particle concentration the hydrodynamic interactions between particles become important and higher-order terms in ϕ have to be considered. The effect of two-sphere hydrodynamic interactions on the suspension viscosity was calculated by Batchelor [15]:

$$\eta_r = 1 + 2.5\phi + 6.2\phi^2 \tag{1.25}$$

This equation is validated to $\phi < 0.1$. For higher particle concentrations multi-particle interactions become imperative and a prediction of viscosity from first principles is still lacking. Numerous phenomenological equations have been introduced to correlate the viscosity of concentrated dispersions to the particle volume fraction. Krieger and Dougherty [16] proposed a semi-empirical equation for the concentration dependence of the viscosity:

$$\eta_r = \left(1 - \frac{\phi}{\phi_{max}}\right)^{-2.5\phi_{max}} \qquad (1.26)$$

where ϕ_{max} is the maximum packing fraction or the volume fraction at which the zero shear viscosity diverges. This equation reduces to the Einstein relation (Equation 1.24) at low particle concentration. Quemada [17] suggested another phenomenological model to predict the $\eta_r(\phi)$ dependence:

$$\eta_r = \left(1 - \frac{\phi}{\phi_{max}}\right)^{-2} \qquad (1.27)$$

This model suits best as $\phi \rightarrow \phi_{max}$. Figure 1.8 shows the volume fraction dependence of relative viscosity, according to the models described above.

The absolute value for the maximum packing fraction ϕ_{max} is determined by the packing geometry, which depends on the particle shape and particle size distribution but not on particle size. The volume fraction at maximum packing has been calculated by theoretical models and different ϕ_{max} values have been found depending on the type of packing. The ϕ_{max} value for hard spheres is often taken as 0.64 [18], which is the value associated with random close packing. However, experiments on colloidal hard sphere dispersions have shown that zero-shear viscosity diverges at the volume fraction of the colloidal glass transition $\phi_g = 0.58$ [19–22]. Above ϕ_g, particle diffusion is restricted to small "cages" formed

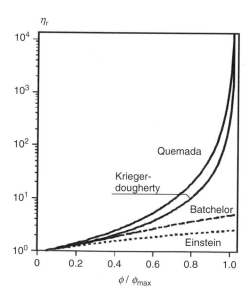

Figure 1.8 Schematic representation of the volume fraction dependence of relative viscosity η_r according to the Einstein, Batchelor, Krieger–Dougherty, and Quemada models.

by the nearest neighbors; correspondingly, the long-time self-diffusion coefficient decreases to zero and the viscosity diverges. The latter two quantities are related to each other by the generalized Stokes–Einstein equation:

$$D = \frac{k_B T}{6 \pi \eta(\phi) a} \tag{1.28}$$

Let us now consider the shear rate dependence of dispersion viscosity in the liquid state. The transition from low shear to high shear plateau referred to as the *shear-thinning region* depends on the balance between Brownian and hydrodynamic forces. The Péclet number Pe is a useful dimensionless quantity to express the relative importance of these two contributions:

$$Pe = \frac{6 \pi a^3 \eta}{k_B T} \dot{\gamma} = \frac{a^2}{D_0} \dot{\gamma} \tag{1.29}$$

where a is the particle size, $k_B T$ is the thermal energy, and $D_0 = D (\phi \to 0)$ is the diffusion coefficient.

The Péclet number is often called the *dimensionless shear rate*; equivalently, the dimensionless shear stress σ_r can be expressed as follows:

$$\sigma_r = \frac{a^3 \sigma}{k_B T} \tag{1.30}$$

The shear thinning region occurs around a characteristic Péclet number $Pe \approx 1$ at which Brownian and hydrodynamic forces are of similar relevance, which strongly depends on the particle size a. A variation of particle size results in a shift of the viscosity/shear rate curve on the $\dot{\gamma}$-axis with a shift factor proportional to the particle radius cubed. Hence a plot of η_r as a function of Péclet number or the dimensionless shear stress σ_r should superimpose for hard sphere colloids of different particle size at a given ϕ. This is illustrated in Figure 1.9a,b using the example of poly(methyl methacrylate) spheres of different size, dispersed in silicone oil , $\eta_{0,r}$ and $\eta_{\infty,r}$ denote the low and high shear limiting values of the relative viscosity.[23].

Figure 1.10a demonstrates schematically the effect of solvent viscosity η_s on the viscosity of hard-sphere dispersions. The Pe number fully accounts for the effect of viscosity of dispersion medium on the shear rate dependence of viscosity and can be used to scale the data onto a master curve (Figure 1.10b) if again the relative viscosity $\eta_r = \eta/\eta_s$.

1.4.1.2 Non-spherical Particles

Particles can deviate from the spherical form by either being axisymmetric or by having an irregular shape. Typically, particles are approximated by prolate or oblate spheroids (Figure 1.11) with a specified axis ratio r_p:

$$r_p = \frac{a}{b} \tag{1.31}$$

where a corresponds to the length of the semi-major axis and b to the length of the semi-minor axis. Some examples of spheroids are shown in Figure 1.11.

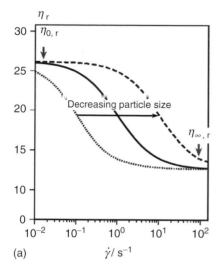

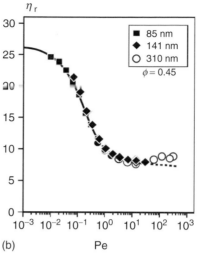

Figure 1.9 Effect of particle size on the shear rate dependence of relative viscosity. (a) Schematic representation of the flow curves of hard sphere dispersion, shifted to high shear rates as the particle size decreases; (b) relative viscosity η_r as a function of Péclet number Pe for sterically stabilized poly(methyl methacrylate) particles of different size. Redrawn from Choi and Krieger [23].

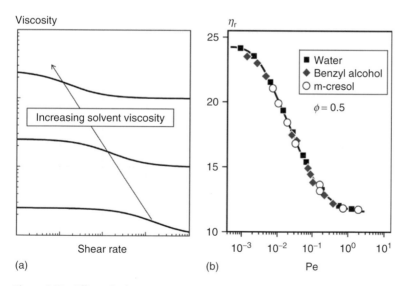

Figure 1.10 Effect of solvent viscosity on the shear rate dependence of relative viscosity. (a) Schematic representation of the flow curves for hard spheres dispersed in solvents with different viscosity; (b) relative viscosity η_r versus Pe number for polystyrene monodispersed spheres in different media. Redrawn from the paper by Krieger [24].

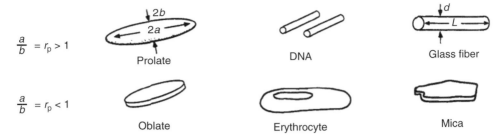

$$\frac{a}{b} = r_p > 1$$

Prolate

DNA

Glass fiber

$$\frac{a}{b} = r_p < 1$$

Oblate

Erythrocyte

Mica

Figure 1.11 Prolate or oblate shaped particles and corresponding examples of typical particles. Taken from Macosko [8]. Copyright © 1994 John Wiley & Sons.

The rheology of suspensions of non-spherical particles is greatly influenced by particle orientation with respect to the flow. The orientation in flowing suspensions is governed by the balance between hydrodynamic forces, which tend to align particles with flow, and Brownian motion randomizing the orientation. The relative importance of each is given by a rotational Péclet number Pe_{rot}:

$$Pe_{rot} = \tau_{rot}\dot{\gamma} \tag{1.32}$$

For disk-like particles with radius b, the rotational relaxation time τ_{rot} is:

$$\tau_{rot}^{-1} = \frac{3k_B T}{32\eta_s b^3} \tag{1.33}$$

and for rod-like particles with length $2a$ such that $r_p \gg 1$:

$$\tau_{rot}^{-1} = \frac{3k_B T(\ln 2r_p - 0.5)}{8\pi\eta_s a^3} \tag{1.34}$$

At low shear rates for small particles and low fluid viscosity $Pe_{rot} \to 0$ and the randomizing effect of Brownian motion dominates. For $Pe_{rot} > 1$ the hydrodynamic forces become enough strong to align the particles with the flow and the suspension shows a considerable shear thinning behavior.

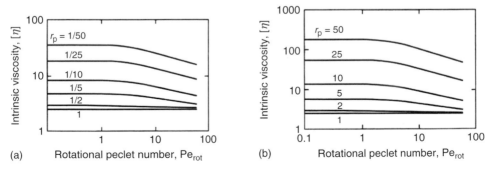

Figure 1.12 Intrinsic viscosity $[\eta]$ as a function of rotational Péclet number Pe_{rot}, calculated for diluted suspensions of (a) disc- and (b) rod-like particles of various aspect ratios [25].

Figure 1.12a,b shows numerical results for the intrinsic viscosity $[\eta]$ as a function of Pe_{rot} for dilute suspensions of disk- and rod-like particles at different aspect ratios [25]. The intrinsic viscosity $[\eta]$ is a dimensionless quantity defined as:

$$[\eta] = \lim_{\phi \to 0} \frac{\eta - \eta_s}{\phi \eta_s} \tag{1.35}$$

It can be seen from Figure 1.12a,b that the zero-shear intrinsic viscosity increases with increasing aspect ratio r_p, which is due to the effective enlargement of the volume inaccessible for other particles. Elongated particles in highly diluted suspensions can rotate freely about their center of gravity and thus occupy a spherical volume with a diameter corresponding to the long dimension of the spheroid. Therefore, particle interactions become relevant beyond a critical volume fraction $\phi^* \ll \phi_{max}$ at which these spheres start to interpenetrate. Hence, particle asymmetry has a strong effect on the concentration dependence of relative viscosity.

In colloidal as well as non-colloidal suspensions axisymmetric particles could be packed more densely than spheres, but the divergence of the zero shear viscosity occurs at lower volume fraction, which decreases with increasing aspect ratio r_p (Figure 1.13).

For anisotropic particles random orientation leads to a higher barrier against flow at low shear rates, that is, to an increase in zero-shear viscosity. However,

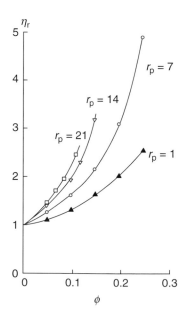

Figure 1.13 Relative viscosity η_r versus particle volume fraction ϕ for non-colloidal glass fiber suspension of various aspect ratios r_p [26]. Taken from Barnes et al. [27].

under shear, these elongated particles can orient in the direction of flow, resulting in a lower high shear viscosity than for spherical particles with equivalent size.

1.4.2
Influence of Colloidal Interactions on Rheology

1.4.2.1 Repulsive Particles

So far we have considered suspensions of hard spheres for which the colloidal or thermodynamic interactions did not play a role. In practice, dispersions are stabilized by repulsive surface forces in order to prevent aggregation. Colloidal interactions such as electrostatic or steric repulsion keep particles far enough apart so that they cannot be attracted by the short-range van der Waals attraction force. This corresponds to an excluded volume that is inaccessible to other particles. The effective volume fraction of the dispersion ϕ_{eff} can be expressed as follows:

$$\phi_{\text{eff}} = \phi \left(\frac{a_{\text{eff}}}{a} \right)^3 \tag{1.36}$$

where a_{eff} is the *effective particle radius* defined as half the distance to which two particle centers can approach each other under the action of colloidal forces. Many rheological features are analogous to those of hard sphere dispersions and can be quantitatively described by mapping the real system onto a hard sphere system with $\phi = \phi_{\text{eff}}$. The effective increase of the volume occupied by the particles causes an increase in the zero-shear viscosity as well as a shift of the liquid to crystalline phase transition and the colloidal glass transition to lower volume fractions ϕ. Note that hard sphere mapping is only valid if the range of repulsive interactions is small compared to the particle radius, which is true for typical commercially or technically relevant dispersions, especially at high particle loading.

Derjaguin–Landau–Verwey–Overbeek (*DLVO*) theory provides a good description of the interactions among electrostatically stabilized colloidal particles (see Chapter 2 in Volume 1 [28]). The strength of the repulsion is given by the surface charge or surface potential and the range of interaction by the so-called Debye length κ^{-1}, which is inversely proportional to the square-root of the ion concentration in the liquid phase. Since the effective volume fraction ϕ_{eff} increases with increasing κ^{-1}, the viscosity of charge-stabilized dispersions depends strongly on the ionic strength of the dispersion medium and diverges at lower volume fraction than predicted for hard spheres. The concentration dependence of the zero-shear viscosity for monodispersed charged polystyrene (PS) latices of different ionic strength and particle size is shown in Figure 1.14a,b [29]. The data on the left-hand side show that the relative zero-shear viscosity $\eta_{0,\text{r}} = \eta_0/\eta_{\text{s}}$ diverges at a volume fraction $\phi_{\text{max,exp}}$ well below that for hard spheres and this experimental maximum volume fraction $\phi_{\text{max,exp}}$ decreases with decreasing ionic strength of the system.

Particle size is also an important parameter that influences ϕ_{eff}. Decreasing the particle radius a, at a constant volume fraction ϕ and constant ionic strength, corresponds to an increase of ϕ_{eff} and, thus, for smaller particles the zero-shear viscosity diverges at a lower particle volume fraction $\phi_{\text{max,exp}}$ [29], when keeping all other conditions the same. The hard sphere mapping concept fully accounts for

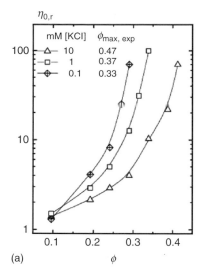

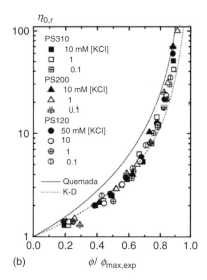

Figure 1.14 (a) Relative zero-shear viscosity $\eta_{0,r}$ versus particle volume fraction ϕ for monodisperse polystyrene particles (PS200), 200 nm in diameter, dispersed in water with concentrations of potassium chloride [KCl]; (b) master curve for all the data including the polystyrene dispersions at different salt concentration and particle size: 120, 200, and 310 nm [29].

the effects of particle size and ionic strength on the volume fraction dependence of viscosity. The zero-shear viscosity data can be collapsed onto a universal master curve by rescaling the volume fraction by $\phi/\phi_{\text{max,exp}}$ (Figure 1.14b). Furthermore, the Quemada (Equation 1.27) and Krieger–Dougherty (Equation 1.26) equations developed for hard sphere dispersions provide a good description of the zero-shear viscosity data for electrostatically interacting systems if ϕ is replaced by ϕ_{eff}.

Electrostatic interactions have a strong impact on the phase behavior of colloidal dispersions and hence on their flow properties. The hard sphere mapping concept can also be applied to categorize different characteristic signatures of the flow curves corresponding to different phase states. Figure 1.15 demonstrates the viscosity as a function of shear rate for an electrostatically stabilized PS/acrylate dispersion at various particle concentrations. Here the phase states typical for hard sphere dispersions (schematically shown in Figure 1.7) can be recognized but shifted to lower particle volume fractions. For repulsively interacting systems the phase diagram may be mapped onto that of a hard-sphere system using the effective radius concept. Accordingly, the transition volume fractions are lower than that for hard-sphere dispersion. In the example presented here, for instance, $\phi = 0.44 = \phi_{\text{lc,exp}}$ corresponds to the liquid/crystalline phase transition occurring at $\phi = 0.50 = \phi_{\text{lc,HS}}$ for hard spheres. Thus all volume fractions in this case can be rescaled as ϕ_{eff}:

$$\phi_{\text{eff}} = \phi \left(\frac{\phi_{\text{lc,HS}}}{\phi_{\text{lc,exp}}} \right) \tag{1.37}$$

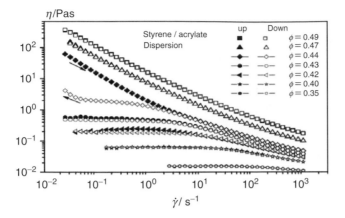

Figure 1.15 Viscosity as a function of increasing (up) and decreasing (down) shear rate for a polystyrene/acrylate dispersion ($a = 35$ nm) measured at various particle volume fractions. The downward arrow indicates the viscosity measurement with increasing the shear rate and the upward arrow indicates the subsequent measurement at gradually decreasing shear rates.

The liquid/crystalline phase transition volume fraction ϕ_{lc} decreases as the range of repulsive interaction increases.

The particle size can also influence the phase behavior of colloidal dispersions. Increasing the particle radius a at constant ϕ and a constant range of the repulsive colloidal interactions corresponds to a decreasing ϕ_{eff}. Thus, dispersions with the same ϕ but different a may exist in different phases. This has a strong impact on

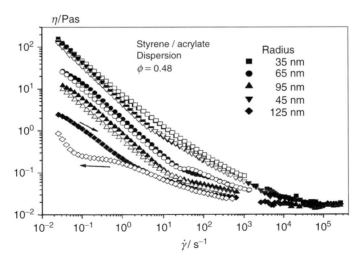

Figure 1.16 Viscosity as a function of increasing (filled symbols) and decreasing (open symbols) shear rate for polystyrene/acrylate dispersions with different particle radii measured at fixed particle volume fraction.

the shear rate dependent viscosity. Figure 1.16 demonstrates the phase transition from the crystalline to the two-phase region upon an increase of particle size of an electrostatically stabilized PS/acrylate dispersion at fixed volume fraction $\phi = 0.48$. At sufficiently high shear rates hydrodynamic interactions become dominant and can overcome the electrostatic repulsive forces so that particles approach each other closer and a_{eff} decreases until the electrostatic contribution is completely suppressed and the particles behave as hard spheres. As a consequence, the viscosity becomes independent of particle size and the flow curves superimpose.

Charged stabilized dispersions show a strong shear thinning behavior until the viscosity is close to that expected for hard spheres, that is, independent of particle size and ionic strength. This is true for the high shear viscosity η_∞ as well as the high frequency viscosity η'_∞. Note that these quantities correspond to different microstructures and η_∞ is always larger than η'_∞. The Cox–Merz rule $\eta(\dot\gamma) = |\eta^*(\omega)|$ for $\dot\gamma = \omega$, which is widely applicable for polymer melts and solutions, can be applied to dispersion rheology only at low ω and/or ϕ.

Figure 1.17 shows the frequency dependence of the elastic modulus G' and viscous modulus G'' for electrostatically stabilized suspensions at three different particle volume fractions. At low volume fraction in the liquid state $G'' \approx \omega$ dominates over $G' \approx \omega^2$, as expected for viscoelastic liquids. In the two-phase region G' and G'' are essentially equal and increase weakly according to $G^* \approx \omega^\alpha$ (power law exponent $\alpha < 1$). In the highly concentrated gel-like or crystalline state $G' \gg G''$ and both moduli are more or less independent of frequency ω.

Let us now consider the rheology of sterically stabilized dispersions. Particle repulsion in sterically stabilized dispersions results from the interactions between polymer chains or surfactant molecules adsorbed or grafted onto the particle surface. The formation of a hairy surface layer gives rise to an increase in the

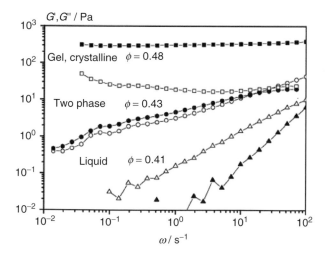

Figure 1.17 G' and G'' as a function of angular frequency ω for a concentrated electrostatically stabilized dispersion at volume fractions around the phase transition region. The filled symbols denote G', open symbols G''.

hydrodynamic particle radius and a dispersions zero-shear viscosity, in a similar way to the case of charged particles. Hence, the rheological behavior of such systems is similar to that of hard spheres with $a_{eff} \approx a + L$. In particular, if the particle radius a is small the stabilizing layer can contribute significantly to the effective volume $\phi_{eff} = \phi(1 + L/a)^3$; and thus give rise to a strong viscosity increase. When using polyelectrolytes or ionic stabilizers with weak functional acid groups, the thickness of the stabilizer layer L depends on the ionic strength and pH of the dispersion medium, which determine the degree of dissociation and range of electrostatic interactions among the functional acid groups. The steric repulsion provided by this surface layer, which is activated and tuned by short-range electrostatic interactions, is called *electrosteric stabilization* and is an important mechanism for stabilization of commercial polymer dispersions. As was the case for charged particles, electrosterically stabilized dispersions show universal scaling independent of ionic strength, pH, or core particle size, but here the data have to be rescaled versus ϕ_{eff} not only for the zero-shear viscosity η_0 but also for the high shear viscosity η_∞ and high frequency viscosity η'_∞. However, the hairy particles show the same η'_∞ as predicted for hard sphere dispersions up to $\phi_{eff} = 0.5$. Beyond this effective volume fraction strong deviations are observed due to the permeability and interpenetration of the stabilizing layers [30].

1.4.2.2 Attractive Particles

Attractive particle interactions either result in large compact aggregates, which rapidly phase separate, or in loose aggregates with fractal structure. Only the latter case is relevant from a rheological point of view. Loose aggregates immobilize water, leading to a larger effective volume fraction ϕ_{eff} and thus to an increase in the zero-shear viscosity. When the shear rate is increased the flocs gradually breakdown and/or align in the flow direction, resulting in a viscosity reduction. Aggregate break-up in dilute dispersions can be estimated by the balance between hydrodynamic forces $F_H = 6\pi\eta_s a^2 \dot{\gamma}$ and the van der Waals force $F_{vdW} = aA_H/12h^2$ (where A_H is the Hamaker constant and h interparticle separation distance). Hence, in the colloidal domain ($a < 1\,\mu m$) very large shear rates are required to break-up the aggregates into primary particles.

The fractal structure of aggregates is characterized by the fractal dimension D_f, which characterizes the mass density of the flocs and is controlled by the aggregation mechanism. The lower the D_f value, the more open the aggregate structure is. Reaction limited- and flow-induced aggregation lead to denser structures, while diffusion limited aggregation results in low D_f values, as confirmed by computer simulation and scattering experiments [31–33]. Above a critical volume fraction fractal aggregates can interconnect, forming a sample-spanning network, which results in a highly elastic gel-like behavior ($G' > G''$) and an apparent yield stress. The rest structure ruptures at a critical stress level and viscosity progressively decreases with increasing applied stress. The shear induced breakdown and recovery of flocs may require a finite amount of time, resulting in thixotropic behavior.

Different flocculation mechanisms in disperse systems can be recognized:

- *Flocculation of charged particles* can be caused by increasing the ionic strength and or lowering the surface charge. Particles can then aggregate in the primary or the secondary minimum of the potential energy. The latter gives rise to fairly weak aggregates and a shear force can easily separate the particles again.
- *Flocculation of sterically stabilized particles* depends on the thickness of the stabilizing layer. Particles aggregate, when the stabilizing layer is not thick enough to screen the van der Waals attraction; as a rule of thumb, the thickness of the stabilizing layer should be $L \approx a/10$. This layer thickness strongly depends on the solvent quality of the continuous phase, and may often be widely tuned by variation of temperature. Systems with an upper or lower critical solution temperature are described in the literature.
- *Depletion flocculation* results from the osmotic pressure induced by the addition of non-adsorbing polymers. Attractive interactions in this case are easily tunable by size and concentration of added polymer.
- *Bridging flocculation* occurs on dissolving high-molecular weight polymers with a strong affinity to particle surface that attach to at least two particles. Strong bridging-flocculated gels may be formed at high particle volume fraction when the particle surface separation is small. Typically, the molecular weight of the polymers is on the order of 10^6 g mol^{-1} so that they can bridge the gap between particles without losing too much conformational entropy.
- *Flocculation by capillary forces*: the addition of small amounts of a secondary fluid, immiscible with the continuous phase of the suspension, causes agglomeration due to the capillary bridges and creates particle networks even at low particle volume fraction.

Investigations of the rheology of strongly flocculated gels are difficult because of the poor reproducibility of sample preparation, sensitivity to shear history, and preparation conditions. On the other hand, weak or reversible flocculation allows for breakup and re-formation of aggregates due to thermal forces and the structure may reach a metastable thermodynamic state.

Rheology of Weakly Flocculated Gels Suspensions in which particles are reversibly captured in a shallow primary or secondary minimum [typically $(-\Psi_{min}/k_B T) < 20$, where Ψ_{min} is the minimum of interaction potential] are classified as weakly flocculated gels. To demonstrate some features of the rheology of these weakly flocculated gels let us consider the results of the investigations of depletion flocculated suspensions and the thermoreversible gelation of sterically interacting particle suspensions. Figure 1.18a shows the shear rate dependence of the relative viscosity of colloidal dispersions of octadecyl grafted silica spheres in benzene ($\phi = 0.367$) at several temperatures [34]. When the temperature is decreased below the theta temperature (316 K) weak aggregates are formed, leading to an increase in viscosity and shear thinning behavior. Buscall *et al.* [35] studied sterically stabilized acrylic copolymer particles dispersed in "white spirit" (mixture of high-boiling hydrocarbons). Adding non-adsorbing polyisobutylene above the critical free polymer concentration for depletion flocculation causes a dramatic increase in viscosity with increasing polymer concentration (Figure 1.18b).

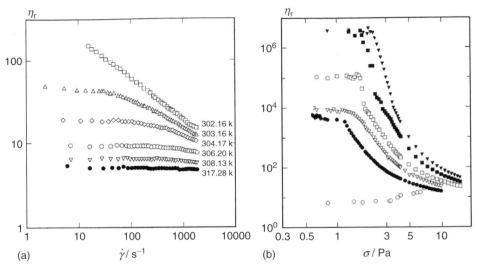

Figure 1.18 (a) Relative shear viscosity versus shear rate for a dispersion of octadecyl grafted silica spheres in benzene ($\phi = 0.367$) at several temperatures [34]; (b) relative viscosity versus shear stress σ for a dispersion of acrylic copolymer particles ($a = 157\,\text{nm}$) grafted with hydroxystearic acid–poly(methyl methacrylate) and dispersed in "white spirit" at volume fraction $\phi = 0.4$ with added polyisobutene ($M_w = 411\,000\,\text{g mol}^{-1}$) of different concentrations in weight per volume: 0.1, 0.4, 0.5, 0.6, 0.85, and 1% (from bottom to top) [35].

Weakly flocculated systems are also characterized by an apparent yield stress. Tadros [1] investigated depletion flocculated aqueous PS dispersions containing free poly(ethylene oxide) (PEO) chains. It was found that the yield stress σ_y increases linearly with increasing PEO concentration ϕ_p and the slope of this linear dependence increases with increasing particle volume fraction ϕ (Figure 1.19). The following scaling relation applies:

$$\sigma_y \sim \phi^p \tag{1.38}$$

where the power-law exponent p depends on the fractal dimension and is around 3 according to experimental investigations, while numerical simulations report higher values: 3.5–4.4, depending on whether the aggregation is slow or rapid.

Figure 1.20 shows the elastic modulus G' of the depletion-flocculated aqueous PS dispersions as a function of the free polymer (PEO) volume fraction ϕ_p at several particle volume fractions. Above the critical free polymer concentration G' increases with increasing ϕ_p since the aggregates grow; G' then reaches a plateau value as soon as a sample-spanning network is formed. Furthermore, it can be seen that at any given ϕ_p the elastic modulus G' increases with increasing particle volume fraction.

Rheology of Strongly Flocculated Gels   Suspensions in which particles are captured in a deep primary or secondary minimum with $(-\Psi_{min}/k_B T) > 20$ are classified as

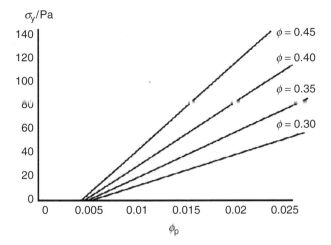

Figure 1.19 Yield stress σ_y versus free polymer (PEO, $M_w = 20\,000$ g mol^{-1}) volume fraction ϕ_p for a polystyrene dispersion at several particle volume fractions ϕ [1].

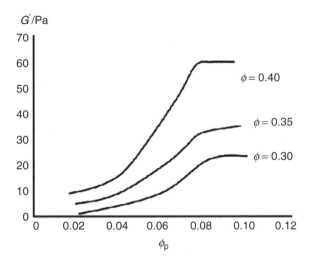

Figure 1.20 Elastic modulus G' versus free polymer (PEO, $M_w = 20\,000$ g mol^{-1}) volume fraction ϕ_p for polystyrene dispersion ($a = 77.5$ nm) at three different particle volume fraction ϕ [1].

strongly flocculated gels. Such systems are not at equilibrium and hence difficult to investigate experimentally. Nevertheless, several studies [36–38] examined the rheological properties of strongly flocculated gels and found some typical trends for these materials. Strongly flocculated gels are highly elastic ($G' \gg G''$) at small amplitudes and have an extremely limited range of viscoelastic response. Above a critical amplitude γ_c the elastic modulus G' rapidly decreases since the flocculated structure breaks down. For strongly flocculated systems γ_c is much lower than for stable dispersions with repulsive interactions or for polymer melts and solutions. The frequency independent elastic modulus G' of strongly flocculated gels is found to be independent of particle size but strongly increases with particle volume fraction ϕ according to the following scaling law:

$$G' \sim \phi^{\alpha} \tag{1.39}$$

where the exponent α varies between 2 and 6 depending on the aggregation conditions. If aggregation is slow (reaction limited) dense structures are formed and gel formation sets in at a higher particle volume fraction and, correspondingly, α is high. Figure 1.21 shows the volume fraction dependence of the G' plateau modulus for a sterically stabilized PS latex dispersion at various concentrations of sodium sulfate (Na_2SO_4) [39]. The stable dispersion shows a strong increase of G' within a narrow concentration range above $\phi = 0.5$, with an exponent $\alpha \approx 30$. At Na_2SO_4 concentrations above the critical flocculation concentration α suddenly decreases and reaches the value of 2.2 at 0.5 M Na_2SO_4, indicating that an open sample-spanning network structure is formed at a particle volume fraction as low as $\phi = 0.35$.

Strongly flocculated dispersions are very sensitive to shear and are characterized by an apparent yield stress. The yield stress σ_y of an aggregated dispersion can be related to the adhesion force F_{adh} between two particles [40]:

$$\sigma_y = \frac{F_{adh}}{a^2} f(\phi) \tag{1.40}$$

The term F_{adh}/a^2 is the stress per particle and for low particle concentration the function $f(\phi)$, referring to the number of particle contacts, can be approximated as the number of binary contacts, that is, $f(\phi) = \phi^2$. The adhesion force is given by the van der Waals attraction, that is, $F_{adh} \approx a$ and thus:

$$\sigma_y \sim \frac{\phi^2}{a} \tag{1.41}$$

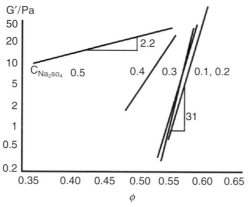

Figure 1.21 Elastic modulus G' versus particle volume fraction ϕ for a sterically stabilized suspension with grafted PEO flocculated by adding Na_2SO_4 [39].

More elaborate models for $f(\phi)$ and the DLVO interaction potential result in the same scaling law and provide good estimates for the absolute value of σ_y and G'. However, various experimental studies [38, 39] have revealed a different scaling for the yield stress of strongly flocculated particulate gels with particle size and volume fraction:

$$\sigma_y \sim \frac{\phi^J}{a^2} \tag{1.42}$$

Capillary Forces in Suspension Rheology   Recently, Koos and Willenbacher [41] reported that the addition of small amounts of a secondary fluid, immiscible with the continuous phase of the suspension, can dramatically change the rheological properties of suspensions. Capillary forces between particles lead to the formation of a sample-spanning network structure resulting in a transition from predominantly viscous to gel-like behavior. This phenomenon is observed for various different fluid/particle systems, independent of whether the primary liquid or the secondary immiscible liquid preferentially wet the solid particles. When the secondary fluid creates isolated capillary bridges between particles the observed gel-like state is termed the *"pendular" state*, analogous to the pendular state in wet granular media (see Chapter 2, in Volume 1 [28]). Even if the second, immiscible fluid does not preferentially wet the solid particles it can still attach to the particles and cause agglomeration due to the negative curvature of the solid/liquid interface. This state is analogous to the capillary state in wet granular media close to the saturation limit. Figure 1.22 shows two examples demonstrating the effect of the fraction of

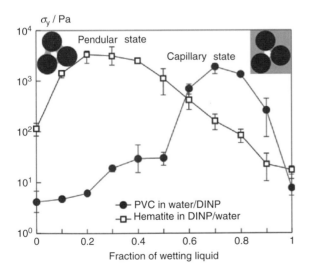

Figure 1.22 Yield stress versus fraction of wetting liquid *S*. For the aqueous PVC dispersion with addition of DINP the yield stress shows a maximum in the capillary state. On adding water to the suspension of hematite particles in DINP the yield stress shows a maximum in the pendular state.

wetting liquid on the yield stress for both the pendular and the capillary state. The increase in yield stress is greatest in the capillary state for the aqueous poly(vinyl chloride) (PVC) dispersion with diisononyl phthalate (DINP) as a secondary fluid. In contrast, the maximum in the yield stress for the dispersion of hematite particles in DINP is in the pendular state where water is the secondary fluid.

SAOS measurements of suspension in the capillary state clearly demonstrate the transition between the weakly elastic, predominantly viscous to highly elastic, gel-like behavior with increasing amount of secondary fluid. Figure 1.23 shows the frequency dependence of the complex shear modulus G^* for hydrophobically modified calcium carbonate ($CaCO_3$) particles suspended in a silicone oil with different amounts of added water as a secondary fluid. Without the secondary fluid the magnitude of the complex shear modulus $|G^*|$ increases with increasing frequency, whereas on addition of only 0.2% wt. water the complex shear modulus G^* becomes frequency independent. This transition in the rheological properties of a suspension upon adding small amount of a secondary fluid is directly evident from the images shown in Figure 1.24. Note that this phenomenon has been observed at a particle volume fraction as low as about 10%.

This phenomenon has important potential technical applications. The formation of a strong sample-spanning network prevents sedimentation. Furthermore, it changes the rheological properties of the system, which is a reversible process and may be tuned by temperature or addition of surfactant. Another field of application is to use such suspensions as precursors for porous materials. The strong capillary forces prevent the collapse of the network structure upon removal of the liquid phase. A solid PVC foam has been already produced under laboratory conditions, using PVC particles ($\phi = 0.2$) dispersed in water, with DINP as a secondary fluid [41].

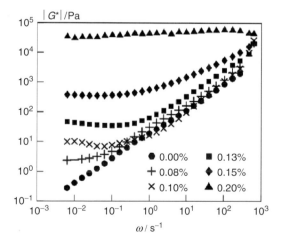

Figure 1.23 Magnitude of complex shear modulus $|G^*|$ versus frequency ω for hydrophobically modified $CaCO_3$ particles ($a = 800$ nm, $\phi = 0.173$) dispersed in a silicone oil, with the addition of various amounts of water.

Fluidization of Highly Concentrated Dispersions Highly concentrated dispersions with a particle volume fraction above the colloidal glass transition ϕ_g behave as gel-like materials with finite plateau modulus G_0. A classical method to keep highly concentrated dispersions fluid and to minimize their viscosity is to shift the maximum packing fraction by mixing of particles of different size (Section 1.4.3). However, in this section we will consider an alternative concept of fluidizing dense colloidal dispersions, based on the so-called re-entry glass transition in colloidal dispersions [42–45].

Weak attractive interactions, for example, introduced by the depletion effect of non-adsorbing polymers dissolved in the continuous phase, can shift the colloidal glass transition ϕ_g to significantly higher values (up to $\phi \approx 0.7$), which can be used to make freely flowing but highly concentrated dispersions. Figure 1.25 shows the viscosity reduction upon addition of a non-adsorbing polymer to an aqueous dispersion of a hard-sphere like polystyrene–(butyl acrylate), P(S-BA), dispersion at a particle volume fraction above the colloidal glass transition [46]. It can be seen

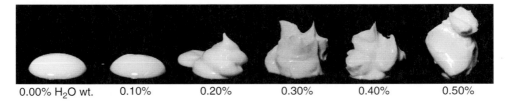

| 0.00% H₂O wt. | 0.10% | 0.20% | 0.30% | 0.40% | 0.50% |

Figure 1.24 Transition from weakly elastic, predominantly viscous to highly elastic, gel-like behavior with increasing amount of water added to a suspension of hydrophobically modified $CaCO_3$ ($a = 800$ nm, $\phi = 0.11$) in DINP.

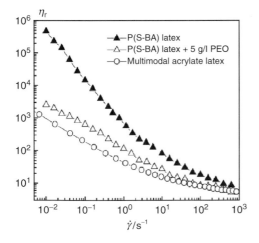

Figure 1.25 Relative viscosity as a function of shear rate for an aqueous polystyrene–(butyl acrylate) P(S-BA) dispersion at $\phi = 0.64$ with and without added PEO ($M_w = 20\,000\,\mathrm{g\,l^{-1}}$) in comparison with a commercial polymer dispersion (acrylate latex) with a broad multimodal size distribution [46].

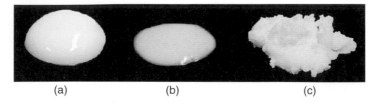

Figure 1.26 Changes in the texture of an aqueous P(S-BA) latex (a) at $\phi = 0.64$ upon addition of different concentrations of PEO ($M_w = 4000\,g\,mol^{-1}$): (b) 5 and (c) $10\,g\,l^{-1}$.

that the low-shear viscosity decreases by two orders of magnitude upon addition of non-adsorbing polymer and the effect is comparable to that resulting from broad multimodal particle size distribution. The fluidization of an aqueous latex dispersion due to added non-adsorbing polymer is also evident in Figure 1.26, which shows images of the suspension with different polymer concentration placed on a glass plate. On adding different amounts of PEO to the aqueous P(S-BA) dispersion the texture of the sample changes from gel-like, due to the particle caging at this concentration (repulsive glass), to fluid like and again to gel-like but now due to particle bonding (attractive glass).

1.4.3
Effect of Particle Size Distribution

Numerous experimental studies have been performed using bimodal and multimodal model systems and various phenomenological models have been developed to describe the effect of particle size distribution on viscosity. Typically, a significant viscosity reduction due to mixing particles with different size is observed at particle volume fractions $\phi > 0.5$ and the effect increases with increasing ϕ. For bimodal systems the viscosity at a given particle loading goes through a pronounced minimum at a relative fraction of small particles $\xi_s \approx 0.3$. This viscosity reduction phenomenon is observed for dispersions of non-Brownian as well as Brownian hard spheres. Typical examples are presented in Figure 1.27a,b. Viscosity reduction has been observed at particle size ratios as low as $\chi = 1.7$ and for hard sphere suspensions the effect increases with increasing $\chi = a_{large}/a_{small}$. This is no longer true if repulsive colloidal interactions become relevant. At a fixed size of the large particles an increasing χ value corresponds to a decreasing size of small particles a_{small} and if the range of the repulsive interactions is constant this corresponds to an increasing ϕ_{eff}. As a consequence the viscosity goes through a minimum and then increases again if the size ratio χ is increased at a constant total particle concentration and a fixed fraction of small particles. This is shown schematically in Figure 1.28. Willenbacher and coworkers [47, 48] have investigated this phenomenon intensively using a large number of polymer dispersions with different particle size ratio and different range of repulsive interaction. They could show that for typical commercial dispersions with short-range repulsive interactions the viscosity reduction effect is most pronounced at a size ratio $\chi = 4-5$. Furthermore,

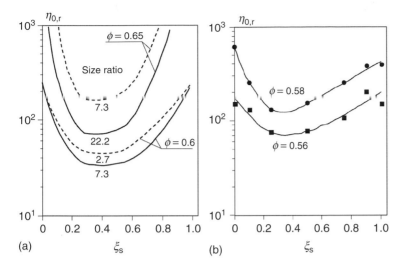

Figure 1.27 Relative zero-shear viscosity $\eta_{0,r}$ versus small particle volume fraction ξ_s: (a) for a suspension of non-Brownian hard spheres at different size ratios (b) For a suspension of Brownian particles with size ratio $\chi = 1.7$ at different particle concentration ϕ (0.58 and 0.56). Redrawn from Rodriguez *et al.* [50].

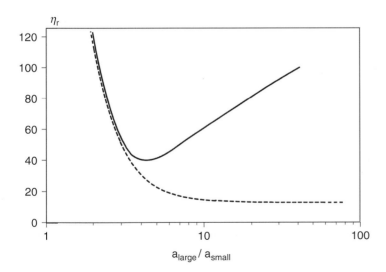

Figure 1.28 Schematic drawing of relative viscosity as a function of particle size ratio calculated according to Equation 1.43 for large particle radius $a_{large} = 400$ nm, total particle concentration $\phi = 0.6$, and small particle volume fraction $\xi_s = 0.25$. The dashed line shows the results for $\varepsilon = 2$, that is, hard sphere dispersions, and the solid line represents the results for ε as a function of average particle size. Adapted from Dames *et al.* [47].

they proposed a generalized Quemada model:

$$\eta = \tilde{\eta}\left(1 - \frac{\phi}{\phi_{max}}\right)^{-\varepsilon} \tag{1.43}$$

with a shear rate dependent pre-factor $\tilde{\eta}$ and they could show that ϕ_{max} can be calculated solely from the particle size distribution according to a phenomenological model derived from a large set of data for non-colloidal hard sphere packing [51], and the colloidal interactions are parameterized by the exponent $\varepsilon \geq 2$. The exponent ε is equal to 2 in the hard sphere limit and increases with decreasing mean particle size. This is attributed to the fact that colloidal interactions among particles become more important as the mean particle separation diminishes and viscosity diverges at lower volume fractions than expected for hard spheres.

The phenomenon of viscosity reduction due to bi- or multimodal particle size distribution is often attributed to an optimized packing that fits small particles into the interstitial volume between the large particles. Along these lines Farris [52] has developed a model for bimodal dispersions with size ratio $\chi > 10$, treating the small particles together with the solvent as a homogeneous fluid with an effective viscosity and assuming that small and large particles do not interact. This model predicts a viscosity minimum at a small particle fraction $\xi_s = 0.27$, which is in good agreement with many experimental observations. But, on the other hand, a minimum value of $\chi_c = 6.46$ [53] is required to fit a small particle into the interstitial volume within a tetrahedron of large particles and for $\chi = \chi_c$ this packing concept corresponds to a fraction of small particles $\xi_s < 0.01$, which is by far not sufficient to induce a viscosity reduction. However, a small particle volume fraction of $\xi_s \approx 0.3$, which is needed to induce a significant viscosity reduction, corresponds to a number ratio $N_{small}/N_{large} \approx 100$ at a size ratio around χ_c. These considerations demonstrate that simple packing considerations are not sufficient to explain the observed phenomena. Accordingly, the formation of ordered superlattice structures or phase separation effects have also been discussed, but a satisfying theory explaining the effect of particle size distribution on viscosity is still lacking.

1.4.4
Shear Thickening

Shear thickening describes the phenomenon of increasing viscosity with increasing shear rate or shear stress. This phenomenon has been observed for a wide variety of colloidal and non-colloidal particle suspensions. Shear thickening becomes important at high shear rates and occurs beyond a critical volume fraction (Figure 1.29a). The thickening effect increases with particle loading and depends on particle size, particle size distribution, and interactions among particles [54].

Early rheological and light scattering results [56, 57] suggested that the shear thickening phenomenon is due to a shear induced order–disorder transition and the shear thinning observed at intermediate shear rates is attributed to the formation of a layered structure. Repulsive interactions are assumed to stabilize this layered structure. At sufficiently high shear rates spatial fluctuations of particle

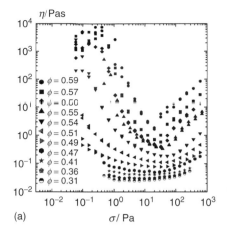

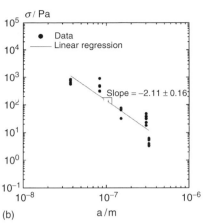

Figure 1.29 Shear thickening of charge stabilized silica dispersions with various particle radii (75, 167, 300, 600, and 1000 nm) and particle volume fraction ranging between 0.31 and 0.59 [55]. (a) Viscosity versus shear stress; (b) critical shear stress σ_c versus particle radius a. The line fits the power law dependence $\sigma_c \approx a^{-2}$. Adapted from Maranzano and Wagner [55].

position destabilize the ordered flow, which results in a strong increase in viscosity. The onset of shear thickening is related to a critical shear rate, above which the hydrodynamic lubrication forces exceed the repulsive colloidal forces [58]. However, comprehensive rheo-optical and small-angle neutron scattering (SANS) experiments [59–62] have revealed that the shear thickening phenomenon may or may not be accompanied by an order–disorder transition but this transition is not a necessary condition. Instead, these investigations clearly revealed that shear thickening is due to the formation of so-called hydroclusters, which form under the action of hydrodynamic forces pushing particles together and instantaneously disintegrate upon cessation of flow. The formation of clusters shows up in turbidity and flow birefringence and has also been confirmed by stress jump experiments [63] as well as Stokesian dynamics simulations of hard sphere dispersion flow [64–66]. Furthermore, Chow and Zukoski [67] investigated the shear thickening behavior of electrostatically stabilized particles in very thin rheometer gaps and found that the critical shear rate for shear thickening increases with increasing the gap size, indicating formation of gap-spanning clusters. The increase in viscosity is attributed to the anisotropic shape of the clusters and the enhanced effective particle volume fraction due to trapped solvent. The hydroclusters can collide with each other and thus "jam" the flow, leading to discontinuous shear thickening at a critical shear stress. If the particle volume fraction is not high enough, hydrocluster formation does not lead to jamming and the shear thickening effect is less pronounced. The formation of hydroclusters is controlled by the balance of hydrodynamic force needed to push particles together and the repulsive thermodynamic forces. Accordingly, a critical stress σ_c for the onset of shear thickening is predicted that scales as $\sigma_c \approx a^{-2}$ for electrostatically stabilized systems, which is consistent with experimental results [55, 58, 61] (Figure 1.29b). This scaling has also been observed for sterically stabilized dispersions [68]. Note that σ_c is almost independent of particle volume fraction ϕ, while the corresponding critical shear rate $\dot{\gamma}_c = \sigma_c / \eta(\phi)$ decreases with increasing ϕ.

Shear thickening can be suppressed or shifted to higher critical stresses by a broad particle size distribution [54]. It has been shown that for bimodal mixtures with size ratio $\chi \approx 3$ the critical shear stress σ_c increases with increasing fraction of small particles ξ_s [55, 61]. Particle shape also has a strong influence on the shear thickening behavior. Beazley [69] demonstrated that anisotropic clay suspensions exhibit shear thickening behavior at lower volume fractions and the effect increases with increasing aspect ratio. Bergstrom [70] investigated aqueous suspensions of rod-shaped silicon carbide whiskers with aspect ratio $r_p \approx 10$ and reported shear thickening behavior at volume fraction as low as 17%. More recently, Egres and Wagner [71] investigated systematically the effect of particle anisotropy on shear thickening using a poly(ethylene glycol) based suspensions of acicular precipitated calcium carbonate (PCC) particles with aspect ratio varying between 2 and 7. Two important results have been pointed out: the critical volume fraction for the onset of shear thickening decreases with increasing aspect ratio but the critical shear stress σ_c is independent of the aspect ratio and follows the scaling laws proposed for hard sphere dispersions with a size corresponding to the minor axis dimension.

1.5
Rheology of Emulsions

The rheology of emulsions exhibits many qualitative analogs to the rheology of solid spherical particle dispersions. Differences arise from the deformability of liquid drops, which is especially relevant at high shear rates and/or high volume fraction of the disperse phase. However, even at low shear rates and low droplet concentrations the relative viscosity of emulsions differs from that of solid sphere dispersions. This is due to circulation of the flow inside the droplets, which leads to deformation of the external streamlines around the fluid spheres such that the flow is less disturbed and viscous dissipation is lower [72]. The degree of this effect depends on the viscosity ratio M:

$$M = \frac{\eta_d}{\eta_s} \tag{1.44}$$

where η_d is the viscosity of the droplet liquid. For high droplet viscosity the viscosity ratio M approaches infinity and the distortion of the stream lines approaches that of rigid spheres. This effect is measurable even in very dilute emulsions and is captured by the Taylor equation [73]:

$$\eta = \eta_s \left[1 + \left(\frac{1 + 2.5M}{1 + M} \right) \phi \right] \tag{1.45}$$

which reduces to the Einstein equation (Equation 1.24) for $M \to \infty$. Taylor's hydrodynamic theory assumes no deformation of droplets, which is satisfied at low enough shear rates. In typical oil-in-water (O/W) emulsions the interfacial tension Γ is high enough to counteract the effect of hydrodynamic forces and leads to fast shape relaxation. The droplet relaxation time τ_d is given by:

$$\tau_d = \frac{a\eta_s}{\Gamma} \tag{1.46}$$

and droplet deformation is not relevant for emulsion rheology as long as $\dot{\gamma} < \tau_d^{-1}$. The balance between surface tension and shear forces is often expressed by the dimensionless capillary number (Ca):

$$Ca = \frac{a\eta_s\dot{\gamma}}{\Gamma} \tag{1.47}$$

Droplet deformation and rupture occur at Ca > 1. A closer look at the phenomenon reveals that the critical Ca at which droplet rupture occurs depends strongly on the viscosity ratio M and can vary by orders of magnitude [74, 75]. Flow kinematics also plays a role and, generally, droplet rupture is easier in elongational than in shear flow.

Experimental results on model emulsions of different viscosity ratios M, reported by Nawab and Mason [76] demonstrated excellent agreement with Taylor's hydrodynamic theory (Figure 1.30). Nawab and Mason pointed out that in some cases adsorbed surfactant layers can reduce the internal circulations and thereby cause an increase of intrinsic viscosity to the rigid sphere limit.

With increasing concentration above the Einstein limit, hydrodynamic interactions become significant and Taylor's equation cannot describe the volume fraction–viscosity dependence. Pal [77] has proposed a phenomenological viscosity equation for concentrated emulsions that takes into account the effect of viscosity ratio M and reduces to the generalized Krieger–Dougherty equation (Equation 1.26)

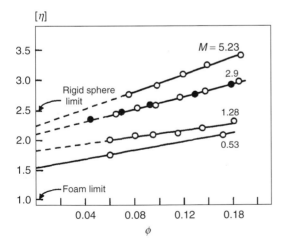

Figure 1.30 Intrinsic viscosity $[\eta]$ versus droplet volume fraction ϕ for monodisperse emulsions of butyl benzoate oil droplets in different water solutions in order to vary the viscosity ratio M [76]. Taken from Macosko [8]. Copyright © 1994 John Wiley & Sons.

when $M \to \infty$:

$$\eta_r \left[\frac{2\eta_r + 5M}{2 + 5M} \right]^{3/2} = \left[1 - \frac{\phi}{\phi_{max}} \right]^{-2.5\phi_{max}} \tag{1.48}$$

This equation as based on a large set of experimental data for emulsions covering a broad range of droplet volume fractions ϕ and viscosity ratios M.

The effect of dispersed phase volume fraction on rheology of typical technical emulsions is less severe in comparison to colloidal dispersions. Since droplet size is usually in the micron range technically relevant shear rates correspond to very high Pe numbers and the measured viscosity data correspond to the upper Newtonian plateau regime. Viscosity is further reduced at high droplet volume fractions due to the usually broad droplet size distributions. As a consequence such emulsions behave as Newtonian fluids up to volume fractions close to dense packing ($\phi \approx 0.6$) [78]. Emulsions with an average droplet radius in the range of several 100 nm exhibit a flow behavior resembling very much that of colloidal hard sphere suspensions. Note that increasing the volume fraction of the dispersed phase does not necessarily result in a monotonic increase in viscosity. At a critical droplet volume fraction, phase inversion may occur that is accompanied by a drastic drop in viscosity. However, emulsions are usually stabilized by surfactants adsorbed onto the droplet surface that prevent the coalescence of droplets at contact.

Repulsive and attractive colloidal interactions as well as droplet deformation and rupture during flow can cause a deviation from the hard sphere behavior of emulsions. The effect of repulsive droplet interactions due to surface charge or adsorbed polymer can be captured by hard sphere mapping ($\phi \to \phi_{eff}$) similar to that for suspensions of repulsive solid particles. Attractive droplet interactions lead to flocculation and gelation analogously to attractive particle suspensions. Emulsion rheology can be tuned over a wide range by adding thickeners to the continuous phase or by excess surfactant providing self-assembling gel-like structure to the continuous phase, which is particularly relevant for stabilization against creaming.

Emulsions can exhibit distinct viscoelastic properties even if both constituents are Newtonian fluids due to the contribution of the interfacial tension, which opposes droplet deformation. This is particularly important for polymer blends, where the viscosity of both components is high and deformed interfaces relax slowly. Various models have been established to describe the complex shear modulus G^* of emulsions. When both phases are Newtonian the Oldroyd model [79, 80] suits:

$$G^* = i\omega\eta_s \left(\frac{1 + \frac{3}{2}\phi\frac{E}{D}}{1 - \phi\frac{E}{D}} \right) \tag{1.49}$$

with:

$$E = 2i\omega(\eta_d - \eta_s)(19\eta_d + 16\eta_s) + \frac{8\Gamma}{a}(5\eta_d + 2\eta_s)$$
$$D = i\omega(2\eta_d + 3\eta_s)(19\eta_d + 16\eta_s) + \frac{40\Gamma}{a}(\eta_d + \eta_s) \tag{1.50}$$

For emulsions with viscosity ratio $M \to \infty$, droplets behave like solid particles and the droplet relaxation time is so short that the ratio E/D reduces to:

$$\frac{E}{D} = \frac{0.4 + M}{1 + M} \qquad (1.51)$$

In the dilute limit with $\phi \to 0$, Equation 1.49 simplifies to:

$$G^* = i\omega\eta_s \left(1 + \frac{5}{2}\phi\frac{E}{D}\right) \qquad (1.52)$$

For emulsions where both continuous and dispersed phase are viscoelastic with frequency dependent complex moduli G_s^* and G_d^*, respectively, the Palierne [81] model provides a good description for the complex modulus G^* of the emulsion:

$$G^* = G_s^* \left(\frac{1 + \frac{3}{2}\phi\frac{E}{D}}{1 - \phi\frac{E}{D}}\right) \qquad (1.53)$$

with:

$$E = 2(G_d^* - G_s^*)(19G_d^* + 16G_s^*) + \frac{8\Gamma}{a}(5G_d^* + 2G_s^*)$$

$$D = (2G_d^* + 3G_s^*)(19G_d^* + 16G_s^*) + \frac{40\Gamma}{a}(G_d^* + G_s^*) \qquad (1.54)$$

Kitade *et al.* [82] investigated the viscoelastic properties of polymer blends consisting of polydimethylsiloxane (PDMS) and polyisoprene and demonstrated that the

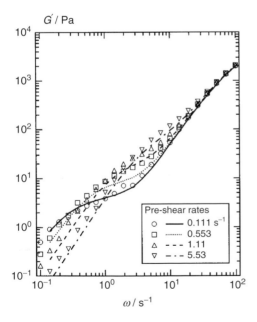

Figure 1.31 Comparison of the Palierne model (lines) with measured $G'(\omega)$ dependence for a blend of 11% polyisoprene ($\eta_0 = 60.9$ Pa s) in PDMS ($\eta_0 = 73.7$ Pa s) with $\Gamma = 3.2$ mN m^{-1}, pre-sheared at four different shear rates [82].

experimentally determined frequency dependence of G' is in agreement with the Palierne model (Figure 1.31). The contribution of the interfacial term results in a pronounced shoulder in the $G'(\omega)$ curve in the low frequency range. Figure 1.31 shows that with increasing pre-shear rate, which corresponds to a decrease of the average droplet size [82], the "shoulder" in the $G'(\omega)$ dependence shifts to higher frequencies. This is due to an increased interfacial area and hence a more pronounced interfacial contribution for smaller droplets. In the high frequency limit the interfacial terms can be ignored and G' is determined only by the viscoelasticity of the dispersion medium. Then, if $G_d^*/G_s^* \approx 1$, the Palierne emulsion model further simplifies to:

$$G^* \approx (1 - \phi)G_s^* + \phi G_d^* \tag{1.55}$$

Emulsions exhibit unique flow properties that are not observable in suspensions when a critical volume fraction ϕ_c is exceeded. For colloidal systems ϕ_c may be associated with the glass transition and for non-Brownian systems with the volume fraction of close packing. At volume fractions $\phi > \phi_c$ dispersions of solid particles can no longer flow. In contrast, emulsions still flow even at $\phi > \phi_c$ since droplets start to deform and take a polyhedral shape. Such emulsions exhibit an apparent yield stress, strong shear thinning, and pronounced elasticity. In the linear viscoelastic regime the storage modulus G' is much larger than G'' and essentially independent of frequency; this G' value is known as the *plateau modulus* G_0. Steady shear flow curves are usually well described by the Herschel–Bulkley model:

$$\sigma = \sigma_y + k\dot{\gamma} \tag{1.56}$$

where k is the consistency parameter and n the power law index. The apparent yield stress, the degree of shear thinning (here expressed in terms of n), and the plateau modulus increase with increasing volume fraction of internal phases and decreasing droplet size. A thermodynamic model developed by Princen [83] related the droplet compression to the osmotic pressure in the system, which increases with increasing droplet volume fraction ϕ. When the osmotic pressure exceeds the Laplace pressure Γ/a droplets start to deform and pack more tightly with increasing ϕ. The elasticity of the system then arises from the surface tension acting to resist the deformation. The plateau modulus G_0 gradually develops when ϕ_c is approached and then increases linearly with effective volume fraction [84]:

$$G_0 = \frac{3\Gamma}{2a}(\phi_{\text{eff}} - \phi_c) \tag{1.57}$$

where ϕ_{eff} accounts for the excluded volume due to repulsive forces. G_0 is also proportional to the Laplace pressure Γ/a and when ϕ_{eff} approaches unity the plateau modulus approaches the limiting value $G_0 \approx \Gamma/2a$.

Densely packed emulsions with $\phi > \phi_c$ are characterized by an apparent yield stress σ_y at which the rest structure breaks down. In oscillatory shear measurements, yielding occurs at a critical deformation amplitude, called the *yield strain* $\gamma_y = \sigma_y/G_0$. For highly concentrated emulsions this yield strain increases linearly with increasing droplet volume fraction [85]:

$$\gamma_y \sim (\phi_{\text{eff}} - \phi_c) \tag{1.58}$$

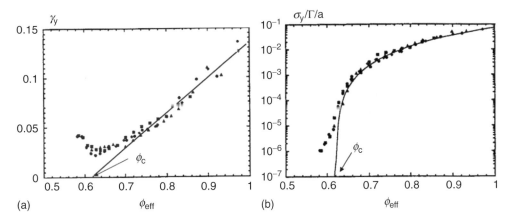

Figure 1.32 (a) Yield strain γ_y versus effective volume fraction ϕ_{eff}; solid line corresponds to Equation 1.58; (b) apparent yield stress σ_y scaled by the Laplace pressure (Γ/a) versus ϕ_{eff} for monodispersed emulsions with droplet size (nm): $a = 250$ (circles), 370 (triangles), 530 (squares), and 740 (diamonds); solid line corresponds to Equation 1.57 [85].

Figure 1.32a demonstrates this linear dependence for monodispersed emulsions having different droplet size. Obviously, the volume fraction dependence of γ_y is independent of droplet size and γ_y reaches its minimum at ϕ_c. The yield stress σ_y can be approximately expressed as $\sigma_y = G_0 \gamma_y$ and together with Equation 1.57 this yields the following relationship:

$$\sigma_y = \frac{3}{2} \frac{\Gamma}{a} (\phi_{eff} - \phi_c)^2 \tag{1.59}$$

which nicely fits the experimental data in Figure 1.32b. For $\phi_{eff} \approx 1$ Equation 1.59 roughly reduces to:

$$\sigma_y(\phi_{eff} = 1) \approx 0.1 \frac{\Gamma}{a} \tag{1.60}$$

These experimental findings are also captured by the Princen–Kiss model [86]:

$$\sigma_y = \frac{\Gamma}{a} \phi^{\frac{1}{3}} Y(\phi) \tag{1.61}$$

This model is based on the affine deformation of a hexagonal structure and $Y(\phi)$ can be expressed in analytical form for two-dimensional systems; however, for three-dimensional emulsions $Y(\phi)$ is an empirical function:

$$Y(\phi) = -0.080 - 0.114 lg(1 - \phi) \tag{1.62}$$

Also distinct in a mathematical sense, the absolute numerical values of the terms $\phi^{1/3}(\phi)$ and $(\phi_{eff} - \phi_c)$ are not very different and ϕ_{eff} does not differ much from ϕ if the layer immobilized by the surfactant is small compared to the droplet size, as for many technically relevant emulsions. Equations 1.59 and 1.61 include the linear relationship between σ_y and the Laplace pressure Γ/a; if the particle size a is known, measuring σ_y or preferentially G_0, since it is accessible with high accuracy

(Equation 1.57), is a valuable tool for determining the interfacial tension Γ, which is otherwise often hard to access.

Highly concentrated emulsions often do not exhibit uniform deformation even in simple shear flow, instead they show shear banding, which can be very irregular in the sense that the plane of deformation changes its position or that the width of the deformed region changes with time [87–90].

References

1. Tadros, T.F. (2010) *Rheology of Dispersions – Principles and Application*, Wiley-VCH Verlag GmbH, Weinheim.
2. James, D.F. and Walters, K. (1993) A critical appraisal of available methods of extensional properties of mobile systems in techniques, in *Rheological Measurements* (ed. A.A. Collyer), Elsevier, London.
3. Willenbacher, N. and Hingmann, R. (1994) Shear and elongational flow properties of fluid S1 from rotational, capillary, and opposed jet rheometry. *J. Non-Newtonian Fluid Mech.*, **52**, 163–176.
4. McKinley, G.H. and Sridhar, T. (2002) Filament-stretching rheometry of complex fluids. *Annu. Rev. Fluid Mech.*, **34**, 375–415.
5. McKinley, G.H. (2005) Visco-elasto-capillary thinning and break-up of complex fluids. *Rheol. Rev.*, **3**, 1–48.
6. Niedzwiedz, K., Buggisch, H., and Willenbacher, N. (2010) Extensional rheology of concentrated emulsions as probed by capillary breakup elongational rheometry (CaBER). *Rheol. Acta*, **49**, 1103–1116.
7. Giesekus, H. and Langer, G. (1977) Die bestimmung der wahren fliesskurven nicht-Newtonscher flüssigkeiten und plastischer stoffe mit der methode der repräsentativen viskosität. *Rheol. Acta*, **16**, 1–22.
8. Macosko, C.W. (1994) *Rheology Principles, Measurements, and Applications*, John Wiley & Sons, Inc., New York.
9. Schümmer, P. and Worthoff, R.H. (1978) An elementary method for the evaluation of the flow curve. *Chem. Eng. Sci.*, **33** (6), 759–763.
10. Bagley, E.B. (1957) End corrections in the capillary flow of polyethylene. *J. Appl. Phys.*, **28**, 624–627.
11. Mooney, M. (1931) Explicit formulas for slip and fluidity. *J. Rheol.*, **2**, 210–222.
12. Pusey, P.N. and van Megen, W. (1986) Phase behaviour of concentrated suspensions of nearly hard colloidal spheres. *Nature*, **320**, 340–342.
13. Einstein, A. (1906) Eine neue bestimmung der moleküldimensionen. *Ann. Phys.*, **19**, 289–306.
14. Einstein, A. (1911) Berichtigung zu meiner arbeit: eine neue bestimmung der moleküldimensionen. *Ann. Phys.*, **34**, 591–592.
15. Batchelor, G.K. (1977) The effect of Brownian motion on the bulk stress in a suspension of spherical particles. *J. Fluid Mech.*, **83**, 97–117.
16. Krieger, I.M. and Dougherty, T.J. (1959) A mechanism for non-Newtonian flow in suspensions of rigid spheres. *Trans. Soc. Rheol.*, **3**, 137–152.
17. Quemada, D. (1977) Rheology of concentrated disperse systems and minimum energy dissipation principle I. Viscosity-concentration relationship. *Rheol. Acta*, **16**, 82–94.
18. Buscall, R., D'Haene, P., and Mewis, J. (1994) Maximum density for flow of dispersions of near monodisperse spherical-particles. *Langmuir*, **10** (5), 1439–1441.
19. van Megen, W., Pusey, P.N., and Bartlett, P. (1990) Phase behavior of dispersions of hard spherical particles. *Phase Transitions*, **21** (2–4), 207–227.
20. van Megen, W. and Underwood, S.M. (1994) Glass transition in colloidal hard spheres: measurement and mode-coupling-theory analysis of the coherent

intermediate scattering function. *Phys. Rev. E*, **49**, 4206–4220.

21. Meeker, S.P., Poon, W.C.K., and Pusey, P.N. (1997) Concentration dependence of the low-shear viscosity of suspensions of hard-sphere colloids. *Phys. Rev. E*, **55**, 5718–5722.

22. Pham, C. D., Russel, W.D., Chang, Z., and Zhu, J. (1996) Phase transition, equation of state, and limiting shear viscosities of hard sphere dispersions. *Phys. Rev. E*, **54**, 6633–6645.

23. Choi, G.N. and Krieger, I.M. (1986) Rheological studies on sterically stabilized model dispersions of uniform colloidal spheres 2. Steady-shear viscosity. *J. Colloid Interface Sci.*, **113** (1), 101–113.

24. Krieger, I.M. (1972) Rheology of monodisperse latices. *Adv. Colloid Interface Sci.*, **3**, 111–136.

25. Brenner, H. (1974) Rheology of a dilute suspension of axisymmetric Brownian particles. *Int. J. Multiphase Flow*, **1** (2), 195–341.

26. Giesekus, H. (1983) in *Physical Properties of Foods* (eds R. Jowitt *et al.*), Applied Science Publishers, London, pp. 205–220.

27. Barnes, H.A., Hutton, J.F., and Walters, K. (eds) (1989) *An Introduction to Rheology*, Elsevier Science, Amsterdam.

28. Bröckel, U., Meier, W., and Wagner, G. (eds) (2007) *Product Design and Engineering*: Volume 1, Basics and Technologies, Wiley-VCH Verlag GmbH, Weinheim.

29. Horn, F.M., Richtering, W., Bergenholtz, J., Willenbacher, N., and Wagner, N.J. (2000) Hydrodynamic and colloidal interactions in concentrated charge-stabilized polymer dispersions. *J. Colloid Interface Sci.*, **225**, 166–178.

30. Fritz, G., Schädler, V., Willenbacher, N., and Wagner, N.J. (2002) Electrosteric stabilization of colloidal dispersions. *Langmuir*, **18** (16), 6381–6390.

31. Wetz, D.A. and Oliveria, M. (1984) Fractal structures formed by kinetic aggregation of aqueous gold colloids. *Phys. Rev. Lett.*, **52** (16), 1433–1436.

32. Weitz, D.A., Huang, J.S., Lin, M.Y., and Sung, J. (1985) Limits of the fractal dimensions for irreversible kinetic aggregation of gold colloids. *Phys. Rev. Lett.*, **54** (13), 1416–1419.

33. Sonntag, R.C. and Russel, W.B. (1986) Structure and breakup of flocs subjected to fluid stresses: I. Shear experiments. *J. Colloid Interface Sci.*, **113** (2), 399–413.

34. Woutersen, A.T.J.M. and de Kruif, C.G. (1991) The rheology of adhesive hard-sphere dispersions. *J. Chem. Phys.*, **94** (8), 5739–5750.

35. Buscall, R., McGowan, J.I., and Morton-Jones, A.J. (1993) The rheology of concentrated dispersions of weakly-attracting colloidal particles with and without wall slip. *J. Rheol.*, **37** (4), 621–641.

36. Buscall, R., Mills, P.D.A., and Yates, G.E. (1986) Viscoelastic properties of strongly flocculated polystyrene latex dispersions. *Colloids Surf.*, **18** (2–4), 341–358.

37. Buscall, R., McGowan, I.J., Mills, P.D.A., Stewart, R.F., Sutton, D., White, L.R., and Yates, G.E. (1987) The rheology of strongly flocculated suspensions. *J. Non-Newtonian Fluid Mech.*, **24** (2), 183–202.

38. Buscall, R., Mills, P.D.A., Goodwin, J.W., and Lawson, D.W. (1988) Scaling behavior of the rheology of aggregate networks formed from colloidal particles. *J. Chem. Soc., Faraday Trans.*, **84** (12), 4249–4260.

39. Tadros, T.F. (1996) Correlation of viscoelastic properties of stable and flocculated suspensions with their interparticle interactions. *Adv. Colloid Interface Sci.*, **68**, 97–200.

40. Larson, R.G. (1999) *The Structure and Rheology of Complex Fluids*, Oxford University Press, New York.

41. Koos, E. and Willenbacher, N. (2011) Capillary forces in suspension rheology. *Science*, **331** (6019), 897–900.

42. Pham, K.N., Puertas, A.N., Bergenholtz, J., Egelhaaf, S.U., Moussaidl, A., Pusey, P.N., Schofield, A.B., Cates, M.E., Fuchs, M., and Poon, W.C.K. (2002) Multiple glassy state in a simple model system. *Science*, **296** (5565), 104–106.

43. Pham, K.N., Egelhaaf, S.U., Pusey, P.N., and Poon, W.C.K. (2004)

Glasses in hard spheres with short-range attraction. *Phys. Rev. E*, **69** (1), 011503.1–011503.13.

44. Eckert, T. and Bartsch, E. (2002) Re-entrant glass transition in a colloid-polymer mixture with depletion attractions. *Phys. Rev. Lett.*, **89** (12), 125701–125704.

45. Eckert, T. and Bartsch, E. (2004) Glass transition dynamics of hard sphere like microgel colloids with short-ranged attractions. *J. Phys. Condens. Matter*, **16**, S4937–S4950.

46. Willenbacher, N., Vesaratchanon, J.S., Thorwarth, O., and Bartsch, E. (2011) An alternative route to highly concentrated, freely flowing colloidal dispersions. *Soft Matter*, **7**, 5777–5788.

47. Dames, B., Morrison, B.R., and Willenbacher, N. (2001) An empirical model predicting the viscosity of highly concentrated, bimodal dispersions with colloidal interactions. *Rheol. Acta*, **40** (5), 434–440.

48. Willenbacher, N., Börger, L., Urban, D., and Varela de la Rosa, L. (2003) Tailoring PSA-dispersion rheology for high-speed coating. *Adhesives Sealants Ind.*, **10** (9), 25–35.

49. Chong, J.S., Christiansen, E.B., and Baer, A.D. (1971) Rheology of concentrated suspensions. *J. Appl. Polym. Sci.*, **15** (8), 2007–2021.

50. Rodriguez, B.E., Kaler, E.W., and Wolfe, M.S. (1992) Binary mixtures of monodisperse latex dispersions 2. Viscosity. *Langmuir*, **8** (10), 2382–2389.

51. Sudduth, R.D. (1993) A generalized model to predict the viscosity of solutions with suspended particles I. *J. Appl. Polym. Sci.*, **48** (1), 25–36.

52. Farris, R.J. (1968) Prediction of the viscosity of multimodal suspensions from unimodal viscosity data. *Trans. Soc. Rheol.*, **12** (2), 281–301.

53. McGeary, R.K. (1961) Mechanical packing of spherical particles. *J. Am. Ceram. Soc.*, **44** (10), 513–522.

54. Barnes, H.A. (1989) Shear-thickening ("Dilatancy") in suspensions of nonagregating solid particles dispersed in Newtonian liquids. *J. Rheol.*, **33** (2), 329–366.

55. Maranzano, B.J. and Wagner, N.J. (2001) The effect of particle size on reversible shear thickening of concentrated colloidal dispersions. *J. Chem. Phys.*, **114** (23), 10514–10527.

56. Hoffman, R.L. (1972) Discontinuous and dilatant viscosity behavior in concentrated suspensions I. Observation of a flow instability. *Trans. Soc. Rheol.*, **16** (1), 155–173.

57. Hoffman, R.L. (1974) Discontinuous and dilatant viscosity behavior in concentrated suspensions II. Theory and experimental tests. *J. Chem. Phys.*, **46** (3), 491–506.

58. Boersma, W.H., Laven, J., and Stein, H.N. (1990) Shear thickening (dilatancy) in concentrated suspensions. *AIChE J.*, **36** (3), 321–332.

59. Laun, H.M., Bung, R., and Schmidt, F. (1991) Rheology of extremely shear thickening polymer dispersions (passively viscosity switching fluids). *J. Rheol.*, **35** (6), 999–1034.

60. Laun, H.M., Bung, R., Hess, S., Loose, W., Hahn, K., Hadicke, E., Hingmann, R., Schmidt, F., and Lindner, P. (1992) Rheological and small angle neutron scattering investigation of shear-induced particle structures of concentrated polymer dispersions. *J. Rheol.*, **36** (4), 743–787.

61. Bender, J. and Wagner, N.J. (1996) Reversible shear thickening in monodisperse and bidisperse colloidal dispersions. *J. Rheol.*, **40** (5), 899–916.

62. Bender, J.W. and Wagner, N.J. (1995) Optical measurement of the contributions of colloidal forces to the rheology of concentrated suspension. *J. Colloid Interface Sci.*, **172** (1), 171–184.

63. Kaffashi, B., O'Brien, V.T., Mackay, M.E., and Underwood, S.M. (1997) Elastic-like and viscous-like components of the shear viscosity for nearly hard sphere, Brownian suspensions. *J. Colloid Interface Sci.*, **181** (1), 22–28.

64. Brady, J.F. and Bossis, G. (1988) Stokesian dynamics. *Ann. Rev. Fluid Mech.*, **20**, 111–157.

65. Brady, J.F. and Bossis, G. (1989) The rheology of Brownian suspensions. *J. Chem. Phys.*, **91** (3), 1866–1874.

66. Phung, T.N., Brady, J.F., and Bossis, G. (1996) Stokesian dynamics simulation of Brownian suspensions. *J. Fluid Mech.*, **313**, 181–207.

67. Chow, M.K. and Zukoski, C.F. (1995) Gap size and shear history dependencies in shear thickening of a suspension ordered at rest. *J. Rheol.*, **39** (1), 15–32.

68. Krishnamurthy, L.N. and Wagner, N.J. (2005) Shear thickening in polymer stabilized colloidal dispersions. *J. Rheol.*, **49** (6), 1347–1360.

69. Beazley, K.M. (1980) Industrial aqueous suspensions, in *Rheometry: Industrial Applications* (ed. K. Walters), Research Studies Press, Chichester.

70. Bergstrom, L. (1998) Shear thinning and shear thickening of concentrated ceramic suspensions. *Colloids Surf., A*, **133**, 151–155.

71. Egres, R.G. and Wagner, N.J. (2005) The rheology and microstructure of acicular precipitated calcium carbonate colloidal suspensions through the shear thickening transition. *J. Rheol.*, **49** (3), 719–746.

72. Bartok, W. and Mason, S.G. (1958) Particle motions in sheared suspensions: VII. Internal circulation in fluid droplets (theoretical). *J. Colloid Interface Sci.*, **13** (4), 293–307.

73. Taylor, G.I. (1932) The viscosity of a fluid containing small drops of another fluid. *Proc. R. Soc. London, Ser. A*, **138** (834), 41–48.

74. Grace, H.P. (1982) Dispersion phenomena in high viscosity immiscible fluid systems and application of static mixers as dispersion devices in such systems. *Chem. Eng. Commun.*, **14** (3), 225–277.

75. Zhao, X. (2007) Drop breakup in dilute Newtonian emulsions in simple shear flow: new drop breakup mechanisms. *J. Rheol.*, **51** (3), 367–392.

76. Nawab, M.A. and Mason, S.G. (1985) The viscosity of dilute emulsions. *Trans. Faraday Soc.*, **54**, 1712–1723.

77. Pal, R. (2001) Novel viscosity equations for emulsions of two immiscible liquids. *J. Rheol.*, **45** (2), 509–520.

78. Pal, R. (2000) Shear viscosity behavior of emulsions of two immiscible liquids. *J. Colloid Interface Sci.*, **225** (2), 359–366.

79. Oldroyd, J.G. (1953) The elastic and viscous properties of emulsions and suspensions. *Proc. R. Soc. London, Ser. A*, **218** (1132), 122–132.

80. Oldroyd, J.G. (1955) The effect of interfacial stabilizing films on the elastic and viscous properties of emulsions. *Proc. R. Soc. London, Ser. A*, **232** (1191), 567–577.

81. Palierne, J.F. (1990) Linear rheology of viscoelastic emulsions with interfacial tension. *Rheol. Acta*, **29** (3), 204–214.

82. Kitade, S., Ichikawa, A., Imura, M., Takahashi, Y., and Noda, I. (1997) Rheological properties and domain structures of immiscible polymer blends under steady and oscillatory shear flows. *J. Rheol.*, **41** (5), 1039–1060.

83. Princen, H.M. (1986) Osmotic pressure of foams and highly concentrated emulsions. 1. Theoretical considerations. *Langmuir*, **2** (4), 519–524.

84. Mason, T.G., Lacasse, M.-D., Grest, S.G., Levine, D., Bibette, J., and Weitz, D.A. (1997) Osmotic pressure and viscoelastic shear moduli of concentrated emulsions. *Phys. Rev. E*, **56** (3), 3150–3166.

85. Mason, T.G., Bibette, J., and Weitz, D.A. (1996) Yielding and flow of monodisperse emulsion. *J. Colloid Interface Sci.*, **179** (2), 439–448.

86. Princen, H.M. and Kiss, A.D. (1986) Rheology of foams and highly concentrated emulsions: 3. Static shear modulus. *J Colloid Interface Sci.*, **112** (2), 427–437.

87. Bécu, L., Grondin, P., Colin, A., and Manneville, S. (2004) How does a concentrated emulsion flow? Yielding, local rheology, and wall slip. *Colloids Surf., A*, **263** (1–3), 146–152.

88. Fall, A., Paredes, J., and Bonn, D. (2010) Yielding and shear banding in soft glassy materials. *Phys. Rev. Lett.*, **105** (22), 225502.

89. Callaghan, P.T. (2008) Rheo-NMR and shear banding. *Rheol. Acta*, **47** (3), 243–255.

90. Coussot, P. and Ovarlez, G. (2010) Physical origin of shear-banding in jammed systems. *Eur. Phys. J. E*, **33** (3), 183–188.

2
Rheology of Cosmetic Emulsions
Rüdiger Brummer

2.1
Introduction

The main purpose of cosmetic products is to supply the skin with lipids and moisture. In the field of medicine the purpose can also be to supply active ingredients that must be applied sufficiently diluted in a cream to diseased skin areas. The main components, however, are always water and oil. Since water and oil are hardly miscible, other ingredients are needed to make them mix. These may be emulsifiers or surfactants that ensure the stability of oil droplets dispersed in water, or vice versa, or they may be polymer molecules that stabilize emulsions by forming a three-dimensional network in which oil droplets can become interspersed.

The following categories of currently manufactured cosmetic products were defined by Brandau [1]:

- ointments
- creams
- gels
- lotions.

Ointments are spreadable, non-transparent formulations at room temperature that are virtually water-free. They comprise only a minor part of the cosmetic products. Creams differ from ointments in that they consist of fat-like substances, water, and usually emulsifiers. Creams can in turn be sub-classified by the emulsion type. In lipophilic creams water is the dispersed and oil is the continuous phase. This type of emulsion is abbreviated as W/O. Conversely, hydrophilic creams have oil as the dispersed and water as the continuous phase and are called oil-in-water (O/W) type emulsions. Amphiphilic creams have both lipophilic and hydrophilic properties. Gels are spreadable, transparent formulations at room temperature, whereas lotions are free flowing creams (mainly of the O/W type) at room temperature. The droplet diameter of the disperse phase usually ranges from 1 to 5 μm. Other possible emulsion types are W/O/W and O/W/O formulations. The oil droplets in a W/O/W emulsion are emulsified in water and the water droplets in turn are emulsified in the oil droplets. The size of the oil droplets ranges

Product Design and Engineering: Formulation of Gels and Pastes, First Edition.
Edited by Ulrich Bröckel, Willi Meier, and Gerhard Wagner.
© 2013 Wiley-VCH Verlag GmbH & Co. KGaA. Published 2013 by Wiley-VCH Verlag GmbH & Co. KGaA.

from 5 to 10 μm and that of the water droplets from approximately 1 to 2 μm. The opposite is true of O/W/O emulsions.

Emulsions are "thermodynamically metastable" systems exposed to physical, chemical, and microbiological influences during manufacture, transport, storage, and use that can produce visible changes in the emulsion. Such changes can be caused by the temperature, exposure to light, external pressure, and so on. These variables affect the solubility product and this can result in crystallization. If interaction of the ingredients with each other or with the packaging material occurs, this can result in instabilities due to chemical reactions. Yeast, bacteria, and molds affect the microbiological stability of the product.

Rheological measurements will be presented that can be used to characterize cosmetic products such as creams, lotions, and gels. These are plastic materials characterized by non-Newtonian flow behavior. The onset of flow is product-specific and differs significantly for lotions and creams. On the basis of the critical shear stress at the yield point, the emulsion type can be determined for creams as well as lotions. The onset of flow of W/O emulsions is observed at a considerably lower shear stress than with an O/W emulsion. Gels do not have a characteristic yield point but can be distinguished by a critical shear rate. The recovery time after loading below the yield point is not a product-specific characteristic for creams, lotions, and gels but is crucial for the reproducibility of measuring results.

Cosmetic cleansing products containing surfactants are characterized by Newtonian flow behavior. In this product group no recovery takes place after shearing. When subjected to periodic, usually sinusoidal, deformation hydrogels show typical polymer characteristics. At low frequencies they behave like a fluid and at high frequencies like an elastic solid.

2.2
Chemistry of Cosmetic Emulsions

2.2.1
Modern Emulsifiers

Modern emulsifiers [2] are mainly surfactant additives that reduce surface tension. They include foaming agents, defoamers, wetting agents, detergents, and solubilizers. Very different emulsion structures can be achieved depending on the emulsifiers used and their concentration. Consequently, various applications are possible. A bar of ordinary soap consists almost entirely of a pure emulsifier that can absorb fats when combined with water. Consequently, ordinary soap is used to cleanse the skin, that is, to remove fatty impurities, and also excess sebum. However, the same emulsifier can be mixed with emollient oils, water, and water-soluble skincare substances to make O/W creams. These skincare creams have long been known as "*stearate creams*" and today are occasionally still found in the skin protection sector, for example, as products with a high content of free stearic acid. This emulsifier has been replaced mainly by pure synthetic emulsifiers that offer several advantages in terms of their performance characteristics.

2.2.2
Skin Care and Cleansing

Emulsions look milky-white like natural milk and are incorporated in cleansing creams as well as skincare cream (semisolid) and lotions (liquid). The oil droplets in these O/W emulsions are about $1-20\,\mu m$ ($0.001-0.020\,mm$) in size. Conversely, emulsions may contain water droplets (W/O creams) or even be multiple systems [3] (W/O/W and O/W/O). O/W creams usually supply more moisture and W/O creams more lipids.

The smaller the mean droplet size, the more transparent the products are. Emulsions with a droplet size distribution between 10 and 50 nm are called *microemulsions.* They are fully transparent and distinguished by a relatively high emulsifier content.

2.2.3
Microemulsions

Microemulsions [4] are used for different purposes. A high emulsifier content has a strong influence on the skin barrier, resulting in very fast penetration or permeation of active ingredients through the skin. This is especially advantageous in the pharmaceuticals sector for drug therapies. In the skincare sector this proves to be more of a disadvantage because emulsifiers severely disturb the integrity of the skin barrier layers. In cosmetics, microemulsions are used mainly for skin cleansing, for example, as oil-containing cleansing gels, shower gels, and bubble baths.

Microemulsions in the narrower sense are systems with a high surfactant content that actually are not emulsions because the water and oil phases can no longer be discerned even under an electron microscope. They are occasionally used for transdermal drug formulations but due to their emulsifier side effects are no longer of much importance. In contrast, no clear distinction is made between two- and one-phase systems in the area of skin cleansers.

2.2.4
Emulsifier-Free Products

Whereas emulsions and microemulsions are based on a more classical concept and contain largely synthetic emulsifiers, nanoemulsions are based on a markedly physiological concept. The particles in nanoemulsions are smaller than those in microemulsions, having a diameter from 0.05 to $>0.1\,\mu m$. Nanoemulsions do not contain typical emulsifiers but, rather, pure, natural phosphatidylcholine. Phosphatidylcholine, which is obtained from lecithin, is the essential building block of all natural cell membranes. Unfortunately, the INCI (International Nomenclature of Cosmetic Ingredients) name [5] for phosphatidylcholine is lecithin, which makes it impossible for the non-professional to distinguish between the two on the package label. Phosphatidylcholine dispersions spontaneously form bilayer membranes like those of the cell membranes, the barrier layers of the skin, and liposomes. Using high-pressure technology it is possible to force phosphatidylcholine to also form simple membranes that can enclose oil droplets, making conventional emulsifiers superfluous. Conditions are achieved that resemble those found in the body's own fat transport system, the chylomicrons.

Phosphatidylcholine can be completely metabolized and additionally provides the skin with two essential substances: linoleic acid and choline. Therefore, phosphatidylcholine actually has little in common with conventional emulsifiers, and the term *nanoemulsion* was quickly supplemented or replaced by terms like *nanodispersion, nanoparticle, or nanoparts*. Nanoemulsions are used, for example, for intravenous fat nutrition. An analogous use of conventional emulsifiers for this purpose would quickly result in destruction of the blood and blood vessels.

In the cosmetics sector, nanoparticles belong to the group of emulsifier-free products. They have one distinct advantage: Whereas emulsifiers are usually stored unchanged in the skin and tend to promote washout of the skin's own lipids with the next skin cleansing, phosphatidylcholine shows just the opposite effect. It has an almost magical attraction for lipids into the skin. This is also true of the donkey milk (known of Cleopata) and observed with balneological products as well. Owing to the high production costs, nanoparticles are incorporated in higher amounts only in special products such as products for elderly and problem skin as well as products for supportive preventive care.

Cold cream is possibly the oldest emulsifier-free cream and DMS (DMS = dermal membrane structure) cream the newest. DMS cream is not classified as an emulsion as no droplet structures can be seen under a normal light microscope. Lamellar structures like those typical of the barrier layers of the skin become visible only under an electron microscope. DMS cream is made of a phosphatidylcholine that contains esters of the palmitic acid and stearic acid predominating in the horny layer rather than of linoleic acid. Interestingly, they have properties similar to those of ceramides. They anchor themselves in the barrier layers of the skin and like ceramides are very resistant to exogenous substances acting on the skin. Consequently, ceramides, DMS, liposomes, and nanoparticles are compatible with each other in nearly any ratio. DMS creams cannot be produced using the common emulsification methods although they do not differ from emulsions in their appearance or use. DMS creams are suitable for extremely sensitive and problem skin because they do not disrupt the skin barrier.

2.2.5
Production of Emulsions

At first glance making an emulsion seems to be a simple process. When two immiscible liquids are dispersed by stirring vigorously an emulsion is obtained briefly. If the two liquids are water and oil, either a W/O or an O/W emulsion will be formed depending on the amounts of each liquid used. Because the free energy of the emulsion system [6] is higher than that of the two liquids, the phases will separate with release of energy. To stabilize these systems for longer periods emulsifiers must be incorporated that delay phase separation into the thermodynamically more stable starting liquids until the emulsion has been used as intended.

The emulsion production process can be divided into three basic steps:

1) pre-emulsification
2) fine emulsification
3) stabilization.

In the pre-emulsification step the water and oil phases are combined at an elevated temperature with stirring, forming a raw emulsion (premix) with large droplets. These are deformed in the subsequent fine emulsification step by external shear forces and their size reduced when a critical deformation is exceeded. The newly formed interface is then protected by emulsifiers against coalescence in the stabilization step.

2.2.6
Processes Occurring During Emulsification

The emulsification process entails the breakup of droplets and wetting of the newly formed interface, which is no longer completely covered by emulsifier molecules immediately after size reduction. Adsorption of more surfactant molecules takes time and depends on the interfacial wetting kinetics of the emulsifier system used. The coverage density influences not only the interfacial tension and hence the energy needed for particle size reduction but also the stability of the droplets generated [7].

Insufficiently stabilized droplets can coalesce upon impact with other droplets if the contact time is long enough. For coalescence to take place, the continuous phase between colliding droplets must be displaced to a critical film thickness (film drainage). Coalescence can be prevented if the repulsive forces between droplets are sufficiently high. These repulsive forces are exerted by the adsorbed emulsifier molecules. Spreading of emulsifier molecules unevenly distributed on the droplet surface (Gibbs–Marangoni effect) [8] slows film drainage and stabilizes the droplets even if the interface is not completely covered [9].

Droplet size reduction as well as coalescence of broken up but not yet completely stabilized droplets determine the emulsification results and the dispersity of the emulsion formed.

2.2.7
Serrated Disc Disperser

Droplet size reduction requires normal and/or tangential tensions at the interface between the internal and external phase. Droplets are broken up when local deforming forces exceed form-retaining interfacial forces for a long enough period.

This requires dissipation of large amounts of energy in the dispersing zone of an emulsifier machine.

The serrated disc disperser consists of a rotor–stator system constructed of coaxially intermeshing discs with slots. The width of the gap between the rotor and stator is of the order of magnitude of millimeters. The emulsion, which is placed in the middle of the disperser, is accelerated by the centrifugal force of the moving rotor and decelerated by the stator. The shear forces arising are generally thought to be responsible for droplet size reduction [10]. Serrated disc dispersers are usually self-propelling due to the way their flow is guided.

After dispersion, the emulsion droplets pass, usually, with a laminar flow through pipes where they can collide. If the phase interface is not sufficiently stabilized by adsorbed emulsifier molecules and the contact time is long enough for the continuous phase between the droplets to be displaced, the droplets will coalesce. The resistance to coalescence immediately after the droplets are broken up is called the *short-term stability* [11]. The short-term stability of emulsions is influenced not only by the adsorption kinetics of the emulsifiers but also by the coalescence probability of the droplets. The latter is determined by the contact time and interparticulate interactions.

When two droplets collide the continuous phase between them is displaced (film drainage), that is, the film ruptures once a critical thickness is reached.

The critical film thickness for emulsions is of the order of magnitude of 1–100 nm [12]. If the film of the continuous phase ruptures, the droplets will coalesce spontaneously. Chesters showed that the coalescence probability depends on the critical film thickness, the viscosity ratio of the dispersed and continuous phase, the droplet radius, and the Weber number [13]. The author also showed that the coalescence probability in laminar flow is higher than in turbulent flow, which means that the droplets in pipelines downstream from the dispersion zone are the most susceptible to coalescence.

Besides the destabilizing mechanisms associated with incomplete coverage of interfaces, there is also a stabilizing effect referred to as *"self-healing"* of the interfacial film. The Gibbs–Marangoni effect produces an increase in emulsifier concentration in the contact zone between two incompletely covered droplets. The pressure in the contact zone increases as the concentration gradient levels off and the droplets are pushed apart [14].

2.3
Rheological Measurements

2.3.1
Stationary Flow Behavior

For rheological characterization of cosmetic emulsions various measurement methods and instruments are used. To detect the apparent yield stress [15], stressed-controlled rheometers are used because with this type of instrument the stress is preset and the onset of flow can be determined more or less accurately depending on the quality of the angle resolution of the instrument. Characterization of creaming conditions requires measurements at high shear rates with a torsional rheometer or for processing at extremely high shear rates with a capillary viscometer. The measurement systems used are also important and will depend on the product to be measured. Watery fluids need a "cylinder system," viscous samples the "cone-plate system," and all other solid materials the "parallel plate system."

Measurements for a new, unknown sample start with the so-called positive ramp test (Figure 2.1). The applied shear stress is increased continuously with time at a constant temperature. This test is also referred to as a *stress–time ramp*. In this experiment an integral viscosity is obtained as a first estimation of the range in which the actual measurements should be performed. A step test is then performed as the second experiment (Figure 2.2). In contrast to the first experiment, here the shear stress is kept constant for a set period of time and then increased in several steps. The result obtained is the so-called stationary viscosity value.

Next the material properties are determined after sudden application (Figure 2.3) or removal (Figure 2.4) of a load.

After determining the integral viscosity in a positive ramp test as shown in Figure 2.1 it is of course also possible to do a negative ramp test (Figure 2.5) to see whether the sample had been changed by the original positive ramp test. Last but not least,

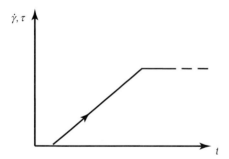

Figure 2.1 Positive ramp test.

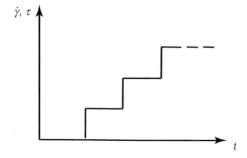

Figure 2.2 Step test.

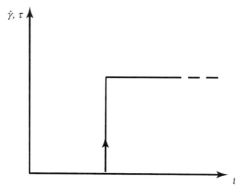

Figure 2.3 Load jump.

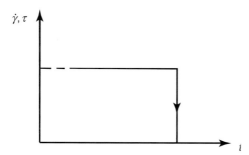

Figure 2.4 Release jump.

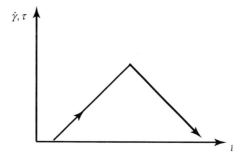

Figure 2.5 Positive and negative ramp.

the user is free to choose any combination of these tests (Figure 2.6) and even to change temperatures.

2.3.2
Stress Ramp Test

The positive ramp or stress ramp test is a simple, quick test that can be performed with a stress-controlled rheometer. The shear stress is increased continuously

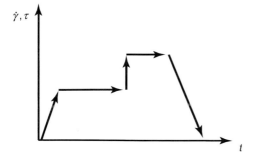

Figure 2.6 Combination test.

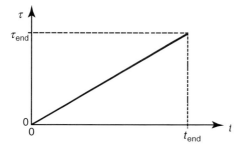

Figure 2.7 Determination of the apparent yield stress.

within a set time (Figure 2.7), the resulting torque is measured at discrete time intervals and from this the viscosity calculated. Programming the right stress ramp is crucial. Is the ramp test with 20 Pa min^{-1} the best or is 100 Pa min^{-1} the better one?

Assuming that the sample has a yield stress, also known as the *yield point*, is this product-specific and should it be determined accurately? A small stress ramp should be programmed for thinner liquids and a larger one for thicker creams. An appropriate number of measurements must be performed to determine if the whole product range can be measured with a single stress ramp.

After programming the stress ramp and performing the measurement the next step is to interpret the resulting curve. Although there are several models [16] available for interpreting measuring curves, they often give rise to relatively large errors because ideal mathematical conditions usually do not exist. Differences already arise from the way in which the results are plotted. If the shear stress is plotted against the shear rate, the curve shown in Figure 2.8 is obtained. Initially, the force or shear stress applied to the sample does not cause any detectable deformation. In other words, the system remains at rest.

No deformation becomes visible until a critical shear stress has been reached, and a shear rate is determined. Using the Herschel–Bulkley relationship:

$$\tau = \tau_0 + K \cdot \dot{\gamma}^n \tag{2.1}$$

the coefficients can be determined and the critical shear stress for the yield stress calculated (corresponding to Figure 2.8: C1 $= \tau_0$, C2 $= K$, and C3 $= n$).

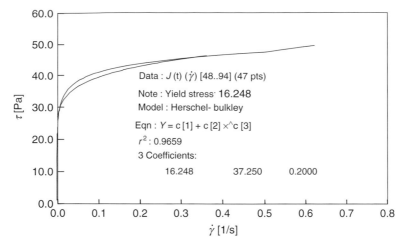

Figure 2.8 Interpretation of the yield stress according to the Herschel–Bulkley model (see text for details).

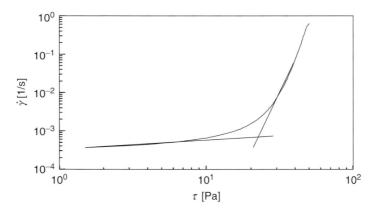

Figure 2.9 Interpretation of the yield stress according to the tangential method.

In the tangent method, the abscissa and ordinate are interchanged in the plot (Figure 2.9). It can be readily seen that the curve shows two linear regions of different slope and a transition region. The intersection of the two tangents to the apparent linear regions is interpreted as the yield stress. This method is strongly dependent on the choice of measuring points for each tangent.

A third method for plotting the results [17] of a stress ramp test is the double logarithmic plot of the viscosity versus the shear stress.

In practice, plotting the viscosity as a function of shear stress gives good results. The viscosity initially increases to a maximum (Figure 2.10) and then decreases again. The stress corresponding to the viscosity maximum is called the *critical shear stress* (τ_{critical}). This is the stress needed to cause the system to flow. Note that the calculated viscosity is an integral of the force over time. Therefore, it is

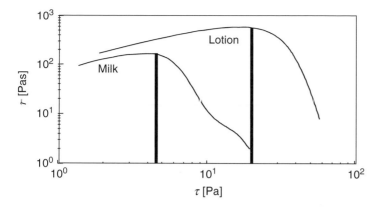

Figure 2.10 Apparent shear stress of two products.

the shear stress rather than the viscosity that is the more important parameter for determining the apparent yield stress. In this way, discrete critical shear stresses can be assigned to specific products.

2.3.3
Newtonian Flow Behavior

Products containing surfactants usually exhibit Newtonian flow behavior (Figure 2.11). This is typically detected in the lower shear rate range [18]. If flow behavior is Newtonian, the viscosity at constant temperature is independent of the applied stress or the velocity gradient. The value obtained is the mean of the viscosities over the velocity gradient.

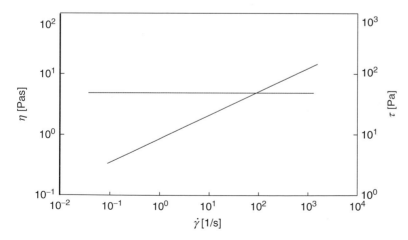

Figure 2.11 Newtonian flow of a surfactant-containing product.

2.3.4
Creep and Creep Recovery Test

The creep test is a simple, quick test for obtaining initial information on the viscoelastic properties of a sample from viscosity-relevant (as opposed to oscillation measurements) data. In this experiment (Figure 2.12) a constant force (shear stress) is applied to the sample at time t_0 and removed again at time t_1. The recovery up to time t_2 is recorded.

The sample responds initially to the force applied at t_0 with deformation. In other words, it starts to creep. At t_1 (after removal of the force), the sample recovers again. There are three different types of creep and creep recovery curves.

2.3.5
Ideal Elastic Behavior

The first case is an ideal elastic body as exemplified by a spring. If a force is applied to the spring it responds with a deformation but returns to its original state after the force is removed (Figure 2.13).

If the force τ is doubled, the deformation γ also doubles. Ideally, 100% of the energy stored in a spring will be recovered. A body with these properties is also known as a *Hookean body*.

2.3.6
Ideal Viscous Behavior

The second case is water as an example of an ideal fluid. The force τ applied to the fluid causes a linear deformation γ over time. In other words, the sample begins to flow.

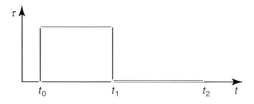

Figure 2.12 Creep and creep recovery test.

Figure 2.13 Creep and creep recovery of an ideal elastic body.

If the force is removed from the sample (Figure 2.14), the deformation attained at this time (in our example t_1) is fully retained. The model in this case is the dashpot model according to Newton.

2.3.7
Real Viscoelastic Behavior

A real body is both viscous and elastic. This means that when a force is applied at the time t_0 deformation begins to take place much more slowly and, if you wait long enough (until t_1), the curve will approach a constant slope.

When the force is removed, part of the energy stored in the body will be released. The result is a recovery of the elastic part γ_e and a permanent deformation (Figure 2.15) of the viscous part γ_v. A viscoelastic solid will therefore recover after a time lag but it will do so almost completely.

2.3.8
Steady Flow Curve

The best way to measure the viscosity of a non-Newtonian sample at constant temperature and known shear rate is to program a time test. Since both the measuring instrument and the sample need a finite time to reach constant conditions (Figure 2.16), it is necessary to wait a certain time before measured values are obtained that can be evaluated.

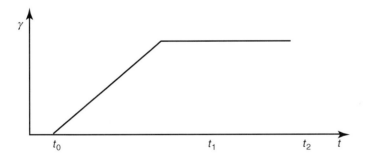

Figure 2.14 Creep and creep recovery of an ideal viscous body.

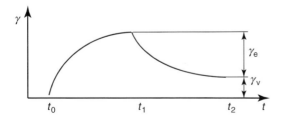

Figure 2.15 Creep and creep recovery of a real viscoelastic body.

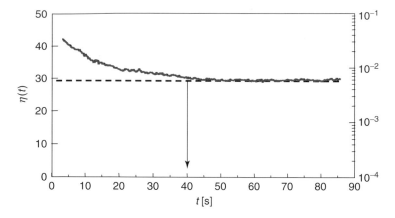

Figure 2.16 Approach to equilibrium in stationary measurements.

In the extreme example shown in Figure 2.16 the viscosity increases within the first few seconds at 25 °C and a shear rate of 0.001 s^{-1}. After passing through a maximum, steady state conditions are not reached until after about 40 s and the viscosity measurement can begin. If several shear rates are applied sequentially in a step test (Figure 2.2), the following results are obtained.

In the first segment of Figure 2.17 at a shear rate of $\gamma = 0.01$ s^{-1} steady state conditions had still not been achieved after 150 s. At $\gamma = 0.1$ s^{-1} constant values could not be measured until after 75 s. At $\gamma = 1$ s^{-1} this was already the case after 25 s. Consequently, the greater the shear rate the sooner steady state conditions will be reached. The lower the shear rate needed for a measurement the longer the measuring time will be.

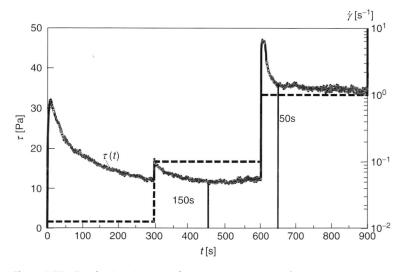

Figure 2.17 Steady viscosity curve for non-Newtonian samples.

Notice that, to measure the stationary viscosity at higher temperatures with low shear rates, the relatively long measuring time is a considerable disadvantage because the sample can begin to dry out during the measurement.

It is extremely important to keep as many boundary conditions as possible constant when measuring the viscosity because the viscosity depends on many factors, including:

- shear rate
- time
- temperature
- density
- solids content
- molecular weight
- history, and so on.

This makes it relatively difficult to determine why the viscosity changes during a measurement. Rheological measurements record the effect and not the cause. Many measurements and the experience gained from them are needed to be able to interpret the data correctly.

2.4
Dynamic Mechanical Tests (Oscillation)

Another very important method by which to characterize cosmetic emulsions is oscillation. This type of measurement provides information on the structural properties of materials. There are four different measuring variations:

1) amplitude variation,
2) structure breakdown and build up,
3) time dependence,
4) frequency variation,
5) temperature dependence,
6) combined temperature–time test.

2.4.1
Amplitude Dependence

The amplitude variation, also known as the *strain test*, is performed at constant temperature and frequency (Figure 2.18). Starting at small amplitudes, the strain is increased in discrete steps.

The responses obtained are the two moduli G'' and G', which run nearly parallel in the lower frequency range. At a certain product-specific frequency, the response is no longer linear. The linear range is also called the linear viscoelastic region (LVR).

In the example shown in Figure 2.19 the experiment was run at a frequency of $\omega = 100$ rad s^{-1} and $T = 25\,°$C. The plots of the moduli G' (storage modulus) and G'' (loss modulus) versus frequency are double logarithmic plots. At a strain of

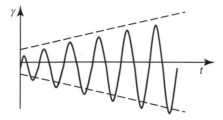

Figure 2.18 Amplitude or strain test at T = constant and ω = constant.

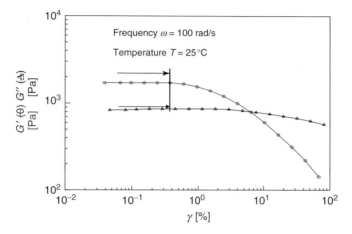

Figure 2.19 Determination of the linear viscoelastic range with the amplitude test.

approximately $\gamma \approx 0.4\%$ (in this example 100% strain corresponds to 0.5 rad) it can be seen that the storage modulus G' deviates from the linear range. Why is it so important to perform the measurement in the linear viscoelastic range?

There are two reasons:

1) Measurements can be compared only if the same boundary conditions were maintained for all measurements. However, if the amplitude leaves the linear range, different results will be obtained in the analysis.
2) The second reason has to do with the measuring instrument. Even the most modern technology has its limits and in this case it is the torque resolution.

A value measured near the torque resolution will be associated with an error of 100%. The minimum accuracy required for the measured value is <10%. This means that the frequency range must be limited as appropriate.

The amplitude test (strain test) is performed as a kind of screening test for every new sample. But there is one more question to be answered: What frequency should you choose? To be absolutely safe this test should be performed at the lowest and highest frequency, namely, at the limits of the frequency test. This procedure must be repeated for every temperature used.

2.4.2
Structure Breakdown and Build-Up

Most cosmetic emulsions are deformed when a small amount is removed. How quickly the structure is restored can be crucial. To determine this [19], the amplitude test must be slightly modified. First the linear viscoelastic range described above must be determined. The test is run for a defined period at the amplitude determined in this test (Figure 2.20) with T and ω constant.

Then the amplitude is suddenly increased, for example, 100-fold and the test is continued at this deformation for about 5 min. Afterwards, the amplitude is returned to the starting amplitude determined in the screening test and the sample observed for a further 30 min.

In Figure 2.21 the storage modulus G' and loss modulus G'' are plotted as a function of time as the response parameters. In the first segment the storage modulus G' is larger than G'' (linear viscoelastic range). After the sudden increase in amplitude it can be seen that G' and G'' respond by decreasing dramatically (nonlinear range).

Strikingly, the relationship is now reversed and G' is smaller than G''. This would mean that the sample flows. To get ketchup out of a bottle you have to shake it vigorously to make it flow. This is in fact exactly what is happening in this example, for the measurement shows us the rheological behavior of ketchup when it is shaken. If the bottle is then left to stand long enough the ketchup again

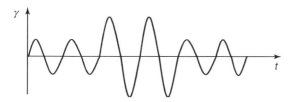

Figure 2.20 Amplitude test for determination of structural breakdown and regeneration.

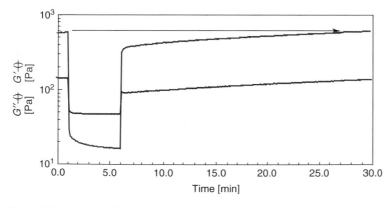

Figure 2.21 Structural breakdown and build up for ketchup.

thickens and will have to be shaken again to make it flow. This is what we see in the plot. After we return the amplitude to a value in the viscoelastic range it takes approximately 20 min for the original state to be achieved.

2.4.3
Time Dependence

Using dynamic-mechanical time-dependence measurements, structural changes can be detected without superposition of shear (as in the static time test). The sample is observed for a certain time at constant temperature, frequency, and amplitude (once again the LVR value) (Figure 2.22).

Figure 2.23 presents the test result as a plot of G' versus time. After about 400 s the absolute value of the storage modulus slowly increases. This can be due to several causes. For instance, the system could crosslink or some of the sample could evaporate, which is more likely for an emulsion.

2.4.4
Frequency Test

After the maximum amplitude has been determined from the LVR, the frequency dependence is studied. Testing usually begins at the highest frequency and is then

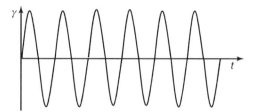

Figure 2.22 Time dependence with T, ω, γ = constant.

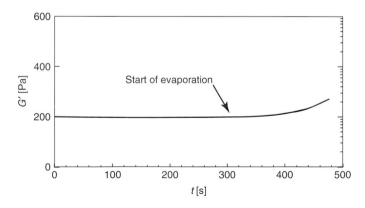

Figure 2.23 Time dependence of the storage modulus with T, ω, γ = constant.

reduced in logarithmic steps. Starting at high frequencies has the advantage that the first measured values can be obtained very quickly because, once again, the following applies:

Low frequencies = long measuring time
High frequencies = short measuring time

Figure 2.24 presents the sinusoidal curve of the frequency test, starting at low frequencies. Obtained as the response, in the double logarithmic plot, are the two moduli G' (storage modulus) and G'' (loss modulus) (Figure 2.25) or $\tan \delta$ as a function of frequency. Alternatively, the complex dynamic viscosity can be plotted versus frequency.

This example clearly shows that the complex dynamic viscosity η^* (Figure 2.25, black, open circles) decreases linearly with increasing frequency in the measured frequency range, as is known from static measurements. The storage modulus G' (blue triangles) is always larger than the loss modulus G'' (red, open squares) in the measurement window. This shows that the sample must be present in the semi-solid state because the loss modulus G'' would always be larger than the storage modulus G' if it were a fluid.

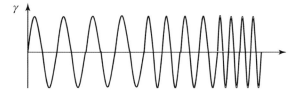

Figure 2.24 Frequency test at constant temperature and amplitude.

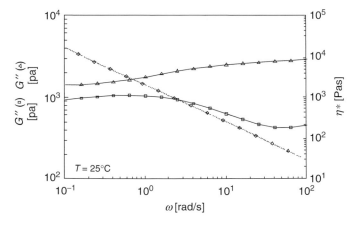

Figure 2.25 Storage/loss modulus and complex dynamic viscosity as a function of frequency.

2.4.5
Temperature Dependence

Often, cosmetic emulsions are exposed to variable temperatures. For example, in summer a hand cream in the glove compartment of a car can easily reach a temperature of $T = 50\,°C$ during the day and drop to room temperature in the garage at night. In winter, the opposite is true, with temperatures dropping to as low as $-20\,°C$. The hand cream is exposed to this stress with no other mechanical forces applied. Therefore, the method of choice is once again a dynamic-mechanical measurement. To measure the influence of temperature [20], first the LVR needs to be determined at the final temperature and attempts made to find a single value that is valid for the entire range.

If this is impossible the experiment must be divided into several parts. The boundary conditions are very simple: All parameters except the temperature are held constant ($\omega, \gamma = $ constant). You have to perform two experiments. Both will start at $T = 25\,°C$. In one case the temperature will be lowered (Figure 2.26) and in the other increased (Figure 2.27).

In Figure 2.26 it can be clearly seen that in the cooling experiment both the storage modulus G' and the loss modulus G'' suddenly increase at a temperature $T = -8\,°C$. This is typical of an O/W emulsion because the external phase – water – freezes. In the heating experiment (Figure 2.27) both moduli decrease at $T = 50\,°C$. This is where the internal structures of the emulsion sample begin to soften. Using these two experiments the temperature range can be determined in which the emulsion will not change during the measurement.

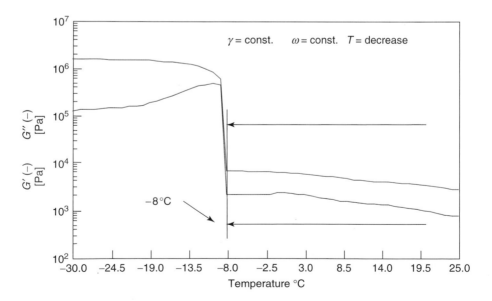

Figure 2.26 Freezing point determination by temperature reduction.

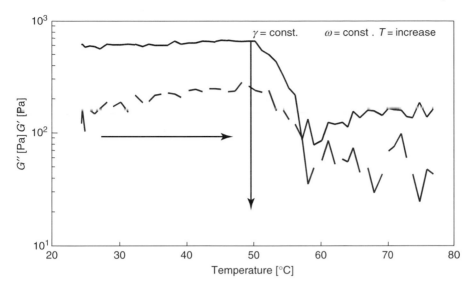

Figure 2.27 Determination of the softening point by temperature elevation.

2.4.6
Combined Temperature–Time Test

If these two tests are repeated at longer and shorter intervals and the absolute values of the storage and loss moduli compared over time, no differences should be detectable for stable emulsions. This is of course a time-consuming procedure. We must also consider whether the moduli G' and G'' will change if the cooling and heating curves are subsequently measured backwards, that is, when heated or cooled, respectively. The test is very simple [21]: Simultaneous negative and positive temperature ramps (Figure 2.28) are programmed at constant frequency

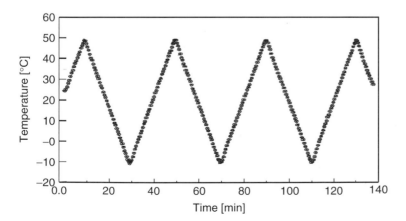

Figure 2.28 Combined temperature–time test; also known as the *cycle test*.

ω and constant amplitude γ. For example, the temperature can be increased from 25 to 50 °C.

The sample is then cooled to -10 °C with a constant cooling rate and subsequently heated again to 25 °C. These ramps are run several times. Once again, the limiting temperatures of the LVR must be determined. The combined temperature–time test is also known as the *cycle test*.

In Figure 2.29 we see the result of a cycle test. The storage modulus G' is always larger than the loss modulus G'' over the entire measurement window. If the storage modulus G' increases so does the loss modulus G'' (but somewhat more). The whole procedure is repeated as often as the temperature ramps are run. In the cold the moduli increase slightly as they approach the freezing point. At higher temperatures a slight softening of the structure begins to become apparent.

This can be seen even more clearly in the plot of the quotient $G''/G' = \tan \delta$ versus temperature (Figure 2.30). It can be easily seen that the temperature cycles do not affect the sample. This sample can be said to be temperature resistant for the temperature range $T = -10$ to $+50$ °C.

Figure 2.31 shows the result of another cycle test. In this case there is a change in the absolute value of the storage modulus G' and the loss modulus G'' at corresponding temperature changes over the four cycles.

At which temperature the mean problem is can be seen more clearly in the $\tan \delta$ plot versus temperature (Figure 2.32).

At higher temperatures the curve is not so clear, but there is no change to be seen over the cycles. In the cold range $-10 < T < +10$ °C you can see hysteresis. This means that the sample will have problems in the cold and, therefore, we can say that this is the behavior of a temperature unstable sample.

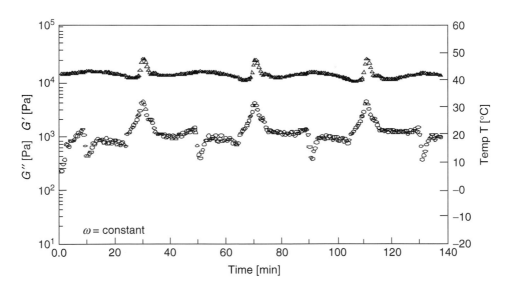

Figure 2.29 Storage and loss moduli (G' and G'', respectively) in a cycle test.

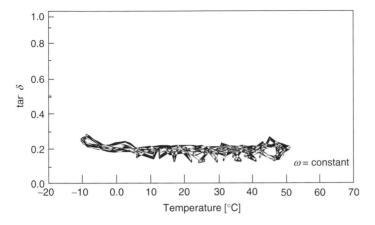

Figure 2.30 Cycle test in the tan δ versus temperature plot.

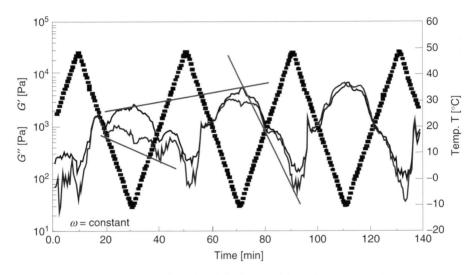

Figure 2.31 Cycle (storage modulus G' and the loss modulus G'' at corresponding temperature changes) test for an unstable formulation.

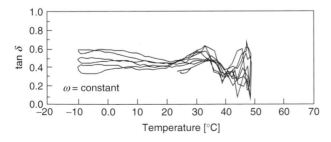

Figure 2.32 Cycle test (tan δ versus temperature) for an unstable formulation.

References

1. Brandau, R. (1983) Definitionen der dermatika unter besonderer berücksichtigung der mehrphasensysteme; arbeitsunterlagen für den fortbildungskursus, Entwicklung von Emulsionen und Cremes, der Arbeitsgemeinschaft für Pharmazeutische Verfahrenstechnik e.V. (APV), Mainz.
2. Junginger, H. (1983) Kolloidchemische betrachtungen an mehrphasensystemen; arbeitsunterlagen für den fortbildungskursus, Entwicklung von Emulsionen und Cremes, der Arbeitsgemeinschaft für Pharmazeutische Verfahrenstechnik e.V. (APV), Mainz.
3. Beiersdorf AG (2002) Eucerin: Das Medizinische Hautpflegeprogramm, Beiersdorf AG.
4. Tadros, Th. (2002) Novel polymeric surfactants in personal care applications - emulsions, nano-emulsions and multiple emulsions. Talk given a conference sponsored by Orafti NON-Food, Aandorestraat 1, B-3300, Tiene, Belgium.
5. Leven, W. (2000) Synonym-INCI-Index, in *Lexikon der Kosmetikinhaltstoffe*, 2 neu bearbeitete und erweiterte Auflage, Govi-Verlag.
6. Brummer, R., Berg, T., and Kulicke, W.M. (1997) *Einfluss des Energieeintrages beim Homogenisieren auf die Struktur von Kosmetischen w/o Emulsionen Posterbeitrag*, Shaker Verlag. ISBN: 3-8265-2908-1.
7. Armbruster, H., Karbstein, H., and Schubert, H. (1991) Herstellung von emulsionen unter berücksichtigung der grenzflächen besetzung kinetik, des emulgators. *Chem. Ing. Tech.*, **63**, 266–267.
8. Exerowa, D. and Kruglyakov, P.M. (1997) *Foam and Foam Films*, Elsevier, Amsterdam.
9. Tadros, T.F., Py, C., Taelman, M.C., and Loll, P. (1995) *SÖFW J.*, **121** (10), 714.
10. Koglin, B., Pawlowski, J., and Schnöring, H. (1981) Kontinuierliches emulgieren mit rotor/stator-maschinen: einfluß der volumenbezogenen dispergierleistung und der verweilzeit auf die emulsionsfeinheit. *Chem. Ing. Tech.*, **53** (8), 641–647.
11. Das, K.P. and Kinsella, J.E. (1991) *Adv. Food Nutr. Res.*, **34**, 81.
12. Blaß, E. (1988) Bildung und koaleszenz von blasen und tropfen. *Chem. Ing. Tech.*, **60** (12), 935–947.
13. Chesters, A.K. (1991) *Trans. IChemE.*, **69A**, 259–270.
14. Van den Tempel, M. (1960) The function of stabilizers during emulsification, in *Proceedings of Third International Congress of Surface Activity, Cologne*, vol. 2, Verlag der Universitätsdruckerei, Mainz GmbH, p. 573.
15. Haag, J. (1992) Praktische Rheologie–Bestimmung der Fließgrenze, Firmenschrift 2/92 der Firma Bohlin Instruments, Mühlacker.
16. Kulicke, W.-M. (1986) *Fließverhalten von Stoffen und Stoffgemischen*, Hüthig & Wepf Verlag, Basel.
17. Brummer, R. and Hamer, G. (1997) Rheological methods to characterize cosmetic products, in *Proceedings of the Second World Congress on Emulsion, Bordeaux*, vol. 2, Theme: 2–4/128. ISBN: 2-86411-106-3.
18. Brummer, R. (2006) *Rheology Essentials of Cosmetic and Food Emulsions*, Springer Laboratory Manuals in Polymer Science, Vol. **XVIII**, Springer, Berlin, pp. 63–79. ISBN: 3-540-25553-2.
19. Metzer, T. (April 1999) *Lohnt sich Rheologie?*, Physica Messtechnik GmbH, Stuttgart.
20. Hetzel, F., Nielsen, J., Wiesner, S., and Brummer R. (2000) Dynamic mechanical freezing points of cosmetic o/w emulsions and their stability at low temperatures. *Appl. Rheol.*, **10** (3), 114–118.
21. Brummer, R., Griebenow, M., Hetzel, F., Schlesiger, V., and Uhlmann, R. (2000) Rheological swing test to predict the temperature stability of cosmetic emulsions, *Proceedings XXIst IFSCC International Congress 2000, Berlin*, Verlag für Chemische Industrie, H. Ziolkowsky GmbH, Augsburg, pp. 476–484.

3
Rheology Modifiers, Thickeners, and Gels

Tharwat F. Tadros

3.1
Introduction

In any formulation (personal care, cosmetic, paint, printing ink, pharmaceutical, or agrochemical) one needs to modify the rheology of the system to achieve the following properties: (i) ease of formulation during manufacturing with minimum use of energy and ease of mixing; (ii) long-term physical stability with absence of creaming or sedimentation and minimum or absence of separation; (iii) ease of application that depends on the formulation type. For most formulations a shear thinning system is required for such requirement. For example, with most cosmetic and personal care applications one requires ease of dispensing of the product from its container, good spreadability, and good skin feel. For paint formulations one requires good coating and lack of sag with the formation of an even coating on the substrate. For pharmaceuticals one requires ease of flow when using injectables (that are applied through a needle with a very small hole), ease of spreading, and formation of a uniform film with topical applications. For agrochemical formulations the flow of the formulation from its large container should be easy with minimum agitation and with ease of dispersion on dilution.

To achieve the above objectives requires the addition of rheology modifiers known as thickeners and gels. These materials must have a specific rheology profile that depends on the type of the formulation and its method of application. In this chapter, I will start with the general classification of thickeners and gels. This is followed by the definition of a "gel" and its rheological behavior. Various types of polymer gels, associative thickeners, and crosslinked gels will be described. The last part of the chapter will deal with gels based on particulate solids and surfactant liquid crystalline phase. For detailed background information the reader is referred to the references [1–6].

Product Design and Engineering: Formulation of Gels and Pastes, First Edition.
Edited by Ulrich Bröckel, Willi Meier, and Gerhard Wagner.

3.2
Classification of Thickeners and Gels

Rheology modifiers can be classified into several categories:

1) thickeners or gels that consist of high molecular weight polymers – these are produced by polymer coil overlap at a particular concentration that depends on the polymer molecular weight and the polymer conformation in solution;
2) associative thickeners that are produced from hydrophobically modified water-soluble polymers obtained by grafting of alkyl chains on the hydrophilic backbone of the polymer – the gels are produced in this case by hydrophobic association of the alkyl chains similar to the formation of surfactant micelles;
3) crosslinked polymers (chemical gels) obtained by chemical reaction of crosslinking agents, thus producing "stiff" gels;
4) gels produced as a result of repulsive interaction, for example, expanded double layers – the gel is produced at a particular concentration of the particles where the double layers begin to overlap;
5) gels produced by attraction of finely divided particulate systems that interact in the continuous phase to form a "three-dimensional" structure;
6) self-structured systems, whereby one induces weak flocculation to produce a "gel" by the particles or droplets – this requires control of the particle size and shape, volume fraction of the dispersion, and depth of the secondary minimum;
7) liquid crystalline structures of surfactants (self-assembly structures) of the hexagonal, cubic, or lamellar phases.

3.3
Definition of a "Gel"

A gel is a "semi-solid" consisting of a "network" in which the solvent is "entrapped." It may be classified as a "liquid-in-solid" dispersion. A gel shows some solid-like properties as well as liquid-like properties, that is, it is a viscoelastic system. Depending on the gel strength, the system may behave as viscoelastic solid or viscoelastic liquid according to the stress applied on the gel. For "strong" gels (such as those produced by "chemical crosslinking") the system may behave as a viscoelastic solid up to high stresses and the gel could also show a significant yield value. For "weaker" gels, for example, those produced by polymer coil overlap or associative thickeners, the system may show viscoelastic liquid-like behavior at lower applied stresses when compared with chemical gels.

3.4
Rheological Behavior of a "Gel"

One of the most practical methods for the investigation of gels is to study their rheological behavior, in particular under low deformation. Three methods may be applied: (i) stress relaxation after sudden application of strain; (ii) strain relaxation after sudden application of stress, referred to as *creep measurements*; (iii) dynamic (oscillatory) techniques whereby a sinusoidal strain or stress with amplitude γ_0 or σ_0 and frequency ω (rad s^{-1}) is applied on the system. By comparing the sine waves of the stress and strain one can obtain the phase angle shift δ. The latter can give an indication of the gel type. For physical gels such as those produced by overlap of the polymer coils, δ is not too small (but still <45°) whereas for "stiff" gels produced by crosslinking δ is usually <20°). Below, a brief description of the three rheological methods (which are equivalent but not identical) is given.

3.4.1
Stress Relaxation (After Sudden Application of Strain)

One of the most useful ways to describe a gel is to consider the relaxation time of the system. Consider a "gel" with the components in some sort of a "three-dimensional" structure. To deform it instantly, a stress is required and energy is stored in the system (high energy structure). To maintain the new shape (constant deformation) the stress required becomes smaller since the components of the "gel" undergo some diffusion, resulting in a lower energy structure being approached (structural or stress relaxation). At long times, deformation becomes permanent with complete relaxation of the structure (new low-energy structure) and viscous flow will occur.

The above behavior is schematically represented in Figure 3.1, which shows the procedure for sudden application of strain. The latter is suddenly increased within a very short time and then kept constant to maintain a constant shear rate. The stress (after sudden application of strain) shows an exponential decay with time (Figure 3.2). This representation is for a viscoelastic liquid (Maxwell element represented by a spring and dashpot in series) with complete relaxation of the springs at infinite time. In other words, the stress approaches zero at infinite time. This behavior is obtained with relatively weak gels such as those produced by overlap of polymer coils (physical gels).

The above exponential decay of the stress can be represented by the following equation:

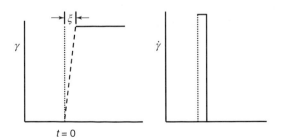

$t = 0$

Figure 3.1 Schematic representation of a strain experiment.

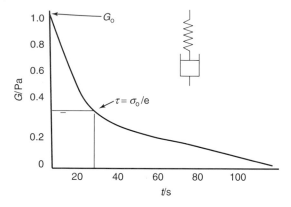

Figure 3.2 Variation of modulus with time for a viscoelastic liquid.

$$\sigma(t) = \sigma_{o} \exp\left(-\frac{t}{\tau}\right) \tag{3.1}$$

where τ is the stress relaxation time.

If the stress is divided by the strain, one obtains the modulus G:

$$G(t) = G_{o} \exp\left(\frac{t}{\tau}\right) \tag{3.2}$$

where G_{o} is the instantaneous modulus (the spring constant).

Many crosslinked gels behave like viscoelastic solids (Kelvin model) with another spring in parallel having an elasticity G_{e}. The modulus does not decay to zero.

The relaxation modulus is given by:

$$G(t) = G_{o} \exp\left(\frac{t}{\tau}\right) + G_{e} \tag{3.3}$$

Figure 3.3 shows the variation of $G(t)$ with time for a viscoelastic solid that is obtained with stronger gels such as those produced by chemical crosslinking of polymers.

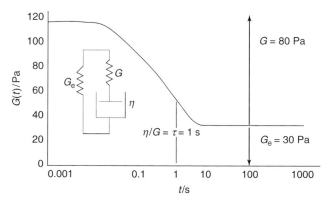

Figure 3.3 Variation of $G(t)$ with t for a viscoelastic solid.

A useful way to distinguish between the various gels is to consider the Deborah number D_e:

$$D_e = \frac{\tau}{t_e} \tag{3.4}$$

For a gel that shows "solid-like" behavior ("three-dimensional structure") D_e is large when compared with a gel that behaves as a viscoelastic liquid.

3.4.2
Constant Stress (Creep) Measurements

In this case a constant stress σ is applied and the strain (deformation) γ or compliance J ($= \gamma / \sigma$, Pa^{-1}) is followed as a function of time. A gel that consists of a strong "three-dimensional" structure (e.g., crosslinked) behaves as a viscoelastic solid (Figure 3.4). This behavior may occur up to high applied stresses. In other words, the critical stress above which significant deformation occurs can be quite high. A weaker gel (produced, for example, by high molecular weight polymers that are physically attached) behaves as a viscoelastic liquid (Figure 3.4). In this case viscoelastic solid behavior only occurs at much lower stresses than that observed with the crosslinked gels.

3.4.3
Dynamic (Oscillatory) Measurements

A sinusoidal strain (or stress) with amplitude γ_o and frequency ω (rad s^{-1}) is applied on the system and the resulting stress (or strain) with amplitude σ_o is simultaneously measured. This is illustrated in Figure 3.5.

For any gel δ is $<90°$ and the smaller the value of δ the stronger the gel. From the amplitudes of stress and strain (σ_o and γ_o) and the phase angle shift δ one can

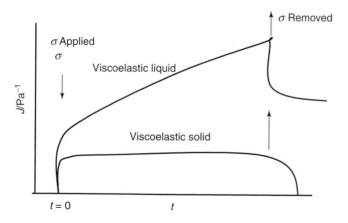

Figure 3.4 Viscoelastic solid and viscoelastic liquid response for gels.

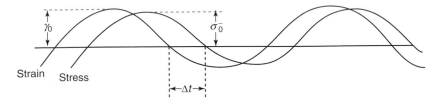

Δt = Time shift for sine waves of stress and strain

$\Delta t \ \omega = \delta$ Phase angle shift

ω = Frequency in radian s^{-1}

$\omega = 2 \, \Pi \, \upsilon$
Perfectly elastic solid $\delta = 0$
Perfectly viscous liquid $\delta = 90°$
Viscoelastic system $0 < \delta < 90°$

Figure 3.5 Strain and stress sine waves for a viscoelastic system.

obtain the following viscoelastic parameters:

$$|G^*| = \frac{\sigma_o}{\gamma_o} \qquad (3.5)$$

Storage (elastic) modulus $G' = |G^*| \cos \delta$ (3.6)

Loss (viscous) modulus $G'' = |G^*| \sin \delta$ (3.7)

$$\text{Tan} \, \delta = \frac{G''}{G'} \qquad (3.8)$$

For gels, $\tan \delta < 1$ and the smaller the value the stronger the gel.

3.5
Classification of Gels

Gels may be conveniently classified into two main categories:

1) gels based on macromolecules (polymer gels);
2) gels based on solid particulate materials.

Numerous examples of gels based on polymers may be identified: Gels produced by overlap or "entanglement" of polymer chains (physical gels), gels produced by association of polymer chains (the so-called associative thickeners), gels produced by physical or chemical crosslinking of polymer chains (sometimes referred to as "*microgel s*"), and so on. The most common particulate gels are those based on "swelling" clays (both aqueous and non-aqueous) and finely divided oxides (e.g., silica gels).

Apart from the above two main classes, gels can also be produced from surfactant liquid crystalline phases: hexagonal, cubic, or lamellar structures. These gels may be produced from single surfactant molecules, usually at high concentrations (>30%). They can also be produced using mixtures of surfactants and other amphiphiles such as long-chain alcohols, for example, mixtures of alcohol ethoxylates with cetyl-, stearyl-, or cetostearyl-alcohol. Gels can also be produced from ionic surfactants by the addition of other ingredients, for example, salts and/or long-chain alcohols.

3.5.1
Polymer Gels

3.5.1.1 Physical Gels Obtained by Chain Overlap

Flexible polymers that produce random coils in solution can produce "gels" at a critical concentration C^*, referred to as the *polymer coil "overlap" concentration*. This picture can be realized if one considers the coil dimensions in solution: considering the polymer chain to be represented by a random walk in three dimensions, one may define two main parameters, namely, the root mean square end-to-end length $\langle r^2 \rangle^{1/2}$ and the root mean square radius of gyration $\langle s^2 \rangle^{1/2}$ (sometimes denoted by R_G). The two are related by:

$$\langle r^2 \rangle^{\frac{1}{2}} = 6^{\frac{1}{2}} \langle S^2 \rangle^{\frac{1}{2}} \tag{3.9}$$

The viscosity of a polymer solution increases gradually with increasing concentration and at a critical concentration, C^*, the polymer coils with a radius of gyration R_G and a hydrodynamic radius R_h (which is higher than R_G due to solvation of the polymer chains) begin to overlap and this shows up as a rapid increase in viscosity. This is illustrated in Figure 3.6, which shows the variation of $\log \eta$ with $\log C$.

In the first part of the curve $\eta \propto C$, whereas in the second part (above C^*) $\eta \propto C^{3.4}$. A schematic representation of polymer coil overlap is given in Figure 3.7, which shows the effect of gradually increasing the polymer concentration. The polymer concentration above C^* is referred to as the *semi-dilute range*.

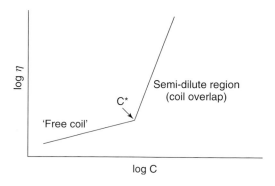

Figure 3.6 Variation of $\log \eta$ with $\log C$.

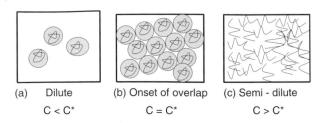

(a)   Dilute   (b) Onset of overlap   (c) Semi - dilute
       $C < C^*$ $C = C^*$ $C > C^*$

Figure 3.7 Cross-over between dilute and semi-dilute solutions: (a) dilute $C < C^*$; (b) onset of overlap $C = C^*$; and (c) semi-dilute $C > C^*$.

C^* is related to R_G and the polymer molecular weight M by:

$$C^* = \frac{3M}{4\pi R_G^3 N_{av}}$$
(3.10)

where N_{av} is the Avogadro constant. As M increases C^* becomes progressively lower. This shows that to produce physical gels at low concentrations by simple polymer coil overlap one has to use high molecular weight polymers.

Another method to reduce the polymer concentration at which chain overlap occurs is to use polymers that form extended chains such as xanthan gum, which produces a conformation in the form of a helical structure with a large axial ratio. These polymers give much higher intrinsic viscosities and they show both rotational and translational diffusion. The relaxation time for the polymer chain is much higher than a corresponding polymer with the same molecular weight but produces random coil conformation.

The above polymers interact at very low concentrations and the overlap concentration can be very low (<0.01%). These polysaccharides are used in many formulations to produce physical gels at very low concentrations.

3.5.1.2 Gels Produced by Associative Thickeners

Associative thickeners are hydrophobically modified polymer molecules whereby alkyl chains (C_{12}–C_{16}) are either randomly grafted on a hydrophilic polymer molecule such as hydroxyethyl cellulose (HEC) or simply grafted at both ends of the hydrophilic chain. An example of hydrophobically modified hydroxyethyl cellulose (HMHEC) is Natrosol Plus (Hercules) which contains 3–4 C_{16} chains randomly grafted onto HEC. Another example of a polymer that contains two alkyl chains at both ends of the molecule is HEUR (Rohm and Haas), which is made of polyethylene oxide (PEO) that is capped at both ends with a linear C_{18} hydrocarbon chain.

The above hydrophobically modified polymers form gels when dissolved in water. Gel formation can occur at relatively lower polymer concentrations than with the unmodified molecule.

The most likely explanation for gel formation is hydrophobic bonding (association) between the alkyl chains in the molecule. This effectively causes an apparent increase in the molecular weight. These associative structures are similar to micelles, except that the aggregation numbers are much smaller.

Figure 3.8 shows the variation of viscosity (measured using a Brookfield at 30 rpm) as a function of alkyl content (C_8, C_{12}, and C_{16}) for HMHEC. The viscosity reaches a maximum at a given alkyl group content and then decreases with increasing alkyl chain length. The viscosity maximum increases with increasing alkyl chain length.

Associative thickeners also interact with surfactant micelles that are present in the formulation. The viscosity of the associative thickeners shows a maximum at a given surfactant concentration that depends on the nature of surfactant. This is shown schematically in Figure 3.9.

The increase in viscosity is attributed to the hydrophobic interaction between the alkyl chains on the backbone of the polymer and the surfactant micelles. Figure 3.10 gives a schematic picture of the interaction between HM polymers and surfactant micelles.

At higher surfactant concentration, the "bridges" between the HM polymer molecules and the micelles are broken (free micelles) and η decreases.

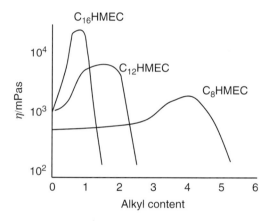

Figure 3.8 Variation of viscosity of 1% HMHEC versus alkyl group content of the polymer.

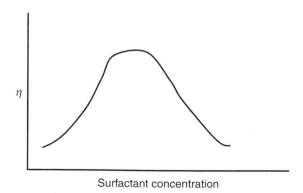

Figure 3.9 Schematic plot of viscosity of HM polymer with surfactant concentration.

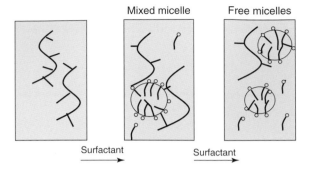

Figure 3.10 Schematic representation of the interaction of polymers with surfactants.

The viscosity of hydrophobically modified polymers shows a rapid increase at a critical concentration, which may be defined as the critical aggregation concentration (CAC), as is illustrated in Figure 3.11 for HMHEC (WSP-D45 from Hercules). The assumption is made that the CAC is equal to the coil overlap concentration C^*.

From a knowledge of C^* and the intrinsic viscosity $[\eta]$ one can obtain the number of chains in each aggregate. For the above example $[\eta] = 4.7$ and $C^*[\eta] = 1$, giving an aggregation number of ~ 4.

At C^* the polymer solution shows non-Newtonian flow (shear thinning behavior) and it exhibits a high viscosity at low shear rates. This is illustrated in Figure 3.12, which shows the variation of apparent viscosity with shear rate (using a constant stress rheometer). Below $\sim 0.1\,\mathrm{s}^{-1}$, a plateau viscosity value $\eta(o)$ (referred to as the *residual* or *zero shear viscosity*) is reached ($\sim 200\,\mathrm{Pa\,s}$).

With increasing polymer concentrations above C^* the zero shear viscosity increases. This is illustrated in Figure 3.13.

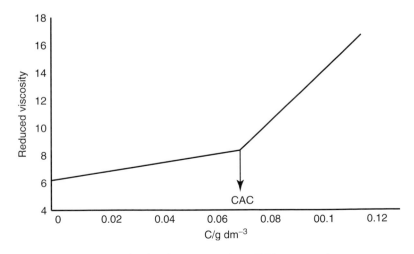

Figure 3.11 Variation of reduced viscosity with HMHEC concentration.

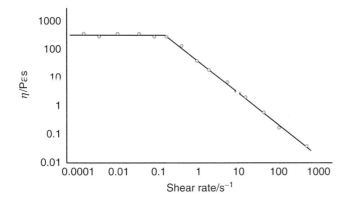

Figure 3.12 Variation of viscosity with shear rate for HMEC WSP-47 at 0.75 g per 100 cm^{-3}.

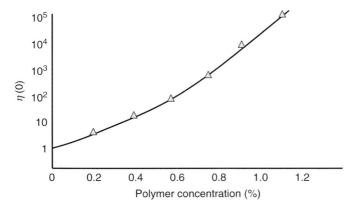

Figure 3.13 Variation of $\eta(0)$ with polymer concentration.

The above hydrophobically modified polymers are viscoelastic. This is illustrated in Figure 3.14 for a 5.25% solution of C$_{18}$ end-capped PEO with $M = 35\,000$, showing the variation of the storage modulus G' and loss modulus G'' with frequency ω (rad s^{-1}). G' increases with increasing frequency and ultimately reaches a plateau at high frequency. G'' (which is higher than G' in the low frequency regime) increases with increasing frequency, reaches a maximum at a characteristic frequency ω^* (at which $G' = G''$) and then decreases to a near zero value in the high frequency regime.

The above variation of G' and G'' with ω is typical for a system that shows Maxwell behavior.

From the cross over point ω^* (at which $G' = G''$) one can obtain the relaxation time τ of the polymer in solution:

$$\tau = \frac{1}{\omega^*} \tag{3.11}$$

For the above polymer $\tau = 8$ s.

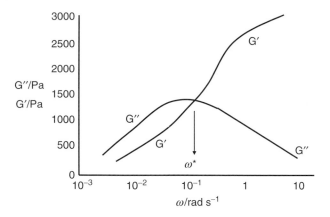

Figure 3.14 Variation of G' and G'' with frequency for 5.24 HM PEO.

The above gels (sometimes referred to as *rheology modifiers*) are used in many formulations to produce the right consistency and also for reduction of sedimentation or creaming of suspensions and emulsions. These hydrophobically modified polymers can also interact with hydrophobic particles in a suspension, forming several other associative structures.

The high frequency modulus, sometimes referred to as the *network modulus,* can be used to obtain the number of "links" in the gel network structure. Using the theory of rubber elasticity, the network modulus G_N is related to the number of elastically effective links N and a factor A that depends on the junction functionality:

$$G_N = ANk_B T \tag{3.12}$$

where k_B is the Boltzmann constant and T is the absolute temperature.

For an end-capped PEO (i.e., HEUR) the junctions should be multifunctional ($A = 1$); for tetra-functional junctions $A = \frac{1}{2}$.

3.5.2
Crosslinked Gels (Chemical Gels)

Many commercially available gels are made using crosslink agents to produce what are sometimes referred to as "*microgels.*" These microgel particles are dispersed in the liquid and they undergo solvent swelling that may also be enhanced by some chemical modification, for example, pH adjustment in aqueous systems.

A typical example of crosslinked gels is that based on poly(acrylic acid), which is commercially sold under the trade name "Carbopol®" (B.F. Goodrich). These microgel particles are dispersed in aqueous solution and on neutralization, with NaOH or ethanolamine, they swell as a result of the ionization of the poly(acrylic acid) chains. The ionization of poly(acrylic acid) occurs when the pH is increased above ~5 and these ionized chains form extended double layers, affording gels at low microgel concentration (mostly less than 1% of the microgel particles).

Another example of a microgel is that based on *N*-isopropyl acrylamide crosslinked with *N,N'*-methylene bisacrylamide (poly-NIPAM). These microgel particles are swollen by temperature changes. Above 35 °C, the crosslinked polymer is in a collapsed state. When the temperature is reduced the crosslinked polymer swells, absorbing water that reaches several orders of magnitudes its volume. These polymer gels are sometimes referred to as *"smart"* colloids and they have several applications in controlled release.

3.6
Particulate Gels

Two main interactions can cause gel formation with particulate materials:

1) long-range repulsions between the particles, for example, using extended electrical double layers or steric repulsion resulting from the presence of adsorbed or grafted surfactant or polymer chains;
2) van der Waals attraction between the particles (flocculation), which can produce three-dimensional gel networks in the continuous phase.

All the above systems produce non-Newtonian systems that show a "yield value" and high viscosity at low shear stresses or shear rates.

Several examples may be quoted to illustrate the above particulate gels:

• **Swellable clays**, for example, sodium montmorillonite (sometimes referred to as *bentonite*) at low electrolyte concentration: These produce gels as a result of the formation of extended double layers. At moderate electrolyte concentrations the clay particles may form association structures as a result of face-to-edge flocculation (see below). Such clays can be modified by interaction with alkyl ammonium salts (cationic surfactants) to produce hydrophobically modified clays sometimes referred to as *organo-clays* or *bentones*. These can be dispersed in non-aqueous media and swollen by addition of polar solvents.
• **Finely divided oxides**, for example, silica, can produce gels by aggregation of the particles to form three-dimensional gel structures.

In many cases, particulate solids are combined with high molecular weight polymers to enhance gel formation, for example, as a result of "bridging" or "depletion flocculation."

3.6.1
Aqueous Clay Gels

Gel formation using swellable clays such as sodium montmorillonite can be understood from the structure of the clay mineral and interparticle interactions in aqueous solutions. These clay minerals are made up of very thin plates (1 nm thick) formed from two layers of tetrahedral silica with one octahedral alumina sheet (between the two silica layers). The charge in a clay mineral is produced by a process referred to as *isomorphic substitution* whereby atoms with high valency are substituted with ions of lower valency (e.g., a Si^{4+} is replaced by an Al^{3+}). This produces a negative charge on the surface of the platelet that is compensated by Na^+ ions. The edges of the clay platelets contain oxide-like material, for example, Al-OH, which can acquire a positive charge at pH < 7 (the isoelectric point of Al-OH is pH 7–9).

Various interactions between clay particles in aqueous solution can be produced depending on the pH and electrolyte concentration. At very low electrolyte concentration, the double layers are extended and gels are produced when these double layers begin to overlap. At intermediate electrolyte concentrations, the double layers at the faces and edges are compressed and interaction may take place by edge-to-face association (making T-junctions, referred to as a "*house of cards structure*"). This also produces a gel in aqueous solution. These two types of interactions are shown schematically in Figure 3.15.

Evidence for the above structures may be obtained using rheological measurements. Figure 3.16 shows the variation of yield value σ_β (for a 3.22% clay dispersion)

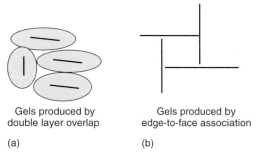

Gels produced by
double layer overlap

Gels produced by
edge-to-face association

(a)   (b)

Figure 3.15 Schematic representation of gel formation in aqueous clay dispersions: gels produced by (a) double layer overlap; (b) edge-to-face association.

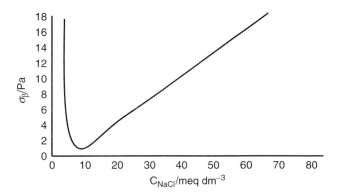

Figure 3.16 Variation of yield value with NaCl concentration for 3.22% sodium montmorillonite dispersions.

as a function of electrolyte concentration. When $C = 0$, the double layers are extended and gel formation is due to double layer overlap (Figure 3.15a). The first addition of NaCl causes compression of the double layers and hence the yield value decreases very rapidly. At intermediate NaCl concentrations, gel formation occurs as a result of face-to-edge association (house of cards structure) and the yield value increases very rapidly with increasing NaCl concentration. If the NaCl concentration is increased further, face-to-face association may occur and the yield value decreases (the gel is destroyed).

3.6.1.1 Organo-clays (Bentones)

These are produced by exchanging the Na^+ ions with alkyl ammonium ions, for example, dodecyl or cetyl trimethyl ammonium ions. In some examples dialkyl ammonium ions are also used. In this case the clay particle surface will be covered with hydrophobic alkyl groups and hence it can be dispersed in organic solvents, for example, hydrocarbon or silicone oils. The exchange is not carried out completely, leaving few hydrophilic groups on the surface. The dispersed organo-clays are then activated by addition of a polar solvent such as propylene carbonate, alcohols, glycols, and so on.

The gel is produced by hydrogen bonding between the polar groups on the surface of the clay and the polar solvent added. Several types of organo-clays are commercially available depending on the application and the type of solvent in which a gel is required. In some cases, organo-clays that are already activated can be supplied.

Organo-clays are applied to "thicken" many personal care products, for example, foundations, non-aqueous creams, nail polish, lipsticks, and so on. The procedure used for dispersion of the organo-clay particles and their subsequent activation is crucial and requires good process control.

3.6.2
Oxide Gels

The most commonly used oxide gels are based on silica. Various forms of silicas can be produced, the most common are referred to as *fumed* and *precipitated silicas*. Fumed silica such as Aerosil® 200 is produced by the reaction of silicon tetrachloride with steam. The surface contains siloxane bonds and isolated silanol groups (referred to as *vicinal*). Precipitated silicas are produced from sodium silicate by acidification. The surface is more populated with silanol groups than fumed silica. It contains geminal OH groups (two attached to the same Si atom). Both fumed and precipitated silicas can produce gels both in aqueous and non-aqueous systems. Gelation results from aggregation of silica particles, thus producing three-dimensional gel networks with a yield value.

In aqueous media, the gel strength depends on the pH and electrolyte concentration. As an illustration Figure 3.17 shows the variation of viscosity and yield value with Aerosil silica (which has been dispersed by sonication) concentration at three different pH values. In all cases, the viscosity and yield value shows a rapid increase above a certain silica concentration that depends on the pH of the system.

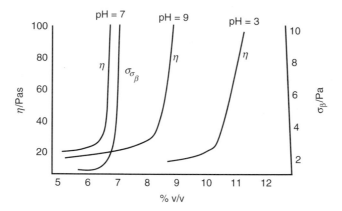

Figure 3.17 Variation of viscosity η and yield value σ_β with Aerosil 200 concentration at three different pH.

At pH 3 (near the isoelectric point of silica), the particles are aggregated (forming flocs) and the increase in viscosity occurs at relatively high silica concentration (>11% v/v). At pH 7, the silica becomes negatively charged and the double layers stabilize the silica particles against aggregation. In this case the particles remain as small units and the viscosity and yield value increases sharply above 7% v/v. At pH 9, some aggregation occurs as a result of the electrolyte released on adjusting the pH; in this case, the viscosity increases at higher concentrations (>9% v/v) when compared with the results at pH 7.

The above results clearly indicate the importance of pH and electrolyte concentration in the gelation of silica. It seems that optimum gel formation occurs at neutral pH.

Silica gels are used in many personal care applications, in particular for control of the rheology of toothpaste. Hydrophilic silica (Aerosil 200) can also be applied to form gels in non-aqueous media. In this case, gel formation is the result of hydrogen bond formation between the particles that produce three-dimensional structures in the non-aqueous medium.

3.6.3
Gels Produced Using Particulate Solids and High Molecular Weight Polymers

In many cases particulate materials are combined with high molecular weight polymers to produce three-dimensional structures by association of the polymer with the particles. Several mechanisms have been suggested for gel formation in these mixtures, for example, bridging by the polymer chains and depletion flocculation.

The above mixtures give more robust gel structures, which in many cases exhibit a smaller temperature dependence than those of the individual components. The optimum composition of these particulate–polymer mixtures can be obtained using rheological measurements.

By measuring the yield value as a function of polymer concentration at a fixed particulate concentration, one can obtain the optimum polymer concentration required. In most cases the yield value reaches a maximum at a given ratio of particulate solid to polymer. This trend may be due to bridging flocculation, which reaches an optimum at a given surface coverage of the particles (usually at 0.25–0.5 surface coverage).

All the above-mentioned gels produce thixotropy, that is, a reversible decrease of viscosity on application of shear (at constant shear rate) and recovery of the viscosity on standing. This thixotropic behavior finds application in many systems in personal care, for example, in creams, toothpastes, and foundations.

One of the most useful techniques for studying thixotropy is to follow the change of modulus with time after application of shear, that is, after subjecting the dispersion to a constant shear rate, oscillatory measurements are carried out at low strains and high frequency and the increase in modulus with time (which is exponential) can be used to characterize the recovery of the gel.

3.7
Rheology Modifiers Based on Surfactant Systems

In dilute solutions surfactants tend to form spherical micelles with aggregation numbers in the range 50–100 units. These micellar solutions are isotropic with low viscosity. At much higher surfactant concentration (>30% depending on the surfactant nature) they produce liquid crystalline phases of the hexagonal (H_1) and lamellar (L_α) phases, which are anisotropic with much higher viscosities. Figures 3.18 and 3.19 show a schematic representation of the hexagonal and lamellar phases, respectively.

Water

Surfactant

Figure 3.18 Schematic representation of hexagonal phase.

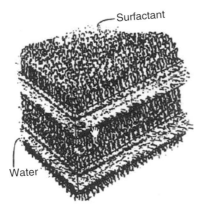

Figure 3.19 Schematic representation of lamellar phase.

These liquid crystalline phases, which are viscoelastic, can be used as rheology modifiers. However, for practical applications such as in shampoos such very high surfactant concentrations are undesirable. One way to increase the viscosity of a surfactant solution at lower concentrations is to add an electrolyte that causes a change from spherical to cylindrical micelles; such micelles can grow in length, and above a critical surfactant volume fraction φ^* these worm-like micelles begin to overlap, forming a "gel" (Figure 3.20).

An alternative method for producing gels in emulsions is to use mixtures of surfactants. By proper choice of the surfactant types (e.g., their hydrophilic–lipophilic balance, HLB) one can produce lamellar liquid crystalline structures that can "wrap" around the oil droplets and extend in solution to form gel networks. These structures (sometimes referred to as *oleosomes*) are schematically shown in Figure 3.21.

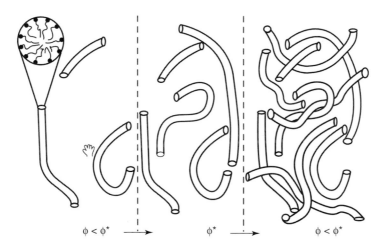

$\phi < \phi^* \longrightarrow \qquad \phi^* \longrightarrow \qquad \phi < \phi^*$

Figure 3.20 Overlap of thread-like micelles.

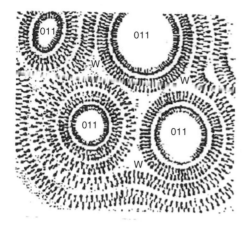

Figure 3.21 Schematic representation of "oleosomes."

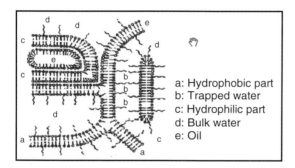

Figure 3.22 Schematic representation of "hydrosomes."

Alternatively, the liquid crystalline structures may produce a "three-dimensional" gel network and the oil droplets become entrapped in the "holes" of the network. These structures are sometimes referred to as *hydrosomes* (Figure 3.22).

The above-mentioned surfactant systems are used in most personal care and cosmetic formulations. Apart from giving the right consistency for application (e.g., good skin feel) they are also effective in the stabilization of emulsions against creaming or sedimentation, flocculation, and coalescence.

Liquid crystalline structures can also influence the delivery of active ingredients, both the lipophilic and hydrophilic types. Since lamellar liquid crystals mimic the skin structure (in particular the stratum corneum) they can offer prolonged hydration potential.

References

1. de Gennes, P.G. (1979) *Scaling Concepts in Polymer Physics*, Cornell University Press, Ithaca.

2. Ferry, J.D. (1980) *Viscoelastic Properties of Polymers*, John Wiley & Sons, Inc., New York.

3. Goddard, E.D. (1999) in *Polymer/Surfactant Interaction* (eds E.D. Goddard and J.V. Gruber), Marcel Dekker, New York, ch. 4 and 5.

4. Van Olphen, H. (1961) *Clay Colloid Chemistry*, John Wiley & Sons, Inc., New York.

5. Heath, D. and Tadros, T.F. (1983) *J. Colloid Interface Sci.*, **93** (307), 320.

6. Tadros, T.F. (2005) *Applied Surfactants*, Wiley-VCH Verlag GmbH, Weinheim.

4

Use of Rheological Measurements for Assessment and Prediction of the Long-Term Assessment of Creaming and Sedimentation

Tharwat F. Tadros

4.1
Introduction

Most formulations undergo creaming or sedimentation as a result of the density difference between disperse phase particles and the medium, unless the particles or droplets are small enough for Brownian motion to overcome gravity [1, 2]. This is illustrated in Figure 4.1 for three cases of suspensions.

The case shown in Figure 4.1a represents the situation when the Brownian diffusion energy (which is in the region of $k_B T$, where k_B is the Boltzmann constant and T is the absolute temperature) is much larger than the gravitational potential energy, which is equal to $(4/3)\pi R^3 \Delta \rho g h$, where R is the particle radius, $\Delta \rho$ is the density difference between the particles and medium, g is the acceleration due to gravity, and h is the height of the container. Under these conditions, the particles become randomly distributed throughout the whole system, and no separation occurs. This situation may occur with nano-suspensions with radii less than 100 nm, particularly if $\Delta \rho$ is not large, say <0.1. In contrast, when $(4/3)\pi R^3 \Delta \rho g h \gg k_B T$, complete sedimentation occurs as is illustrated in Figure 4.1b with suspensions of uniform particles. In such case, the repulsive force necessary to ensure colloid stability enables the particles to move past each other to form a compact layer [1, 2]. As a consequence of the dense packing and small spaces between the particles, such compact sediments (which are technically referred to as *"clays"* or *"cakes"*) are difficult to redisperse. In rheological terms, the close packed sediment undergoes shear thickening, which is referred to as *dilatancy*, that is, a rapid increase in the viscosity with increasing shear rate.

The most practical situation is that represented by Figure 4.1c, whereby a concentration gradient of the particles occurs across the container. The concentration of particles C can be related to that at the bottom of the container C_0 by the following equation:

$$C = C_0 \exp\left(-\frac{mgh}{k_B T}\right) \tag{4.1}$$

Product Design and Engineering: Formulation of Gels and Pastes, First Edition.
Edited by Ulrich Bröckel, Willi Meier, and Gerhard Wagner.

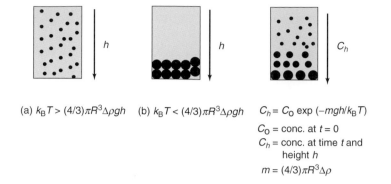

(a) $k_BT > (4/3)\pi R^3 \Delta \rho g h$ (b) $k_BT < (4/3)\pi R^3 \Delta \rho g h$ $C_h = C_0 \exp(-mgh/k_BT)$

C_0 = conc. at $t = 0$
C_h = conc. at time t and
height h
$m = (4/3)\pi R^3 \Delta \rho$

Figure 4.1 Schematic representation of sedimentation of suspensions: (a) $k_BT >$ $(4/3)\pi R^3 \Delta \rho g h$; (b) $k_BT < (4/3)\pi R^3 \Delta \rho g h$; and (c) $C_h = C_0 \exp(-mgh/k_BT)$, $C_0 =$ conc. at $t = 0$, $C_h =$ conc. at time t and height h, $m = (4/3)\pi R^3 \Delta \rho$.

where m is the mass of the particles, which is given by $(4/3)\pi R^3 \Delta \rho$ (R is the particle radius and $\Delta \rho$ is the density difference between particle and medium), g is the acceleration due to gravity, and h is the height of the container.

4.2
Accelerated Tests and Their Limitations

Several tests have been designed to accelerate the process of sedimentation or creaming – the most commonly used methods are based on increasing temperature or subjecting the suspension or emulsion to high g forces (using a high speed centrifuge).

With increasing temperature, the viscosity of the system usually decreases and hence sedimentation or creaming is accelerated. The assumption is usually made that if a suspension or emulsion does not show any sedimentation, creaming, or separation at 50 °C for say one month then the system will show no separation at ambient temperatures for more than one year.

The above method is only valid if the formulation viscosity η follows the Arrhenius equation, which predicts a linear increase in $\ln \eta$ with $(1/T)$ where T is the absolute temperature. Most practical formulations do not follow such plot due to the possible phase changes or flocculation that may occur at high temperatures. With many surfactant systems, such phase changes may result in the formation of liquid crystalline phases that have higher viscosity at high temperatures and hence no separation results at high temperatures, although such changes could occur at ambient conditions.

4.3
Application of High Gravity (g) Force

This method, if carefully applied, may offer a better accelerated technique. It has been particularly applied to emulsions. The assumption is also made here that by increasing the g force the rate of sedimentation or creaming is significantly increased and this could be applied to predict the process from measurements at short time periods.

In a centrifuge, the gravity force is given by:

$$g = \omega^2 x \tag{4.2}$$

where x is the mean distance of the centrifuge tube from the axis of rotation and ω is the angular velocity ($\omega = 2\pi\nu$, where ν is the number of revolutions per second). Note that if the centrifuge tube is not small compared to x then the applied centrifugal field cannot be considered to be uniform over the length of the tube.

Modern analytical ultracentrifuges allow one to follow the separation of emulsions in a quantitative manner. With typical oil-in-water (O/W) emulsions, three layers are generally observed: (i) a clear aqueous phase; (ii) an opaque phase consisting of distorted polyhedral oil droplets; (iii) a clear separated oil phase, resulting from coalescence of the polyhedra.

The degree of emulsion stability may be taken as the volume of the opaque phase remaining after time t. Alternatively, one may use the volume of oil separated at infinite time as an index for stability.

A simple expression may be used to treat the data in a quantitative manner:

$$\frac{t}{V} = \frac{1}{bV_\infty} + \frac{1}{V_\infty} \tag{4.3}$$

where V is the volume of oil separated at time t, V_∞ is the extrapolated volume at infinite time, and b is a constant.

A plot of t/V versus t should give a straight line from which b and V_∞ may be calculated. These two parameters may be taken as indices for emulsion stability.

A more rigorous procedure to study emulsion stability using an ultracentrifuge is to observe the system at various speeds of rotation. At relatively low centrifuge speeds one may observe the expected opaque cream layer. At sufficiently high centrifuge speeds, one may observe a coalesced oil layer and a cream layer that are separated by an extra layer of deformed oil droplets. This deformed layer looks like a "foam," that is, it consists of oil droplets separated by thin aqueous films.

For certain emulsions, one may find that by increasing the centrifuge speed the "foam"/cream layer boundary does not move. Under conditions where there is an equilibrium between the "foam" and the cream layer, one may conclude that there is no barrier to be overcome in forming the foam layer from the cream layer. This implies that in the foam layer the aqueous film separating two oil droplets thins to a "black" film under the action of van der Waals forces. The boundary between the foam layer and the coalesced layer is associated with a force (or pressure) barrier.

One may observe the minimum centrifuge speed that is necessary to produce a visible amount of coalesced oil after say 30 min of centrifugation. This centrifuge speed may be used to calculate the "critical pressure" that needs to be applied to induce coalescence.

4.4
Rheological Techniques for Prediction of Sedimentation or Creaming

Sedimentation or creaming is prevented by the addition of "thickeners" that form a "three-dimensional elastic" network in the continuous phase. If the viscosity of the elastic network, at shear stresses (or shear rates) comparable to those exerted by the particles or droplets, exceeds a certain value then creaming or sedimentation is completely eliminated.

The shear stress, σ_p, exerted by a particle (force area^{-1}) can be calculated simply [3]:

$$\sigma_p = \frac{\frac{4}{3}\pi R^3 \Delta\rho g}{4\pi R^2} = \frac{\Delta\rho\, Rg}{3} \tag{4.4}$$

For a 10 μm radius particle with a density difference $\Delta\rho$ of 0.2 g cm^{-3} the stress is equal to:

$$\sigma_p = \frac{0.2 \times 10^3 \times 10 \times 10^{-6} \times 9.8}{3} \approx 6 \times 10^{-3}\,\mathrm{Pa} \tag{4.5}$$

For smaller particles smaller stresses are exerted.

Thus, to predict creaming or sedimentation, one has to measure the viscosity at very low stresses (or shear rates). These measurements can be carried out using a constant stress rheometer (Carri-Med, Bohlin, Rheometrics, Haake, or Physica).

Usually, one obtains a good correlation between the rate of creaming or sedimentation v and the residual viscosity $\eta(0)$. This is illustrated in Figure 4.2. Above a certain value of $\eta(0)$, v becomes equal to 0. Clearly, to minimize creaming or sedimentation one has to increase $\eta(0)$; an acceptable level for the high shear viscosity η_∞ must be achieved, depending on the application. In some cases, a high $\eta(0)$ may be accompanied by a high η_∞ (which may not be acceptable for application, for example, if spreading of a dispersion on the skin is required). If this is the case, the formulation chemist should look for an alternative thickener.

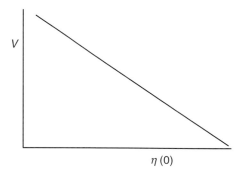

Figure 4.2 Variation of creaming or sedimentation rate with residual viscosity.

4.5
Separation of Formulation ("Syneresis")

Another problem encountered with many dispersions is that of "syneresis," that is, the appearance of a clear liquid film at either the bottom (if creaming is the case) or top (if sedimentation is the case) of the container. Syneresis occurs with most "flocculated" and/or "structured" (i.e., those containing a thickener in the continuous phase) dispersions.

Syneresis may be predicted from measurement of the yield value (using steady state measurements of shear stress as a function of shear rate) as a function of time or by using oscillatory techniques (whereby the storage and loss modulus are measured as a function of strain amplitude and frequency of oscillation).

The oscillatory measurements are perhaps more useful since to prevent separation the bulk modulus of the system should balance the gravity forces given by $h\Delta\rho g$ (where h is the height of the disperse phase, $\Delta\rho$ is the density difference, and g is the acceleration due to gravity).

The bulk modulus is related to the storage modulus G'. A more useful predictive test is to calculate the cohesive energy density of the structure E_c that is given by the following equation [3]:

$$E_c = \int_0^{\gamma_{cr}} G'\gamma\,d\gamma = \frac{1}{2}G'\gamma_{cr}^2 \tag{4.6}$$

The separation of a formulation decreases with increasing E_c. This is illustrated in Figure 4.3, which shows schematically the reduction in % Separation with increasing E_c. The value of E_c that is required to stop complete separation depends on the particle or droplet size distribution, the density difference between the particle or droplet and the medium, as well as the volume fraction φ of the dispersion.

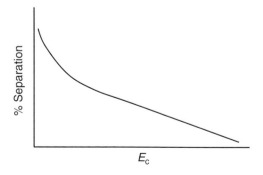

Figure 4.3 Schematic representation of the variation of % separation with E_c.

4.6
Examples of Correlation of Sedimentation or Creaming with Residual (Zero Shear) Viscosity

4.6.1
Model Suspensions of Aqueous Polystyrene Latex

The sedimentation rate is a complex function of the volume fraction φ. This was tested using polystyrene latex suspensions with radius $R = 1.55\,\mu m$ in 10^{-3} mol dm^{-3} NaCl [4].

One may be able to correlate the change in the rate of sedimentation with increasing φ with the viscosity of the suspension as predicted by the Dougherty–Krieger equation [5]:

$$\frac{v}{v_0} \propto \frac{\eta_0}{\eta} \tag{4.7}$$

$$\frac{v}{v_0} = \alpha \frac{\eta_0}{\eta} \tag{4.8}$$

where α is a constant.

By combining Equation 4.8 with:

$$\frac{\eta}{\eta_0} = \left[1 - \left(\frac{\phi}{\phi_p}\right)\right]^{-[\eta]\phi_p} \tag{4.9}$$

one can obtain the following:

$$\frac{v}{v_0} = \left(1 - \frac{\phi}{\phi_p}\right)^{\alpha[\eta]\phi_p} = \left(1 - \frac{\phi}{\phi_p}\right)^{k\phi_p} \tag{4.10}$$

where φ_p is the maximum packing fraction and $[\eta]$ is the intrinsic viscosity.

Equation 4.10 was tested for polystyrene dispersions (Figure 4.4).

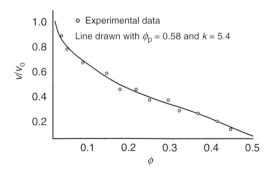

Figure 4.4 Variation of sedimentation rate with volume fraction for polystyrene dispersions.

4.6.2
Sedimentation in Non-Newtonian Liquids

To reduce sedimentation one usually adds high molecular weight material, for example, hydroxyethyl cellulose or xanthan gum (Kelzan®, Keltrol®, or Rhodopol®). Above a critical concentration, C^* such polymer solutions show non-Newtonian flow in aqueous solution. This is illustrated in Figure 4.5, which shows the variation of shear stress and viscosity with shear rate. Figure 4.6 shows the variation of log η with log C to illustrate the onset of free coil overlap.

Before overlap $\eta \propto C$, whereas after overlap $\eta \propto C^{3.4}$. Two limiting Newtonian viscosities are identified: (i) residual (zero shear) viscosity $\eta(0)$ and (ii) Newtonian high shear rate viscosity η_∞. The value of $\eta(0)$ can be several orders of magnitude $(10^3–10^5)$ higher than η_∞ and such a high $\eta(0)$ can significantly reduce creaming or sedimentation.

4.6.3
Role of Thickeners

As mentioned above, thickeners reduce creaming or sedimentation by increasing the residual viscosity $\eta(0)$, which must be measured at stresses comparable to

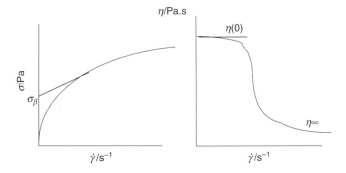

Figure 4.5 Flow behavior of "thickeners."

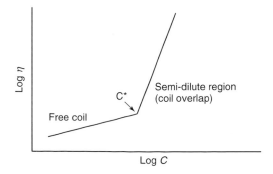

Figure 4.6 Variation of log η with log C.

those exerted by the droplets or particles (mostly less than 0.1 Pa). At such low stresses, $\eta(0)$ increases very rapidly with increasing "thickener" concentration. This rapid increase is not observed at high stresses, which illustrates the need for measurement at low stresses (using constant stress or creep measurements). For example, Figure 4.7 shows the variation of η with applied stress σ for *EHEC*, a thickener that is applied in some formulations [4].

It can be seen that the limiting residual viscosity increases rapidly with increasing EHEC concentration. A plot of sedimentation rate for 1.55 μm PS (polystyrene) latex particles versus $\eta(0)$ is presented in Figure 4.8, which shows an excellent correlation. In this case a value of $\eta(0) \geq 10$ Pa s is sufficient to reduce the rate of sedimentation to 0 [4].

4.6.4
Prediction of Emulsion Creaming

For the above purposes some model emulsions were prepared using mixtures of oils and commercial surfactants. The oil phase of the emulsion consisted of ten

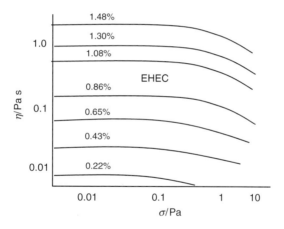

Figure 4.7 Constant stress (creep) measurements for PS latex dispersions as a function of ethyl hydroxyethyl cellulose (*EHEC*) concentration.

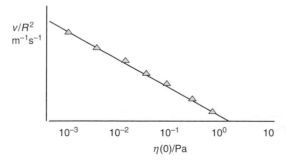

Figure 4.8 Sedimentation rate versus $\eta(0)$.

parts Arlamol™ HD (isohexadecane, supplied by UNIQEMA, ICI), two parts Estol 3603 (caprylic/capric triglyceride supplied by UNIQEMA), one part sunflower oil (Florasen® 90, *Helanthus annus*, supplied by Florateck), and one part avocado oil (*Persea gratissima* supplied by Mosselman) [6].

Two emulsifier systems were used for the preparation of O/W emulsions. The first emulsifier was Synperonic® PEF 127, an A-B-A block copolymer of poly(ethylene oxide), *PEO* (the A chains, about 100 EO units each) and poly(propylene oxide), *PPO* (the B chain, about 55 PO units), supplied by UNIQEMA. The second emulsifier system was Arlatone V-100 (supplied by UNIQEMA), which is a nonionic emulsifier systems made of a blend of Steareth-100 (stearyl alcohol with 100 EO units), Steareth-2 (Stearyl alcohol with 2 EO units), glyceryl stearate citrate, sucrose, and a mixture of two polysaccharides, namely, mannan and xanthan gum (Keltrol F, supplied by Kelco). In some emulsions, xanthan gum was used as a thickener. All emulsions contained a preservative (Nipaguard® BPX).

The rate of creaming and cream volume was measured using graduated cylinders. The creaming rate was assessed by comparing the cream volume V_c with that of the maximum value V_∞ obtained when the emulsion was stored at $55\,^\circ$C. The time $t_{0.3}$ taken to reach a value of $V_c/V_\infty = 0.3$ (i.e., 30% of the maximum rate) was calculated [7].

All rheological measurements were carried out using a Physica *UDS* 200 (Universal Dynamic Spectrometer). A cone-and-plate geometry was used with a cone angle of 2°. The emulsions were also investigated using optical microscopy and image analysis.

Figure 4.9 shows the results for creaming rates obtained at various temperatures, using a 20/80 O/W emulsion stabilized with Synperonic PEF 127. Clearly, $t_{0.3}$ decreases with increasing temperature [6].

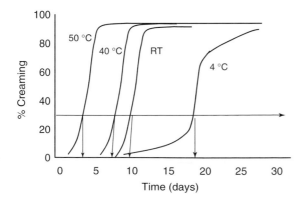

Figure 4.9 Percentage creaming versus time at various temperatures.

4.6.5
Creep Measurements for Prediction of Creaming

The most useful method by which to predict creaming is to use constant stress (creep) measurements. From these measurements one can obtain the residual (zero shear) viscosity $\eta(0)$.

Results were obtained for 20/80% v/v emulsions as a function of Arlatone V-100 concentration. The data are shown in Figure 4.10 after several periods of storage (one day, one week, two weeks, and one month). The $\eta(0)$ showed a large decrease after one day, which could be due to equilibration of the structure. The results after one week, two weeks, and one month are close to each other. There is a significant increase in $\eta(0)$ when the Arlatone® V 100 concentration is increased above 0.8%. The creaming rate of the emulsion also showed a sharp decrease above 0.8% Arlatone V 100, indicating the correlation between $\eta(0)$ and creaming rate.

4.6.6
Oscillatory Measurements for Prediction of Creaming

A very useful method for the prediction of creaming is to measure the cohesive energy density (E_c) as given by Equation 4.6. As an illustration Figure 4.11 shows the variation of E_c with Arlatone V 100 concentration. The results clearly show a rapid increase in E_c above 0.8% Arlatone V 100. The value of E_c seems to show a decrease after storage for two weeks. This may be due to a small increase in droplet size (as a result of some coalescence) that results in a reduction in the cohesive energy density. This small increase in droplet size could not be detected by microscopy since the change was very small.

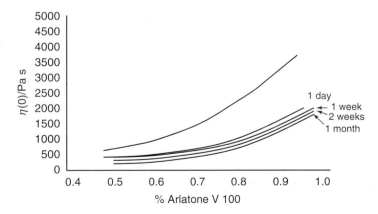

Figure 4.10 Variation of residual viscosity with Arlatone V 100 concentration at various storage times.

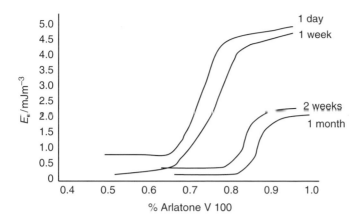

Figure 4.11 Variation of E_c with % Arlatone V 100 in the emulsion.

4.7
Assessment and Prediction of Flocculation Using Rheological Techniques

4.7.1
Stability/Instability of Electrostatically Stabilized Dispersions: Derjaguin–Landau–Verwey–Overbeek (DLVO) Theory

The Derjaguin–Landau–Verwey–Overbeek (*DLVO*) theory [8, 9] is based on a combination of electrostatic repulsion free energy G_{el} with that of van der Waals attraction G_A:

$$G_T = G_{el} + G_A \tag{4.11}$$

This is illustrated in Figure 4.12. G_{el} decays exponentially with h, that is, $G_{el} \to 0$ as h becomes large. G_A is ∞ $1/h$, that is, G_A does not decay to 0 at large h. At long separation distances, $G_A > G_{el}$, resulting in a shallow minimum (secondary minimum) (few $k_B T$ units). At very short distances, $G_A \gg G_{el}$, resulting in a

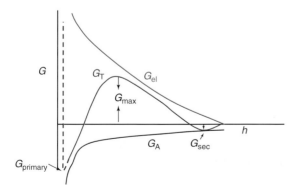

Figure 4.12 Energy–distance curves according to the DLVO theory.

deep primary minimum (several hundred $k_B T$ units). At intermediate distances, $G_{el} > G_A$, resulting in an energy maximum, G_{max}, whose height depends on the surface potential ψ_0 (or the Stern potential ψ_d), particle radius R, and the electrolyte concentration and valency.

At low electrolyte concentrations ($<10^{-2}$ mol dm^{-3} for a $1:1$ electrolyte), G_{max} is high ($>25k_B T$) and this prevents particle aggregation into the primary minimum. The higher the electrolyte concentration (and the higher the valency of the ions), the lower the energy maximum. Under some conditions (depending on electrolyte concentration and particle size), flocculation into the secondary minimum may occur. This flocculation is weak and reversible. By increasing the electrolyte concentration, G_{max} decreases till at a given concentration it vanishes and particle coagulation occurs. This is illustrated in Figure 4.13, which shows the energy–distance curves at various electrolyte concentrations.

Flocculation of colloidal dispersions occurs under conditions where the van der Waals attraction exceeds the repulsive energy. Flocculation can be "weak" or "strong" depending on the magnitude of the attractive energy. Weak flocculation is accompanied by a low energy of attraction (of the order of few $k_B T$ units, where k_B is the Boltzmann constant and T is the absolute temperature) as is the case with the secondary minimum for electrostatically stabilized systems. Strong flocculation (sometimes referred to as *coagulation*) is accompanied by large attractive energies (several hundred $k_B T$ units), as is the case with the primary minimum.

The stability of dispersions containing adsorbed polymeric surfactants arises from two main effects [10]: (i) unfavorable mixing of the stabilizing chains when these are in good solvent conditions – this gives a repulsive energy G_{mix} – and (ii) reduction of configurational entropy on significant overlap characterized by a repulsive energy G_{el}. The combination of G_{mix} and G_{el} with G_A gives the total energy of interaction G_T (assuming there is no contribution from any residual electrostatic interaction), that is:

$$G_T = G_{mix} + G_{el} + G_A \tag{4.12}$$

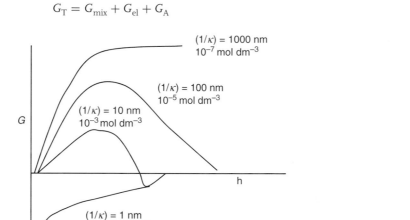

Figure 4.13 Energy–distance curves at various 1 : 1 electrolyte concentrations.

Figure 4.14 shows a schematic representation of the variation of G_{mix}, G_{el}, G_A, and G_T with surface–surface separation distance h.

G_{mix} increases very sharply with decreasing h when $h < 2\delta$; G_{el} increases very sharply with decreasing h when $h < \delta$. G_T versus h shows a minimum, G_{min}, at separation distances comparable to 2δ; when $h < 2\delta$, G_T shows a rapid increase with decreasing h. Unlike the $G_T–h$ curve predicted by DLVO theory (which shows two minima and one energy maximum), the $G_T–h$ for systems that are sterically stabilized show only one minimum, G_{min}, followed by sharp increase in G_T with decreasing h (when $h < 2\delta$). The depth of the minimum depends on the Hamaker constant A, the particle radius R, and the adsorbed layer thickness δ; G_{min} increases with increasing A and R. At a given A and R, G_{min} increases with decreasing δ (i.e., with decreasing molecular weight, M_w, of the stabilizer, This is illustrated in Figure 4.15, which shows the energy–distance curves as a function of δ/R; the larger the value of δ/R, the smaller the value of G_{min}.

Two main types of flocculation of sterically stabilized dispersions may be distinguished: (i) Weak flocculation: this occurs when the thickness of the adsorbed

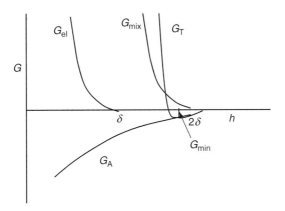

Figure 4.14 Variation of G_{mix}, G_{el}, G_A, and G_T with surface–surface separation distance h.

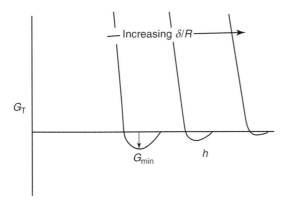

Figure 4.15 Energy–distance curves as a function of δ/R.

layer is small (usually <5 nm), particularly when the particle radius (R) and Hamaker constant (A) are large. (ii) Incipient flocculation: this occurs when the solvency of the medium is reduced to become worse than θ-solvent (i.e., $\chi > 0.5$). This is illustrated in Figure 4.16, where χ was increased from <0.5 (good solvent) to >0.5 (poor solvent). When $\chi > 0.5$, G_{mix} becomes negative (attractive), which when combined with the van der Waals attraction at this separation distance gives a deep minimum causing flocculation. In most cases, there is a correlation between the critical flocculation point and the θ condition of the medium.

4.7.2
Rheological Techniques for Studying Flocculation

Steady state rheological investigations may be used to investigate the state of flocculation of a dispersion. Weakly flocculated dispersions usually show thixotropy, and the change of thixotropy with applied time may be used as an indication of the strength of this weak flocculation [3].

The above methods are only qualitative and one cannot use the results in a quantitative manner. This is due to the possible breakdown of the structure on transferring the formulation to the rheometer and also during the uncontrolled shear experiment. Better techniques to study the flocculation of a formulation are constant stress (creep) or oscillatory measurements. By careful transfer of the sample to the rheometer (with minimum shear) the structure of the flocculated system may be maintained.

4.7.3
Wall Slip

A very important point that must be considered in any rheological measurement is the possibility of "slip" during the measurements [3]. This is particularly the

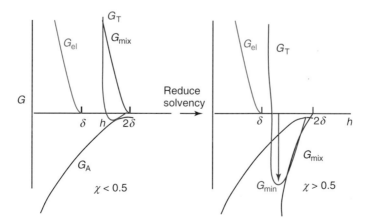

Figure 4.16 Influence of reduction in solvency on the energy–distance curve.

case with highly concentrated dispersions, whereby the flocculated system may form a "plug" in the gap of the platens leaving a thin liquid film at the walls of the concentric cylinder or cone-and-plate geometry. This behavior is caused by some "syneresis" of the formulation in the gap of the concentric cylinder or cone-and-plate. To reduce "slip" one should use roughened walls for the platens. A vane rheometer may also be used.

4.7.4
Steady State Shear Stress–Shear Rate Measurements

This is by far the most commonly used method in many industrial laboratories. Basically, the dispersion is stored at various temperatures and the yield value σ_β and plastic viscosity η_{pl} are measured at various intervals of time. Any flocculation in the formulation should be accompanied by an increase in σ_β and η_{pl}.

A rapid way to study the effect of temperature changes on the flocculation of a formulation is to carry out temperature sweep experiments, running the samples from say 5 to 50 °C. The trend in the variation of σ_β and η_{pl} with temperature can quickly give an indication on the temperature range at which a dispersion remains stable (during that temperature range σ_β and η_{pl} remain constant).

4.7.5
Influence of Ostwald Ripening and Coalescence

If Ostwald ripening and/or coalescence occur simultaneously, σ_β and η_{pl} may change in a complex manner with storage time. Ostwald ripening and/or coalescence result in a shift of the particle size distribution to higher diameters. This has the effect of reducing σ_β and η_{pl}. If flocculation occurs simultaneously (having the effect of increasing these rheological parameters), the net effect may be an increase or decrease of the rheological parameters.

The above trend depends on the extent of flocculation relative to Ostwald ripening and/or coalescence. Therefore, following σ_β and η_{pl} with storage time requires knowledge of Ostwald ripening and/or coalescence. Only in the absence of these latter breakdown processes can one use rheological measurements as a guide for the assessment of flocculation.

4.7.6
Constant Stress (Creep) Experiments

Basically, a constant stress σ is applied on the system and the compliance J (Pa^{-1}) is plotted as a function of time. These experiments are repeated several times, increasing the stress from the smallest possible value (that can be applied by the instrument), increasing the stress in small increments. A set of creep curves are produced at various applied stresses. From the slope of the linear portion of the creep curve (after the system reaches a steady state), the viscosity at each applied stress, η_σ, is calculated. A plot of η_σ versus σ allows one to obtain the limiting (or zero shear) viscosity $\eta(0)$ and the critical stress σ_{cr} (which may be identified with the "true" yield stress of the system) [3]. The values of $\eta(0)$ and σ_{cr} may be used to assess the flocculation of the dispersion on storage.

If flocculation occurs on storage (without any Ostwald ripening or coalescence), the values of $\eta(0)$ and σ_{cr} may show a gradual increase with increasing storage time. As discussed in the section above on steady state measurements, the trend becomes complicated if Ostwald ripening and/or coalescence occur simultaneously [both have the effect of reducing $\eta(0)$ and σ_{cr}].

The above measurements should be supplemented by particle size distribution measurements of the diluted dispersion (making sure that no flocs are present after dilution) to assess the extent of Ostwald ripening and/or coalescence. Another complication may arise from the nature of the flocculation. If the latter occurs in an irregular way (producing strong and tight flocs), $\eta(0)$ may increase, while σ_{cr} may show some decrease and this complicates the analysis of the results. Despite these complications, constant stress measurements may provide valuable information on the state of the dispersion on storage.

Carrying out creep experiments and ensuring that a steady state is reached can be time consuming. One usually carries out a stress sweep experiment, whereby the stress is gradually increased (within a predetermined time period to ensure that one is not too far from reaching the steady state) and plots of η_σ versus σ are established.

The above experiments are carried out at various storage times (say every two weeks) and temperatures. From the change of $\eta(0)$ and σ_{cr} with storage time and temperature, one may obtain information on the degree and the rate of flocculation of the system. Clearly, interpretation of the rheological results requires expert knowledge of rheology and measurement of the particle size distribution as a function of time.

One main problem in carrying the above experiments is sample preparation. When a flocculated dispersion is removed from the container, care should be taken not to cause much disturbance to that structure (minimum shear should be applied on transferring the formulation to the rheometer). It is also advisable to use separate containers for assessment of the flocculation; a relatively large sample is prepared and this is then transferred to several separate containers. Each sample is used separately at a given storage time and temperature. One should be careful in transferring the sample to the rheometer. If any separation occurs in the formulation the sample is gently mixed by placing it on a roller. It is advisable to use the minimum shear possible when transferring the sample from the container to the rheometer (the sample is preferably transferred using a "spoon" or by simple pouring from the container). The experiment should be carried out without an initial pre-shear.

4.7.7
Dynamic (Oscillatory) Measurements

In oscillatory measurements one carries out two sets of experiments [3].

4.7.7.1 Strain Sweep Measurements

In this case, the oscillation is fixed (say at 1 Hz) and the viscoelastic parameters are measured as a function of strain amplitude. G^*, G', and G'' remain virtually constant up to a critical strain value, γ_{cr}. This region is the linear viscoelastic region. Above γ_{cr}, G^* and G' starts to fall, whereas G'' starts to increase; this is the nonlinear region.

The value of γ_{cr} may be identified with the minimum strain above which the "structure" of the dispersion starts to break down (e.g., breakdown of flocs into smaller units and/or breakdown of a "structuring" agent).

From γ_{cr} and G', one can obtain the cohesive energy E_c (J m^{-3}) of the flocculated structure using Equation 4.6. E_c may be used in a quantitative manner as a measure of the extent and strength of the flocculated structure in a dispersion. The higher the value of E_c the more flocculated the structure is.

Clearly, E_c depends on the volume fraction of the dispersion as well as the particle size distribution (which determines the number of contact points in a floc). Therefore, for quantitative comparison between various systems, one has to make sure that the volume fraction of the disperse particles is the same and that the dispersions have very similar particle size distributions. E_c also depends on the strength of the flocculated structure, that is, the energy of attraction between the droplets. This depends on whether the flocculation is in the primary or secondary minimum. Flocculation in the primary minimum is associated with a large attractive energy and this leads to higher values of E_c than those obtained for secondary minimum flocculation (weak flocculation). For a weakly flocculated dispersion, such as the case with secondary minimum flocculation of an electrostatically stabilized system, the deeper the secondary minimum the higher the value of E_c (at any given volume fraction and particle size distribution of the dispersion).

With a sterically stabilized dispersion, weak flocculation can also occur when the thickness of the adsorbed layer decreases. Again, the value of E_c can be used as a measure of the flocculation – the higher the value of E_c, the stronger the flocculation. If incipient flocculation occurs (on reducing the solvency of the medium for the change to worse than θ-condition) a much deeper minimum is observed and this is accompanied by a much larger increase in E_c.

To apply the above analysis, one must have an independent method for assessing the nature of the flocculation. Rheology is a bulk property that can give information on the interparticle interaction (whether repulsive or attractive) and to apply it in a quantitative manner one must know the nature of these interaction forces. However, rheology can be used in a qualitative manner to follow the change of the formulation on storage.

Providing the system does not undergo any Ostwald ripening and/or coalescence, the change of the moduli with time and in particular the change of the linear viscoelastic region may be used as an indication of flocculation. Strong flocculation is usually accompanied by a rapid increase in G' and this may be accompanied by a decrease in the critical strain above which the "structure" breaks down. This may be used as an indication of the formation of "irregular" and tight flocs, which become sensitive to the applied strain. The floc structure will entrap a large amount of the continuous phase and this leads to an apparent increase in the volume fraction of the dispersion and hence an increase in G'.

4.7.7.2 Oscillatory Sweep Measurements

In this case, the strain amplitude is kept constant in the linear viscoelastic region (one usually takes a point far from γ_{cr} but not too low, that is, in the mid-point of the linear viscoelastic region) and measurements are carried out as a function of frequency. Both G^* and G' increase with increasing frequency and, ultimately, above a certain frequency they reach a limiting value and show little dependence on frequency. G'' is higher than G' in the low frequency regime; it also increases with increasing frequency and at a certain characteristic frequency ω^* (that depends on the system) it becomes equal to G' (usually referred to as the *cross-over point*), after which it reaches a maximum and then shows a reduction with a further increase in frequency.

From ω^* one can calculate the relaxation time τ of the system [3]:

$$\tau = \frac{1}{\omega^*} \tag{4.13}$$

The relaxation time may be used as a guide for the state of the dispersion. For a colloidally stable dispersion (at a given particle size distribution), τ increases with increasing volume fraction of the disperse phase, φ. In other words, the cross-over point shifts to lower frequency with increasing φ. For a given dispersion, τ increases with increasing flocculation providing the particle size distribution remains the same (i.e., no Ostwald ripening and/or coalescence).

The value of G' also increases with increasing flocculation, since the aggregation of particles usually results in liquid entrapment and the effective volume fraction of the dispersion shows an apparent increase. With flocculation, the net attraction between the particles also increases and this results in an increase in G'. The latter is determined by the number of contacts between the particles and the strength of each contact (which is determined by the attractive energy).

In practice, one may not obtain the full curve, due to the frequency limit of the instrument and, also, measurement at low frequency is time consuming. Usually, one obtains part of the frequency dependence of G' and G''. In most cases, one has a more elastic than viscous system.

Most disperse systems used in practice are weakly flocculated and they also contain "thickeners" or "structuring" agents to reduce creaming or sedimentation and to acquire the right rheological characteristics for application, for example, in hand creams and lotions. The exact values of G' and G'' required depend on the system and its application. In most cases a compromise has to be made between acquiring the right rheological characteristics for application and the optimum rheological parameters for long-term physical stability. Application of rheological measurements to achieve these conditions requires a great deal of skill and understanding of the factors that affect rheology.

4.8
Examples of Application of Rheology for Assessment and Prediction of Flocculation

4.8.1
Flocculation and Restabilization of Clays Using Cationic Surfactants

Hunter and Nicol [11] studied the flocculation and restabilization of kaolinite suspensions using rheology and zeta potential measurements. Figure 4.17 shows plots of the yield value σ_β and electrophoretic mobility u as a function of cetyl(trimethyl)ammonium bromide (*CTAB* or $C_{16}TAB$) concentration at pH 9. The value of σ_β increases with increasing CTAB concentration, reaching a maximum at the point where the mobility reaches zero (the isoelectric point, *i.e.p.*, of the clay) and then decreases with further increase in CTAB concentration. This trend can be explained on the basis of flocculation and restabilization of the clay suspension.

Initial addition of CTAB causes a reduction in the negative surface charge of the clay (by adsorption of CTA^+ on the negative sites of the clay). This is accompanied by reduction in the negative mobility of the clay. When complete neutralization of the clay particles occurs (at the i.e.p.) maximum flocculation of the clay suspension takes place and this is accompanied by a maximum in σ_β. On further increasing CTAB concentration, further adsorption of CTA^+ occurs, resulting in charge reversal and restabilization of the clay suspension. This is accompanied by a reduction in σ_β.

4.8.2
Flocculation of Sterically Stabilized Dispersions

Neville and Hunter [12] studied the flocculation of poly(methyl methacrylate) (*PMMA*) latex stabilized with PEO. Flocculation was induced by addition of electrolyte and/or an increase of temperature. Figure 4.18 shows the variation of σ_β with increasing temperature at constant electrolyte concentration.

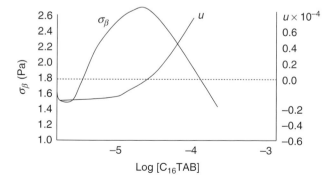

Figure 4.17 Variation of yield value σ_β and electrophoretic mobility u with $C_{16}TAB$ (CTAB) concentration.

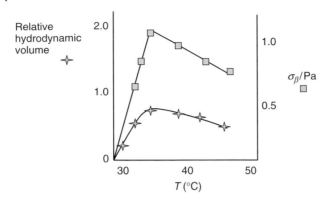

Figure 4.18 Variation of σ_β and hydrodynamic volume with temperature.

It can be seen that σ_β increases with increasing temperature, reaching a maximum at the critical flocculation temperature (*CFT*) and then decreasing with further increase in temperature. The initial increase is due to flocculation of the latex with increasing temperature, as result of a reduction of solvency of the PEO chains with increasing temperature. The reduction in σ_β after the CFT is due to the reduction in the hydrodynamic volume of the dispersion.

4.8.3
Flocculation of Sterically Stabilized Emulsions

Emulsions were prepared using an ABA block copolymer of PEO-PPO-PEO (Synperonic F127). Flocculation was induced by addition of NaCl [6]. Figure 4.19 shows the variation of the yield value, calculated using the Herschel–Bulkley model, as a function of NaCl concentration at various storage times. In the absence of NaCl, the yield value did not change with storage time over a period of one month, indicating an absence of flocculation. In the presence of NaCl, the yield value

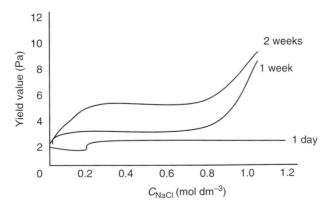

Figure 4.19 Variation of yield value with NaCl concentration.

increased with increasing storage time and this increase was very significant when the NaCl concentration was increased above 0.8 mol dm^{-3}.

The above increase in yield value indicated flocculation of the emulsion and this was confirmed by optical microscopy. The smaller increase in yield value below 0.8 mol dm^{-3} NaCl is indicative of weak flocculation, which could be confirmed by redispersion of the emulsion by gentle shaking. Above 0.8 mol dm^{-3} NaCl, the flocculation was strong and irreversible. In this case, the solvency of the medium for the PEO chains becomes poor, resulting in incipient flocculation.

Further evidence of flocculation was also obtained from dynamic (oscillatory) measurements. Figure 4.20 shows the variation of G' with NaCl concentration at various storage times. Below 0.8 mol dm^{-3} NaCl, G' shows a modest increase with storage time over a period of two weeks, indicating weak flocculation. Above 0.8 mol dm^{-3} NaCl, G' shows a rapid increase with increasing storage time, indicating strong flocculation. This strong (incipient) flocculation is due to the reduction of solvency of PEO chains (worse than θ-solvent) resulting in strong attraction between the droplets, which are difficult to redisperse.

4.9
Assessment and Prediction of Emulsion Coalescence Using Rheological Techniques

4.9.1
Introduction

The driving force of emulsion coalescence is the thinning and disruption of the liquid film between the droplets [1]. When two emulsion droplets come into contact, say in a cream layer or a floc, or even during Brownian collision, the liquid film between them undergoes some fluctuation in thickness; the thinnest part of the film will have the highest van der Waals attraction and this is the region where coalescence starts. Alternatively, the surfaces of the emulsion droplets may undergo

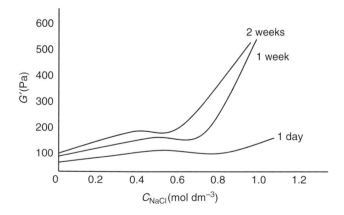

Figure 4.20 Variation of G' with NaCl concentration.

fluctuation, producing waves, which may grow in amplitude; the strongest van der Waals attraction is at the apices of these fluctuations and coalescence occurs by further growth of the fluctuation. One may define a critical film thickness below which coalescence occurs.

4.9.2
Rate of Coalescence

The rate of coalescence is determined by the rate at which the film thins and this usually follows first-order kinetics:

$$N = N_0 \exp(-kt) \tag{4.14}$$

where N is the number of droplets after time t, N_0 is the number at zero time, and k is the rate constant of coalescence.

Alternatively, one can measure the average droplet diameter d as a function of time:

$$d = d_0 \exp(kt) \tag{4.15}$$

Providing the emulsion does not undergo any flocculation, the coalescence rate can be simply measured by following the number of droplets or average diameter as a function of time. A given volume of the emulsion is carefully diluted into the isotone solution of a Coulter Counter and the number of droplets is measured. The average diameter can be obtained using laser diffraction methods (e.g., using a Master Sizer). By following this procedure at various time periods, one can obtain the coalescence rate constant k.

Usually, one plots $\log N$ or $\log d$ versus t and the slope of the line in the initial period gives the rate of coalescence k. Clearly, the higher the value of k is the greater the coalescence of the emulsion. An accelerated test may be used by subjecting the system to higher temperatures; usually, the rate of coalescence increases with increasing temperature (although this is not always the case). One should be careful in the dilution procedure, particularly if the oil is significantly soluble (say greater than 10 ppm) in the isotone solution or in the tank of the Master Sizer. In this case, one should saturate the solution with the oil before diluting the concentrated emulsion for droplet counting or sizing.

4.9.3
Rheological Techniques

4.9.3.1 Viscosity Measurements

In the absence of any flocculation, coalescence of an emulsion results in reduction of its viscosity. At any given volume fraction of oil, an increase in droplet size results in a reduction in viscosity; this is particularly the case with concentrated emulsions. Thus, by following the decrease in emulsion viscosity with time one may obtain information on its coalescence. However, one should be careful in applying simple viscosity measurements, particularly if flocculation occurs simultaneously (which results in an increase in the viscosity). It is possible in principle to predict the extent of viscosity reduction on storage, if one combines the results of droplet size analysis (or droplet number) as a function of time with the reduction in viscosity in the first few weeks [6].

Freshly prepared emulsions with various droplet sizes are prepared (by controlling the speed of the stirrer used for emulsification). The emulsifier concentration in these experiments should be kept constant and care should be taken that excess emulsifier is not present in the continuous phase. The viscosity of these freshly prepared emulsions is plotted versus the average droplet diameter. A master curve is then produced that relates the emulsion viscosity to the average droplet size; the viscosity decreases monotonically with increase in the average droplet size.

Using a Coulter Counter or Master Sizer, one can determine the rate of coalescence by plotting the log of the average droplet diameter versus time in the first few weeks. This allows one to predict the average droplet diameter over a longer period (say 6–12 months). The predicted average droplet diameter is used to obtain the viscosity that is reached on storage by applying the master curve of viscosity versus average drop size.

The above procedure is quite useful for setting the limit of viscosity that may be reached on storage as a result of coalescence. With many creams, the viscosity of the system is not allowed to drop below an acceptable limit (which is important for application). The limit that may be reached after one year storage may be predicted from the viscosity and rate constant measurements over the first few weeks.

4.9.3.2 Measurement of Yield Value as a Function of Time

Since the yield value σ_β of an emulsion depends on the number of contacts between the droplets, any coalescence should be accompanied by a reduction in the yield value. This trend is only observed if no flocculation occurs (this causes an increase in σ_β).

The above change was recently measured using O/W emulsions stabilized with an A-B-A block copolymer of PEO (A) and PPO (B) [Synperonic NPE 127 (UNIQEMA)]. Emulsions (60 : 40 O/W) were prepared using 0.5, 1.0, 1.5, 2.0, 3, 4, and 5% emulsifier [6]. Figure 4.21 shows the variation of droplet size with time at various Synperonic PEF 127 concentrations. At emulsifier concentrations >2% there is no change of droplet size with time, indicating an absence of coalescence. Below 2% the droplet size increased with time, indicating coalescence.

4.9.3.3 Measurement of Storage Modulus G' as a Function of Time

This is perhaps the most sensitive method for predicting coalescence. G' is a measure of the contact points of the emulsion droplets as well as their strength. Providing no flocculation occurs (which results in an increase in G'), any reduction in G' on storage indicates coalescence.

The above trend was confirmed using the emulsions described above. The emulsions containing less than 3% Synperonic PEF 127 showed a rapid reduction in G' when compared with those containing >3%, which showed virtually no change in G' over a two-week period [6]. This is illustrated in Figure 4.22.

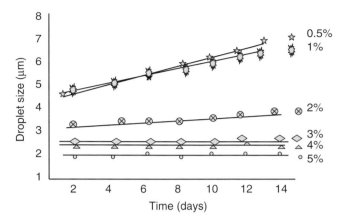

Figure 4.21 Variation of droplet size with time at various Synperonic PEF 127 concentrations.

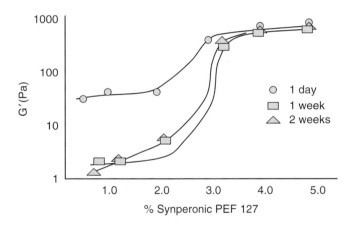

Figure 4.22 Variation of G' with Synperonic PEF 127 concentration at various storage times.

4.9.4
Correlation between Elastic Modulus and Coalescence

The correlation between emulsion elastic modulus and coalescence rate can be easily represented if one calculates the relative decrease in G' after two weeks [6]:

$$\text{Relative decrease of } G' = \left(\frac{G_{\text{initial}} - G_{\text{after 2 weeks}}}{G_{\text{initial}}} \right) \times 100 \qquad (4.16)$$

Figure 4.23 shows the variation of the relative decrease of G' and relative increase in droplet size with Synperonic PEF127 concentration. The correlation between the relative decrease in G' and relative increase in droplet size as a result of coalescence is now very clear.

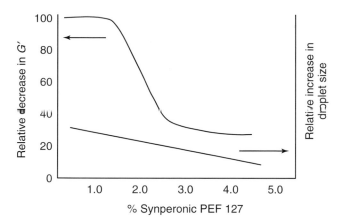

Figure 4.23 Correlation of relative decrease in *G'* with relative increase in droplet size.

4.9.5
Cohesive Energy *E*c

The cohesive energy E_c is the most sensitive parameter for the assessment of coalescence. In its definition given in Equation 4.6, γ_{cr} is the critical strain above which the linear response (where G' is independent of the applied strain) changes to a nonlinear response. Any coalescence results in a decrease in the number of contact points and causes a reduction in E_c.

Using the above-mentioned emulsions E_c was found to decrease with increasing droplet size (as a result of coalescence). At and above 3% Synperonic PE 127, E_c remained virtually constant, indicating an absence of coalescence. Figure 4.24 shows the variation of relative decrease of E_c with relative increase in droplet size; the correlation is clear [6].

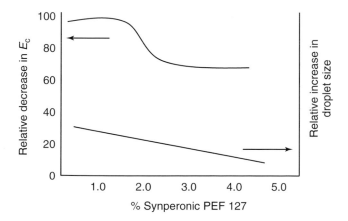

Figure 4.24 Correlation of relative decrease in E_c with relative increase in droplet size.

References

1. Tadros, T. (2005) *Applied Surfactants*, Wiley-VCH Verlag GmbH, Weinheim.
2. Tadros, T. (ed.) (1987) *Solid/Liquid Dispersions*, Academic Press, London.
3. Tadros, T. (2010) *Rheology of Dispersions*, Wiley-VCH Verlag GmbH, Weinheim.
4. Buscall, R., Goodwin, J.W., Ottewill, R.H., and Tadros, T.F. (1982) *J. Colloid Interface Sci.*, **85**, 78.
5. Krieger, I.M. (1972) *Adv. Colloid Interface Sci.*, **3**, 111.
6. Tadros, T. (2004) *Adv. Colloid Interface Sci.*, **227**, 108–109.
7. Salager, J.L. (2000) *Pharmaceutical Emulsions and Suspensions*, Marcel Dekker, New York.
8. Derjaguin, B.V. and Landau, L. (1941) *Acta Physicochem. USSR*, **14**, 633.
9. Verwey, E.J.W. and Overbeek, J.T.G. (1948) *Theory of Stability of Lyophobic Colloids*, Elsevier, Amsterdam.
10. Napper, D.H. (1983) *Polymeric Stabilization of Colloidal Dispersions*, Academic Press, London.
11. Hunter, R.J. and Nicol, S.K. (1968) *J. Colloid Interface Sci.*, **28**, 250.
12. Neville, P.C. and Hunter, R.J. (1974) *J. Colloid Interface Sci.*, **49**, 204.

5

Prediction of Thermophysical Properties of Liquid Formulated Products

Michele Mattei, Elisa Conte, Georgios M. Kontogeorgis, and Rafiqul Gani

5.1
Introduction

In chemical product design, the objective is to find molecules or mixtures of molecules with desired sets of target properties [1]. Examples of single chemical (molecular type) liquid products are solvents and process fluids, while mixtures of different chemicals are typically known as *liquid formulated (chemical) products*. In the latter type of products, the liquid formulations (or blends) may be a single liquid phase (e.g., an insect repellent lotion) or an emulsion (e.g., a detergent). In both types of liquid products, the chemical (molecular type) responsible for providing the main function of the product is usually known as the *active ingredient*. For example, in sunscreen lotions the active ingredient(s) provides protection from the sun's ultraviolet rays, in insect repellents they repel the insects, while as solvents the active ingredient dissolves other chemicals. Since, in single chemical liquid products, the active ingredient is present as a liquid for the range of conditions where it would be applied, additional chemicals do not need to be added. In liquid formulated products, however, the active ingredient is usually a solid in the standard state and, therefore, additional chemicals (such as solvents) are needed to dissolve and deliver it. The role of the solvents, which could be regarded as a single chemical product, is to deliver the active ingredient(s) and then leave the product application site. Other chemicals are usually added to the liquid formulations to provide further enhancements of the product quality (smell, color, stability, etc.). A large variety of household, health-care, and cosmetic products are chemical-based liquid formulated products.

The desired set of target properties for both types of liquid products represent the functions (or needs) of the product. Therefore, properties play a very important role in the design and development of chemicals based liquid products. It is through these properties that the stability of the liquid product, the evaporation of the solvent on application of the product, the spread-ability of the product, the environmental impact of the product, and so on are verified. A common approach to designing these liquid products is to measure the needed properties for each candidate product. While this approach is reliable, it is also time consuming,

Product Design and Engineering: Formulation of Gels and Pastes, First Edition.
Edited by Ulrich Bröckel, Willi Meier, and Gerhard Wagner.
© 2013 Wiley-VCH Verlag GmbH & Co. KGaA. Published 2013 by Wiley-VCH Verlag GmbH & Co. KGaA.

expensive, and does not allow the consideration of all potential product candidates. In a hybrid approach, where model-based techniques are used to estimate the desired set of properties, a set of promising candidates are identified through a model-based approach in the first (evaluation) stage. In the second (verification) stage, the properties of selected product candidates are verified and the product formulations are further improved by experiments [2, 3]. In both cases, a database of collected experimental data may also be used if the required property data are available in the database.

The objective of this chapter is to give an overview of the models, methods, and tools that may be used for the estimation of liquid formulated products. First, a classification of the products is given and the thermophysical properties needed to represent their functions are listed. For each property, a collection of the available models are presented according to the property type and the model type. Notably, however, the property models considered or highlighted in this chapter are only examples and are not necessarily the best and most accurate for the corresponding property.

5.2
Classification of Products, Properties and Models

5.2.1
Classification of Products

The basic distinction between liquid formulated products is whether they are microstructured or non-microstructured. Non-microstructured products are homogeneous liquid products, characterized by a single liquid (homogeneous) phase; examples are sunscreen and insect repellent lotions where oil-soluble ingredients are dissolved in organic solvents. On the other hand, the class of microstructured products can be divided into different subclasses, each of them characterized by two non-miscible phases, one dispersed in the other. Table 5.1 gives a list of microstructured formulated products [4].

This chapter covers only the non-microstructured liquid formulated products and emulsions, which form one of the most common types of microstructured liquid formulated products.

The selection of the most appropriate chemicals for each type of active ingredient as well as the final choice of the overall composition in the formulation require the evaluation of a wide range of thermophysical properties related to the candidate product formulations. For this reason, property models play a fundamental role in the design of formulated products. For example, density, dynamic viscosity, and surface tension are bulk functional properties that are usually needed for the design of almost all liquid formulated products together with the test for liquid stability of non-microstructured products or the test for emulsion stability for microstructured products. Table 5.2 gives a list of the relations between properties and product functions developed on the basis of their use in different liquid formulated products.

Table 5.1 Classes of microstructured formulated products [4].

Class of product	Dispersed phase	Continuous phase
Liquid aerosol	Liquid	Gas
Solid aerosol	Solid	Gas
Foam	Gas	Liquid
Emulsion	Liquid	Liquid
Sol, suspension	Solid	Liquid
Solid foam	Gas	Solid
Solid emulsion	Liquid	Solid
Solid suspension	Solid	Solid

Table 5.2 Product function–target properties relations.

Product function	Target properties	Symbols
Conductivity	Dielectric constant	ε
Cost	Cost	C
Drying time	Evaporation time	T_{90}
Flammability	Open cup flash point	T_f
Foam-ability (emulsions only)	Critical micelle concentration, surface tension	CMC, σ
Solubility	Hansen and Hildebrand solubility parameters	$\delta_D, \delta_P, \delta_H, \delta_T$
	Hydrophilic–lipophilic balance (surfactants only)	HLB
Skin-irritancy	Hansen and Hildebrand solubility parameters	$\delta_D, \delta_P, \delta_H, \delta_T$
Spray-ability	Density, dynamic viscosity, surface tension	ρ, η, σ
Spread-ability	Kinematic viscosity, density	υ, ρ
Stability	Gibbs energy change of mixing	ΔG^{mix}
	Critical micelle concentration, cloud point, Krafft temperature (surfactants only)	CMC, T_c T_K
Toxicity	Toxicity parameter	LC_{50}

5.2.2
Classification of Properties

The set of desired target properties representing the product functions may be classified in terms of:

- **Primary properties** – These are single value properties of the pure compound. Every molecule is characterized by a single value of these properties. These are the critical properties, the normal boiling point, the normal melting point, and many more. Measured values of these properties can usually be found in databases of chemicals. From a modeling point of view, they are also classified as those dependent only on the molecular structure of the compound.

- **Secondary properties** – From a modeling point of view, these properties of pure compounds are dependent on the molecular structure as well as other properties of the compound. For example, the density or heat of vaporization at the normal boiling point may be calculated from knowledge of the critical properties and the normal boiling point. In some cases, such as the enthalpy of vaporization at the boiling point, a secondary property may be converted into a primary property.

- **Functional properties (pure compound or mixture)** – These are properties that depend on temperature, pressure, and/or mixture composition. Those that depend only on temperature and/or pressure are related to pure compound, while those that depend also on mixture composition, may be further classified as bulk-properties or compound properties in mixtures:

  - **Functional pure compound properties** – These are properties, such as vapor pressure, density, heat of vaporization, and so on, of the pure compound that depend on the temperature and/or pressure. From a modeling point of view, the pressure effect is usually neglected and the temperature effect is modeled through regressed correlations. Details of the regressed correlations can be found for properties and compounds in databases of chemicals.

  - **Functional bulk properties** – These properties are functions of mixture compositions as well as temperature and/or pressure (or a defined mixture state). They represent the bulk property of the mixture, for example, the density or viscosity of the liquid mixture (or formulation). From a modeling point of view, estimation of these properties requires the corresponding pure compound properties of the involved compounds and a mixing rule to take into account the composition effect.

  - **Functional compound properties in mixtures** – These are phase equilibrium related properties of the compounds present in a mixture. For example, the activity or fugacity coefficients of each compound present in the mixture. From a modeling point of view, they may require primary and secondary as well as other functional properties.

- **Performance related properties** – These properties are related to the performance of the product – such as the evaporation rate of the solvent and the stability of the liquid or emulsion. From a modeling point of view, they may require the above property models embedded into a process model. For example, the use of activity coefficients (functional compound properties in a mixture) within a liquid-phase stability test algorithm [2].

5.2.3
Classification of Property Models

The models used to estimate the properties may be classified, for each class of properties, into those that are predictive by nature and those that are not. For example, estimating properties only from molecular structural information involves predictive models, such as the group contribution (*GC*) based models, while estimating properties from compound-specific coefficients involves the use of correlations that are not predictive by nature. In liquid formulated product design, both types of models are needed. During the evaluation of candidate products, the models need to be predictive and computationally fast and cheap, while during the verification of a small number of candidates correlated models may be used, if the correlation coefficients are available. During the evaluation stage the models need to be, at least, qualitatively correct, while during the verification stage the models also need to be quantitatively correct.

The property models may also be classified in terms of mechanistic (e.g., applying quantum mechanical techniques) or correlative (e.g., data-based regression of correlations) or hybrid (e.g., GC-based models that are predictive but require the use of regressed parameters). Truly predictive models are mechanistic models but they are usually computationally expensive and, as yet, their application range is limited. A vast variety of hybrid models have, however, been developed, ranging from simple and easy to use GC-based models [5, 6] to more complex "theoretical" models [7–9]. Other models that are not predictive with respect to the molecules present in the mixture but can be extrapolated in terms of temperature, pressure, and composition are the well-known equations of state [10, 11] and activity coefficient models [6, 12, 13].

Table 5.3 lists a collection of pure compound (primary and secondary) properties typically encountered in liquid formulated product design, while Tables 5.4 and 5.5 list a collection of mixture properties (functional bulk and compound in mixture) typically encountered in the design of homogeneous and emulsion-based liquid products, respectively. The properties and models mentioned in Tables 5.3–5.5 are described in the following sections (Sections 5.3–5.6).

Table 5.3 Target pure compound properties, symbols, and models for their prediction.

Properties	Symbols	Models to use
Cloud point (surfactants only)	T_c	M&G GC$^+$ method
Cost	C	Correlation
Critical micelle concentration (surfactants only)	CMC	M&G GC$^+$ method
Density	ρ	GC-method; equations of state
Dielectric constant	ε	Correlation
Dynamic viscosity	η	M&G GC$^+$ method
Evaporation time	T_{90}	Correlation
Hansen solubility parameters	$\delta_D, \delta_P, \delta_H$	M&G GC$^+$ method
Hildebrand solubility parameter	δ_T	M&G GC$^+$ method
Hydrophilic–lipophilic balance (surfactants only)	HLB	Definition
Kinematic viscosity	υ	Definition
Krafft temperature (surfactants only)	T_K	QSPR method
Open cup flash point	T_f	M&G GC$^+$ method
Surface tension	σ	M&G GC$^+$ method
Toxicity parameter	LC$_{50}$	GC-based method

Table 5.4 Target properties, symbols, and models to use for homogeneous liquid mixtures.

Properties	Symbols	Models to use
Cost	C	Linear mixing rule
Gibbs energy change of mixing	ΔG^{mix}	Activity models
Density	ρ	Linear mixing rule on molar volume
Dielectric constant	ε	Linear mixing rule
Dynamic viscosity	η	Linear mixing rule; GC(UNIFAC)-based method
Evaporation time	T_{90}	GC(UNIFAC)-based method
Hansen solubility parameters	$\delta_{\text{D}}, \delta_{\text{P}}, \delta_{\text{H}}$	Linear mixing rule
Hildebrand solubility parameter	δ_{T}	Linear mixing rule
Open cup flash point	T_f	GC(UNIFAC)-based method
Surface tension	σ	Linear mixing rule; GC(UNIFAC)-based method
Toxicity parameter	LC_{50}	Linear mixing rule

Table 5.5 Target properties, symbols, and models to use for emulsion-based products.

Properties	Symbols	Models to use
Cost	C	Linear mixing rule
Density	ρ	Linear mixing rule on molar volume
Dielectric constant	ε	Linear mixing rule
Dynamic viscosity	η	Dedicated model
Evaporation time	T_{90}	GC(UNIFAC)-based method
Hansen solubility parameters	$\delta_{\text{D}}, \delta_{\text{P}}, \delta_{\text{H}}$	Linear mixing rule
Hildebrand solubility parameter	δ_{T}	Linear mixing rule
Open cup flash point	T_f	GC(UNIFAC)-based method
Surface tension	σ	Linear mixing rule; GC(UNIFAC)-based method; M&G GC$^+$ method (water + surfactant)
Toxicity parameter	LC_{50}	Linear mixing rule

5.3
Pure Compound Property Modeling

The target properties needed to design liquid formulated products require properties of the mixture. However, as given in Tables 5.3–5.5, the calculation of mixture properties also needs pure compound properties of all the constituent chemicals. In this section, the pure compound properties relative to homogeneous liquid formulated products are discussed first, followed by those relative to emulsion-based formulated products.

5.3.1
Homogeneous Formulated Products – Primary and Secondary Properties

5.3.1.1 Cost

Even though cost is not a property it has been correlated as a function of volume [2] and plays an important role in the final selection of a formulated product; that is, when all feasible formulations have been identified, a criterion for selection of the formulation could be the cost. The pure compound cost data may be subject to various uncertainties such as purity and source. Therefore, Conte *et al.* [2] proposed a simple correlation to provide qualitatively correct estimations of the pure compound cost. For different molecular types, the cost (*C*) of a chemical has been found to depend linearly on the molar volume (*V*) (Figure 5.1) [2].

From Figure 5.1, the cost of alcohols and esters (chemicals frequently employed as solvents in liquid formulated products) can be estimated from Equations 5.1 and 5.2, respectively:

$$C = 2.152 \times V - 38.714 \tag{5.1}$$

$$C = 2.356 \times V - 119.000 \tag{5.2}$$

where the units of measure for the cost (*C*) and the molar volume (*V*) are \$ kmol^{-1} and l kmol^{-1}, respectively. Note that this model does not take into account the fluctuation of market prices, and it has been developed only for preliminary selection purposes.

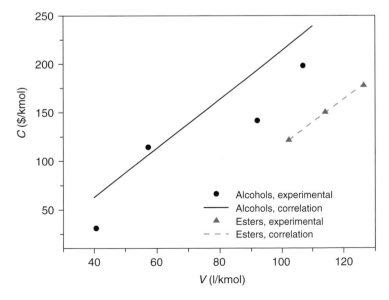

Figure 5.1 Calculated values of cost versus molar volume for some alcohols and esters [2].

5.3.1.2 Density

Liquid density is a key property in liquid product design since it is related to the sizing of equipment, volume per mass of the product, as well as the calculation of other secondary and functional properties. Most prediction methods for saturated liquid densities are based on the corresponding state principle. One of the most popular methods for the prediction of liquid densities is the Rackett equation [14] and its modification by Spencer and Danner [15] and by Yamada and Gunn [16]. The modified Rackett equation by Spencer and Danner [15] is the most commonly used model, which is given by Equation 5.3:

$$1/\rho_S = (R \times T_C/\rho_C) \times Z_{Ra}^{[1+(1-T_R)^{2/7}]} \tag{5.3}$$

where ρ_S is the liquid density, T_C is the critical temperature, ρ_C is the critical density, T_R is the reduced temperature, R the universal gas constant, and Z_{Ra} is the modified constant for the Rackett equation, which needs to be determined from experimental data. Another option is to use the definition of the critical compressibility factor as given by Equation 5.4:

$$Z_C = P_C \times V_C/(R \times T_C) \tag{5.4}$$

where P_C is the critical pressure and V_C is the critical volume.

In addition, GC-based models, which only need molecular structural information of the chemical, have been developed for the prediction of pure compound densities (molar volumes) at the standard state. The GCVOL model [17] and its extensions [18, 19] are good examples of models of this type. Here, density is modeled as the inverse of molar volume, a state variable also found in equations of state; that is, from any equation of state, given two state variables (such as temperature and pressure), the corresponding molar volume (or density) of the fluid can be calculated.

5.3.1.3 Dielectric Constant

The dielectric constant is an important property in liquid product design, as this property is correlated to several other properties, such as solubility and polarizability. The most used method to predict this property is a correlation (see Equation 5.5, where δ_T is the Hildebrand solubility parameter) reported by Horvath *et al.* [20], which is suitable for a broad variety of chemicals:

$$\varepsilon = \left(\frac{\delta_T - 7.5}{0.22}\right) \tag{5.5}$$

Another option by which to calculate the dielectric constant for a compound is through GC-based models where this property is modeled as a primary property [21–23]. In the GC-based method, a chemical is uniquely represented by a set of functional groups. By retrieving the GCs for a corresponding property from the GC table, the property is estimated through Equation 5.6, which is an example of the Marrero–Gani method [5]. Note, however, that the dielectric constant has not been modeled yet by the Marrero–Gani method and a version of GC-based method [24] using only the first summation term on the right-hand side of Equation 5.6

can be found in ICAS-ProPred [25]. A brief overview of the Marrero–Gani group contribution ($M\&G\ GC^+$) method is given in Appendix 5.A:

$$F(\vartheta) = \sum N_i C_i + \gamma \sum M_j D_j + z \sum O_k E_k \qquad (5.6)$$

5.3.1.4 Dynamic Viscosity

Knowledge of this property plays an important role in design issues related to transport of mass and/or energy, and many more properties such as spread-ability and spray-ability (Table 5.2) of the liquid product. Among the models available, the one proposed by Sastri and Rao [26] is well-known for its correlation of dynamic viscosity for different types of organic chemicals:

$$\mu = \mu_{\mathrm{B}} \times P^{-N} \qquad (5.7)$$

where μ is expressed in mPa s^{-1}, P is the pressure in atm, while N and μ_{B} are regressed parameters that are found to vary very little for compounds of similar molecular type. Therefore, these parameters may be evaluated through a GC method and the pressure, when it is above the boiling point, could be estimated from an equation of state or corresponding principles. The Sastri and Rao [26] method can predict the viscosity quite well, but the recently developed M&G GC$^+$ method for prediction of dynamic viscosity at 298 K has been found to perform better [27].

5.3.1.5 Evaporation Time

Evaporation times are needed in the design of liquid formulated products when the solvent is supposed to vaporize out after application of the product. Measured values for the evaporation time are difficult to find and, therefore, correlations based on model systems reported by others are usually employed. Van Wesenbeeck *et al.* [28] proposed a linear correlation between the logarithm of the evaporation rate (ER, in g cm^2 s^{-1}) and the vapor pressure (P^{sat} in Pa) of the solvent:

$$\ln(ER) = 0.865 \times (P^{\mathrm{sat}}) + 12.70 \qquad (5.8)$$

This correlation has been found to estimate the evaporation rates of a pure solvent quite accurately [29], but other correlations have also been proposed [30] and they are reported to give good qualitative accuracy in describing the relation between the vapor pressure of the solvent and its evaporation rate.

Very often, instead of the evaporation rate, the target property is the evaporation time needed for 90% by weight of the solvent to evaporate (T^{90}, in s). This property is also correlated as a function of the vapor pressure of the solvent:

$$\ln(T^{90}) = -0.793 \times \ln(P^{\mathrm{sat}}) + 12.416 \qquad (5.9)$$

The accuracy of this correlation has also been found to be quite satisfactory [2, 3]. Note that the vapor pressures needed for Equations 5.8 and 5.9 can be predicted in various ways (Section 5.3.1.6).

5.3.1.6 Vapor Pressure and Heat of Vaporization

The vapor pressure and the heats of vaporizations of pure compounds are classified as a functional pure compound property, dependent on temperature. This means that measured data of these properties are correlated for each compound as functions of temperature. The regressed coefficients can be found in databases of chemicals and their properties [31–33]. In addition, equations of state, such as the cubic equations of state [10], could be used if the necessary critical properties of the compounds are known. For non-cubic equations of state, the parameters, for example, of CPA [11] and PC-SAFT [34], are estimated based on vapor pressure data. Another option that requires less storage of parameters, needing only the molecular structural information, and is predictive in nature is to use GC-based methods to estimate the regressed parameters of the temperature-dependent functions [35].

To model the vapor pressure and heat of vaporization of lipids, which are finding increasing use in chemicals based liquid products, Ceriani *et al.* [35] developed a GC model using an extensive data bank of lipids of various classes with improved predictive power. The property model for vapor pressure is of the form:

$$\ln P^{vp} = A + \frac{B}{K} + C_{1k} \ln(T)$$

$$A = \sum_k N_k(A_{1k} + MA_{2k}) + (s_o + N_{cs}s_1) + \alpha(f_o + N_c f_1)$$

$$B = \sum_k N_k(B_{1k} + MB_{2k}) + \beta(f_o + N_c f_1)$$

$$C = \sum_k N_k(C_{1k} + MC_{2k}) \tag{5.10}$$

The heat of vaporization is calculated using the following expression:

$$\Delta H^{vap} = 8.3144(-B + CT)\left(1 - \frac{T}{T_c}\right)^{\frac{P^{vp}}{P_c}} \tag{5.11}$$

where N_{cs} is the number of carbon atoms in the alcohol part of esters, N_c is the total number of carbon atoms, T_c and P_c are the critical temperature (in K, as for T) and pressure (in Pa as for P^{vp}), respectively, ΔH^{vap} is in J gmol^{-1}, and A_{1k}, B_{1k}, C_{1k}, A_{2k}, B_{2k}, C_{2k}, α, β, s_o, s_1, f_o, and f_1 are the model parameters whose values are reported by Ceriani *et al.* [35].

5.3.1.7 Solubility Parameters

Solvent selection is one of the major concerns in the early development of many liquid formulated products because of the relation between active ingredients and the additives in the product. Since the active ingredients and/or additives in their standard states may not be in the liquid form, solvents are needed to dissolve them and deliver them to the product application site as well as for storage of the product. The prediction of solid solubility is a major topic of current research and many methods and tools are available for estimation of the solubility of solids in solvents [24, 36, 37]. The solid–solvent systems may be quite complex [38] and the applicability of any single method with acceptable accuracy for a wide range of chemical systems is questionable [39]. A model-based liquid formulated product design technique would require the screening of thousands of candidate solid–solvent mixtures. Therefore, a predictive and easy to apply method is needed. A good review of the model-based estimation of solid solubility is given by Conte *et al.* [27]. In addition, see also Section 5.4.2 where the prediction of solid–liquid equilibrium (*SLE*) is discussed. This section, however, discusses the use of solubility parameters in solvent selection.

Hansen Solubility Parameters   As a first option, it is useful to identify the best solvents without detailed calculations of the solid–solvent solubility. For this purpose, the three Hansen solubility parameters may be used, especially if a predictive model is available [40]. The Hansen solubility parameters (δ_D represents the dispersion parameter, δ_P represents the polar parameter, and δ_H represents the hydrogen bonding parameter) give a qualitative understanding of solvent issues, thereby allowing a fast screening between suitable and unsuitable solvent mixtures, significantly reducing the search space and the computational load for more rigorous quantitative approaches (Section 5.4.2). With the M&G GC$^+$ method (Equation 5.6), each of the three Hansen solubility parameters can be predicted with good accuracy [41].

Hildebrand Solubility Parameter   The Hildebrand solubility parameter [42] is an alternative for fast screening of solvents. It is defined as:

$$\delta_T = (\Delta E_V / V_m)^{0.5} \tag{5.12}$$

where ΔE_V and V_m are the heat of vaporization and the molar volume of the compound, respectively, at a specified temperature. The reliability of this property is, however, questionable since it does not take into account the variations in the contributions to the vaporization energy due to dispersion, polar, and hydrogen-bonding effects. These are considered in the Hansen solubility parameters. The relationship between Hansen and Hildebrand solubility parameters is:

$$\delta_T = (\delta_D^2 + \delta_P^2 + \delta_H^2)^{0.5} \tag{5.13}$$

Like the Hansen solubility parameters, the Hildebrand solubility parameters can also be predicted by the M&G GC$^+$ method [41].

5.3.1.8   Open Cup Flash Point

The open cup flash point is related to the flammability of a chemical and therefore is included as a target property to accommodate safety issues. Since the solvent is supposed to evaporate from the product, the flammability issue is important, as also is the environmental impact (such as toxicity, ozone depletion, etc.). A solvent mixture should have a flash point that is at least higher than the usage temperature of the product, considering that in the formulation the solvent mixture is diluted by active ingredients and additives that are usually not highly flammable. Here also, GC models have been developed to predict this important property with acceptable accuracy [24].The M&G GC$^+$ method can be used (Equation 5.6).

5.3.1.9 Surface Tension

Like dynamic viscosity, surface tension is also a key-property in process and product design since it strongly influences transport phenomena and affects, among other properties, the spread-ability and spray-ability (Table 5.2) of the liquid product. In the past, the C. Orrick and J.H. Erbar (personal communication) method has been used, but these days the more accurate and predictive M&G GC$^+$ method (Equation 5.6) is an alternative worth considering for the prediction of surface tension at 298 K. In addition, models based on the corresponding state principle, or quantitative structure–property relationships (QSPRs), have been proposed [43], but when surface tension values at 298 K are required the GC approach has been found to provide good results with reasonable accuracy for a wide range of chemicals [27].

5.3.1.10 Environmental Health and Safety Related Properties

Recently, Hukkerikar *et al.* [44] developed a series of M&G GC$^+$ method (Equation 5.6) based models for a range of environmental, health, and safety related properties. The following properties were covered: the fathead minnow 96-h LC_{50}, *Daphnia magna* 48-h LC_{50}, oral rat LD_{50}, aqueous solubility, bioconcentration factor, permissible exposure limit (OSHA-TWA), photochemical oxidation potential, global warming potential, ozone depletion potential, acidification potential, emission to urban air (carcinogenic and non-carcinogenic), emission to continental rural air (carcinogenic and non-carcinogenic), emission to continental fresh water (carcinogenic and non-carcinogenic), emission to continental sea water (carcinogenic and non-carcinogenic), emission to continental natural soil (carcinogenic and non-carcinogenic), emission to continental agricultural soil (carcinogenic and non-carcinogenic). In this section, the LC_{50} property is highlighted, which is among the most used target properties in process and product design. The parameter LC_{50} indicates the lethal concentration of a pure chemical or mixture that causes 50% of deaths in a fathead minnow population. Especially in the design of liquid (skin-care or cosmetic) products, or products to be inhaled or ingested, this parameter is used as a measure of the toxicity along with other properties listed above. The model parameters for all the listed properties for the M&G GC$^+$ method (Equation 5.6) are given by Hukkerikar *et al.* [44]. When only first-order groups are used, Equation 5.6 takes the following form for LC_{50}:

$$-\log(LC_{50}) = \sum (n_i \alpha_i) \tag{5.14}$$

where n_i is the number of groups of the type i and α_i is the property contribution of group i.

5.3.2
Emulsified Formulated Products – Primary and Secondary Properties

Emulsified formulated products are characterized by the presence, often in small amounts to ensure the formation of the emulsion, of chemicals classified as surfactants. These chemicals are characterized by an amphiphilic nature, which means that part of them is hydrophilic while another part is hydrophobic or lipophilic. To describe their behavior in relation to two non-miscible phases and the range of temperatures at which they are active, some properties, such as cloud point and critical micelle concentration (*CMC*), which are not covered above, are needed. Notably, however, model-based emulsified formulated product design is not as developed as the homogeneous case and so predictive models for many of the needed properties are lacking. In addition, although many of these properties cannot be strictly considered to be pure compound properties, since they refer to a mixture with water at fixed temperature and sometimes at fixed composition as well, they can be modeled from only the molecular structure of the involved surfactant. Hence these properties are modeled as primary properties and can be estimated (predicted) using the GC-type of methods and correlations.

5.3.2.1 Cloud Point

The *cloud point* is defined as the temperature at which, when heating, a solution of 1% molar surfactant in water becomes cloudy as the surfactant "drops out of" the solution since a separation into two phases occurs. The cloud point is a characteristic of non-ionic poly(ethylene oxide) based surfactants, and it arises because the solubility of the poly(ethylene oxide) chain is affected by hydrogen bonding [45]. If the ability of compounds to solubilize depends on hydrogen bonding in aqueous solutions they are commonly found to exhibit an inverse temperature/solubility relationship [46].

The cloud point is known to increase with increasing surfactant (e.g., ethylene oxide) content and decreases if the hydrophobicity of the surfactant is increased [47]. This observation led to the development of a large number of methods based on knowledge only of the chemical structure of the desired surfactant. The use of QSPR type models [48–50] has also been developed for the prediction of cloud points. An example of models of this type is described by Equation 5.15 [48]:

$$T^C = (-237.30 \mp 76.42) + (95.09 \mp 10.14) \log \ EO - (107.30 \mp 25.60) AMW +$$
$$+ (974.55 \mp 137.26) AAC - (51.09 \mp 12.42) MAXDP - (1337.21 \mp 354.39) PW4 +$$
$$- \ (72.88 \mp 22.18) PJI2 - (25.90 \mp 4.40) IC4 + (85.88 \mp 23.22) Elu \qquad (5.15)$$

where, EO, AMW, AAC, MAXDP, PW4, PJI2, IC4, and Elu are the values of the descriptors that need to be calculated for the pure compound. For this reason, the use of GC-based methods may be more convenient since the descriptors (i.e., the groups) are well known and their parameters are readily available. In addition, the clear dependence of the cloud point and the length of the hydrophobic and the hydrophilic chains suggest that a GC-based approach could provide satisfactory predictions. A new GC model based on the M&G GC$^+$ method (Equation 5.6) has been developed.

5.3.2.2 Critical Micelle Concentration

The CMC of a surfactant indicates the point at which the monolayer adsorption of surfactant molecules at the surface is complete and the surface tension of the mixture is at its minimum. The CMC is dependent on the chemical structure of the surfactant and from experimental tests it has been observed that for a surfactant like alkyl poly(ethylene oxide) the CMC increases with decreasing molecular weight of the hydrophobic moiety and it decreases as the number of ethylene oxide groups decreases. There is considerable interest in monitoring the CMC property as this in practice represents the lowest surfactant concentration needed to form an emulsion and to obtain maximum benefit if surfactant abilities based on surface activity (e.g., wetting and foaming) are needed [51].

The CMC has been modeled in various ways – UNIFAC [6], NRTL [12], and also the PC-SAFT equation of state [34] – but its dependence on the chemical structure together with experimental evidence that micelles are formed only in the presence of water [52] has pushed the development of models for the CMC based only on the molecular structure of the surfactant involved. QSPR models for non-ionic as well as ionic surfactants have been developed recently [53, 54] and their accuracy is quite satisfactory. However, similarly to the cloud points of surfactants, the M&G GC+ method has also been adopted for CMC prediction of non-ionic surfactants with very high accuracy (Figure 5.2). More details on the model and its parameters can be obtained from the corresponding author of this chapter (R. Gani).

5.3.2.3 Hydrophilic–Lipophilic Balance

The hydrophilic–lipophilic balance system (*HLB*) is one of the most common methods by which to correlate surfactant structures with their effectiveness as emulsifiers [55]. The system consists of an arbitrary scale to which HLB numbers are experimentally determined and assigned [56]. The value of the HLB number indicates how the surfactant will behave in a solution and which kind of emulsion would be formed by adding that surfactant to a system of two non-miscible phases. The HLB concept [55–57] works very well for non-ionic surfactants [such as alkyl poly(ethylene oxide)], but it is less successful with ionic surfactants [58].

The original model for estimation of the HLB value of alkyl poly(ethylene oxide) surfactants [56] states that the HLB value is equivalent to the mass (or weight)

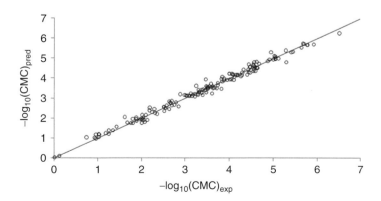

Figure 5.2 Experimental versus predicted values (M&G GC+ method) of the critical micelle concentration.

percentage of poly(ethylene oxide) content of a surfactant (E) divided by 5:

$$HLB = E/5 \tag{5.16}$$

For most polyhydric alcohol fatty acid esters, the HLB value can be calculated by considering the saponification value (S) of the ester and the acid value (A) of the acid, by means of Equation 5.17 [59]:

$$HLB = 20(1 - S/A) \tag{5.17}$$

For surfactants that contain poly(ethylene oxide) chains and polyhydric alcohols (e.g., glycerol and sorbitol) as hydrophilic groups, Equation 5.18 has been proposed:

$$HLB = (E + P)/5 \tag{5.18}$$

where P is the mass (or weight) percentage of polyhydric alcohol content.

In the above methods, the HLB estimated with Equation 5.18 may give errors and therefore predictions based on it should be used with care [60]. Alternative methods that include the hydrophilic–lipophilic effects have also been proposed. For example, the cohesive energy ratio (CER), has been proposed as an alternative to HLB. The CER is calculated as a secondary property from the molar volume and Hansen solubility parameter of both the hydrophilic and lipophilic moieties [61]:

$$CER = \left[V_L \left(\delta_d^2 + 0.25\delta_p^2 + 0.25\delta_h^2 \right)_L \right] / \left[V_H \left(\delta_d^2 + 0.25\delta_p^2 + 0.25\delta_h^2 \right)_H \right] \tag{5.19}$$

A CER value equal to one indicates that the surfactant has the same affinity for both phases and, therefore, no stable emulsion can be formed. A CER value higher than unity indicates, instead, a higher affinity of the surfactant for the oil-phase, eventually leading to a water-in-oil emulsion, while a CER value that is lower than unity has to be chosen if an oil-in-water emulsion is designed.

The critical packing parameter (CPP), a geometric expression that relates the molar volume of the surfactant (V), the lipophilic chain length (L), and also the interfacial area occupied by the head group (A) has been given by Israelachvili et al. [62]:

$$CPP = V_s/(L_S \times A_S) \tag{5.20}$$

A CPP value of $<1/3$ indicates spherical micelles, while a value between $1/3$ and 0.5 indicates rods micelles. In both cases, an oil-in-water emulsion is expected. A CPP value between 0.5 and 1, instead, indicates lamellar micelles or vesicles, which are different association structures assumed by the surfactants, still leading to oil-in-water emulsion. A CPP value higher than 1, on the other hand, results in hexagonal reverse micelles, typical of water-in-oil emulsions. An intermediate value of 1, instead, refers to planar bilayered structures that do not lead to any stable emulsified system.

In addition, the surface-affinity difference (SAD) has been proposed as an equivalent concept to the HLB value, where a reliable equation of state or activity coefficient model is necessary to quantify the activity coefficients needed by the model [63]:

$$\ln(a^O/a^W) = SAD/RT = \ln S - K \times ACN - f(A) + \sigma - \alpha_T \times \Delta T \tag{5.21a}$$

$$= \alpha - \text{EON} + b \times S - K \times \text{ACN} - \varphi(A) + c_T \Delta T \qquad (5.21b)$$

where a^O and a^W are the activity of the surfactant in the oil and in the water phase, respectively. Equation 5.21a is valid for ionic surfactant and Equation 5.21b for non-ionic surfactants. The different terms have been experimentally determined and they are now available for a large number of chemicals [63]. The desired SAD value depends on the structure desired. A value of zero indicates $a^O = a^W$, which means that the affinity of the surfactant towards the two phases is comparable and, therefore, no stable emulsions can be formed (similarly to CER = 1 or CPP = 1). On the other hand, positive SAD values indicate a higher affinity of the surfactant for the oil-phase, so a stable water-in-oil emulsion can be formed, while negative SAD values indicate a higher affinity of the surfactant for the water-phase and, therefore, lead to the formation of stable oil-in-water emulsions.

5.3.2.4 Krafft Temperature

Surfactants forming micelles exhibit unusual solubility behavior as their solubilities show a rapid increase above a certain temperature, known as the *Krafft point* or *Krafft temperature*. This solubility behavior is mostly observed for ionic surfactants as only a few non-ionic surfactants possess a real Krafft temperature [64]. The Krafft temperature depends on a complex three-phase equilibrium condition that is largely determined by the counterion of the surfactant as well as the length of the hydrophobic moiety. The presence of electrolytes does also influence the Krafft temperature, which increases with increasing concentration of electrolytes [65]. These experimental observations make the development of dedicated models for Krafft temperature complicated. So far, the only published reliable method has been presented by Li *et al.* [66], which is a QSPR type of model:

$$T_K = 57.4 - 7.6 \times \text{KS2} - 0.06 H_f + 47.1 \times A \log P98 - 28.1 \times$$
$$A \log P - 36.1 \times \text{IC} + 6.7 \text{nO} \qquad (5.22)$$

where KS2, H_f, P98, P, IC, and nO are descriptors, which need to be known before any property estimation can be made. However, also for this property, GC models based on the M&G GC$^+$ method (Equation 5.6) are feasible.

5.3.2.5 Surface Tension

Surface tension needs a special modeling consideration when referring to the water phase of an emulsion-based formulated product. Here the surface tension needed is not that of the pure surfactant but rather that of the water–surfactant mixture at the CMC. However, based on experimental observations, it has been found that the addition of surfactant to water strongly decreases the surface tension of the mixture until the CMC is reached; then adding surfactant influences only slightly the surface tension of the mixture, which is therefore usually considered constant [67]. Hence, as in modeling of the CMC, this property may also be modeled as a function of only the surfactant molecular structure since all other variables are fixed. Therefore, methods like the M&G GC$^+$ method (Equation 5.6) can also be developed for this property once sufficient data-points with respect to different surfactants have been collected.

5.4
Functional Bulk Property Modeling – Mixture Properties

As listed in Tables 5.2–5.5, many of the target properties are those of the formulated liquid mixture, which are listed also as bulk properties of the mixture. The state of the mixture is liquid and the target properties need to be estimated for given ranges of temperatures, pressures, and/or compositions of the identified chemicals in the formulation.

5.4.1
Bulk Properties Based on Linear Mixing Rule

The simplest model for estimation of the bulk (mixture) properties is to use the corresponding pure compound properties and a linear mixing rule. The question of when to apply the linear mixing rules depends on the type of chemicals present in the mixture. For mixtures with negligible excess properties of mixing the linear mixing rules may be safely used. For others, depending on the accuracy needed and the availability of other models, the linear mixing rules may also be employed, as an initial trial. According to the linear mixing rule, the bulk mixture property is estimated through Equation 5.23, where the mixture property ζ is determined from the pure compound properties and mole fractions of the compounds present in the mixture, at a specific temperature and/or pressure:

$$\zeta = \sum (\zeta_i \times x_i) \tag{5.23}$$

A good indication of mixtures that may show negligible excess properties of mixing can be obtained from plots of excess properties of known binary mixtures (Figure 5.3). Note that for some properties, such as density, even though the linear mixing rule cannot be applied the molar volume has been observed to behave linearly with composition. Thus, it is possible to estimate the molar volume at a fixed temperature and pressure, through the linear mixing rule, and then convert it into the density.

5.4.2
Bulk Properties Based on Nonlinear Mixing Rules

For some bulk properties, calculation methods employing nonlinear mixing rules may already be available. For example, saturation temperatures of the liquid in equilibrium with the vapor (i.e., the bubble point of the liquid mixture) or with the solid (i.e., the saturation solubility temperature) are determined from the corresponding conditions of phase equilibrium [Equations 5.24 and 5.25 indicate the equilibrium conditions that need to be satisfied for vapor–liquid equilibrium

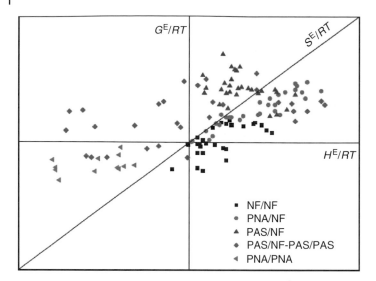

Figure 5.3 Equimolar excess properties (G^E: excess Gibbs energy, H^E: excess enthalpy, and S^E: excess entropy) for more than 130 mixtures at 298.15 K. The mixtures are represented according to the hydrogen bonding classification (*NF*: normal fluid; *PNA*: polar non-associating; *PAS*: polar associating) [68].

(*VLE*) and SLE, respectively]:

$$x_i \gamma_i^L \, P_i^{sat} = y_i \varphi_i^V P \tag{5.24}$$

$$x_i \, \gamma_i^L = \exp\left[\frac{\Delta H_i^{fus}}{RT_{m,i}} \left(\frac{T - T_{m,i}}{T} \right) \right] \tag{5.25}$$

where x_i and y_i represent the molar fraction of compound i in the liquid and vapor phases, respectively, γ_i is the activity coefficient and φ_i the fugacity coefficient of compound i, the superscripts L and V stand for liquid and vapor phases, respectively, P_i^S is the vapor pressure of compound i, ΔH_i^{fus} is the enthalpy of fusion of compound i, $T_{m,t}$ its melting temperature, T the temperature, and P the pressure of the system.

At a given liquid phase composition and pressure, the above equations may be used to calculate the temperatures at which the conditions of equilibrium are satisfied. The pure compound vapor pressures and the normal melting points and enthalpy of fusions are obtained from the models discussed in Section 5.3.1. Estimations of the needed fugacity coefficients and the activity coefficients are discussed in Section 5.5. Figure 5.4 illustrates the work-flow and data-flow involved in the calculations of VLE and SLE. Note that, in addition to property models, databases as well as numerical solvers are needed.

For a given phase of a mixture together with its temperature and pressure, nonlinear mixing models that are also based on GCs have been developed for the estimation of bulk mixture viscosity and surface tension. For surface tension, the

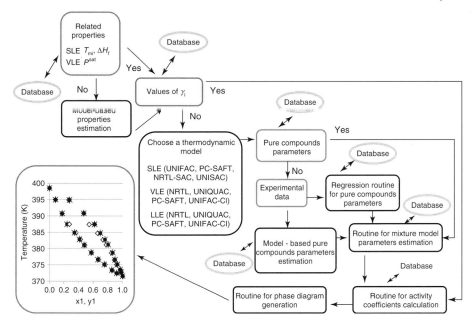

Figure 5.4 Work-flow and data-flow for phase equilibrium calculations.

UNIFAC GC-based model, developed by Suarez *et al.* [69], is quite convenient to use, if the UNIFAC-model with its group parameters are available (Section 5.5):

$$\sigma = \sigma_i + (R \times T/A_i) \times \left[\left(x_{i,s} \times \gamma_{i,b}\right) / \left(x_{i,s} \times \gamma_{i,b}\right) \right] \tag{5.26}$$

This model needs, besides the pure compound surface tension values, the activity coefficients (γ_{ij}) of each compound in the liquid phase and the molar surface area. The activity coefficients can be estimated with an appropriate UNIFAC method [6] or the Paquette equation. The surface area A_i (in $cm^2\ mol^{-1}$) is estimated through Equation 5.27 or 5.28, depending on the available data, where the molar volume V_i is expressed in $cm^3\ mol^{-1}$:

$$A_i = 2.5 \times 10^9 \times \sum v_{k,i} \times Q_k \tag{5.27}$$

$$A_i = 1.021 \times 10^8 \times V_{c,i}^{6/15} \times V_i^{4/15} \tag{5.28}$$

The accuracy of this model is mostly acceptable, but for highly non-ideal or polar mixtures the errors may be large. A useful GC-based model is also available for bulk mixture dynamic viscosity [70]. Here, the model equations are given by:

$$\ln(\eta \times V) = \sum \varphi_i \times \ln(\eta_i \times V_i) + 2\sum \varphi_i\ \ln(x_i/\varphi_i) +$$
$$- \sum [(q_i \times n \times p_i \times \varphi_i)/r_i] \times \sum \theta_{ji} \times \ln(\tau_{ji}) \tag{5.29}$$

Note that except for the pure compound viscosity η_i all other terms in the right-hand side of Equation 5.29 come from the UNIFAC model and therefore, as long as the group parameters are available, this GC-based model can also be used. Good accuracy has been found for the test mixtures reported by Conte *et al.* [2, 3].

Note that the properties discussed above with nonlinear mixing rules are applicable for different temperatures and mixture compositions. To also obtain the same properties as a function of pressure, an equation of state may be used, for example, the CPA [11] or the PC-SAFT [34].

5.5
Functional Compound Properties in Mixtures – Modeling

These are phase equilibrium related properties of each compound present in the mixture. For example, for a two-phase VLE system, these could be the fugacity coefficients of each compound in the vapor phase at a fixed temperature and pressure, and the activity coefficients for the same compounds in the liquid phase at the same condition of temperature and pressure. Given the phase identity, mixture composition, temperature, pressure, and compound identity, the fugacity coefficients and/or activity coefficients are estimated by employing appropriate models. As the equations for most well-known models can be found in standard textbooks they are not given in this chapter.

5.5.1
Fugacity Coefficients

The most well-known models for this property are the cubic equations of state [10, 71, 72], which are, however, not valid for non-ideal liquid mixtures. Other non-cubic equations of state, such as the CPA [11] or the PC-SAFT [34], may also be used to obtain these properties, even for non-ideal mixtures in the liquid phase.

5.5.2
Activity Coefficients

The most well-known models for this property are the molecular interactions based models, such as the Wilson [73], NRTL [12], and UNIQUAC [74] models or the GC-based models such as the different versions of the UNIFAC model [6]. One useful extension of the UNIFAC-based methods is the UNIFAC-CI method, which uses atom connectivity rules to predict the missing group parameters [75, 76]. A brief overview of this method is given in Appendix 5.B.

5.6
Performance Related Property Modeling

Several properties that may not be employed as target properties but are, nevertheless, important for evaluation of the performance of liquid formulated products are discussed in this section. For example, in homogeneous liquid formulated products, the liquid product must be a stable liquid during storage and at application. The flash point temperatures, 90% evaporation rate, and stability of the emulsion are examples of performance-related properties.

5.6.1
Prediction of Liquid Phase Stability

The test of liquid-phase stability is determined from the stability criterion, which is based on the ΔG^{mix} (Gibbs energy of mixing):

$$\frac{\Delta G^{mix}}{RT} < 0 \tag{5.30}$$

where ΔG^{mix} is obtained through a suitable activity coefficient model, such as given by Equations 5.31a and 5.31b:

$$\frac{\Delta G^{mix}}{RT} = \frac{G^E}{RT} + \sum x_i \times \ln(x_i) \tag{5.31a}$$

$$\frac{G^E}{RT} = \sum x_i \times \ln(\gamma_i) \tag{5.31b}$$

The correct estimation of ΔG^{mix} is important to ensure the stability of the liquid formulated product being designed. Depending on the chemical system, different models may be used for the activity coefficients needed by Equation 5.31b. The most common models used to calculate excess Gibbs energies (G^E) for organic chemical systems are UNIFAC [6], UNIQUAC [72], and NRTL [12]. For systems involving polymers, the GC-Flory [77] and UNIFAC-FV [78] models have been used. The accuracy of these models can be improved by fine-tuning the model parameters, if experimentally measured data is available.

5.6.2
Flash Point

For many liquid formulated mixtures, as discussed above, the linear mixing rules or ideal thermodynamics may not be applicable. Estimations of the open-cup flash points (a measure of flammability) and the evaporation rates are two examples of product performance related properties that may also require models suitable for non-ideal mixtures depending on the chemicals present in the mixture.

Mathematical models for predicting the open cup flash point have been developed for miscible mixtures [79] and non-miscible mixtures [80]:

$$1 = \sum (x_i \times P_i^{sat})/P_{i,fp}^{sat} \tag{5.32}$$

$$1 = \sum 10^a \times x_i \times [1642 - (T_{i,fp} + 230)]/ \ [1642 - (T_{mix,fp} + 230)] \tag{5.33}$$

$$a = \frac{m_i(T_{i,fp} - T_{mix,fp})}{T_{i,fp} \times T_{mix,fp}} \tag{5.34}$$

where m_i is a parameter that may be estimated from the literature [80], $T_{i,fp}$ is the flash point of each component i, while $T_{mix,fp}$ is the flash point of the mixture. The results provided, once accurate flash point values of the pure compound thermophysical properties are available, confirm that these models can be reliably applied for process-product design issues.

5.6.3
Performance Properties of Liquid (Emulsion) Formulated Products

When considering emulsified formulated products it is necessary to note that, even though from a macroscopic point of view the product may appear to be homogeneous, at the microscopic level emulsions are actually formed by two well-defined non-miscible phases that are kept together by the emulsifiers. Therefore, standard bulk (mixture) property models may not be suitable for estimating bulk thermophysical properties of emulsions. In this section, the target (bulk) properties that are also performance related properties of the liquid (emulsion) product are discussed. First, those properties for which two distinct values that characterize the formulated product are presented; then, those properties for which a linear mixing rule may be suitable are discussed; and, finally, we present those properties for which special models need to be developed that consider the specific nature of the formulated (emulsion) product.

5.6.3.1 Two Distinct Values
The dual nature of emulsified products, where two non-miscible phases coexist in a macroscopically homogeneous mixture, does not help in specifying two distinct values (one per phase) for a thermophysical property of interest. Among the properties listed in Table 5.2, evaporation time is a good example of this family of properties. The emulsified form does not influence the vapor pressure of the two phases; consequently, for the evaporation time (a function of the vapor pressure) two values need to be specified to characterize an emulsified product. As a result of this, the two phases of an emulsified formulated product are subject to different evaporation rates and if the difference between the values of the two phases is large enough it is possible that one of the two phases would evaporate completely before the other one, destroying the emulsion and leading to a homogeneous formulated product. In addition, in the case of solid solubility calculations, the solubility parameters (both Hansen and Hildebrand) need to be considered separately for an emulsified formulated product; that is, once they have been determined relative to each phase, the emulsion is characterized by two distinct values for each of the solubility parameters, identifying each of the two co-existing phases.

5.6.3.2 Linear Mixing Rule
A linear mixing rule is said to be suitable only when the excess properties of the mixture are not significant. When considering an emulsion, since the two phases have to be non-miscible by definition, it is hard to expect the excess properties of the whole product to be negligible. However, many properties of interest can be

estimated through simple averaging schemes such as:

$$\zeta = \zeta_{\text{oil-phase}} x_{\text{oil-phase}} + \zeta_{\text{water-phase}} \; x_{\text{water-phase}} \tag{5.35}$$

Among the properties listed in Table 5.2, the following are commonly estimated, in relation to emulsified formulated properties, with a linear mixing rule: cost, density (based on the molar volume), open cup flash point, and toxicity parameters.

5.6.3.3 Dedicated Models

Some relevant thermophysical properties need special versions of models when applied to emulsions, for example, the dynamic viscosity and the surface tension.

Dynamic Viscosity The viscosity of emulsions depends not only on the viscosity of the two phases and the ratio between them but it is also highly dependent on the droplet size distribution both in terms of average value and in terms of dispersion of the distribution. Moreover, emulsions show highly non-ideal behavior, far from the assumption of Newtonian fluids. Depending on the formulation variables, emulsions can behave both as pseudo-plastic and as pseudo-dilatant fluids. However, strong modeling efforts have been made in recent years to develop reliable models and now they are widely used, especially for the production of creams and pastes even though, due to the high uncertainty implicit in the viscosity of emulsions, an experimental refinement of the modeled value is usually required. For diluted emulsions (dispersed phase volume fraction: $\varphi < 0.62$), two equations are currently used, namely, those of Krieger and Dougherty (Equation 5.36) [81] and Pal (Equations 5.37 and 5.38) [82]:

$$\mu_e = \mu_c (1 - \varphi/\varphi_C)^{-k^* \varphi_C} \tag{5.36}$$

$$\mu_e = \mu_c \times \left[1 + (5k + 2) \; \varphi/2 \times (k + 1) + (5k + 2)^2 \varphi^2 / 10 \times (k + 1)^2 \right]$$
$$\times \left[\frac{1 + \lambda_1 \times \lambda_2 \times N_{\text{Ca}}^2}{1 + \lambda_1^2 \times N_{\text{Ca}}^2} \right] \tag{5.37}$$

$$\lambda_1, \lambda_2 = f(k, \varphi) \tag{5.38}$$

where k is the oil-water viscosity ratio, φ the dispersed phase volume fraction, and N_{Ca} is the capillarity dimensionless number, describing the correlation between the viscosity and the droplet size distribution. When dealing with concentrated emulsions ($\varphi > 0.74$), use of the following model [83] is proposed:

$$\mu_e = \tau_0/\dot{\gamma} + 32(\varphi - 0.73) \times \mu_c \times N_{\text{Ca}}^{0.5} \tag{5.39}$$

where γ is the shear rate and τ_0 the yield value.

Surface Tension The surface tension of emulsions is extremely relevant since the use of this particular kind of formulated product is extremely widespread in the field of detergency. For the cleaning mechanism to happen, wetting between the detergent and the surface to be cleaned needs to be ensured [84], which means that the surface tension of the fluid needs to be equal to or lower than the critical surface tension of the surface to be cleaned. Surfactants, besides having many other effects, are responsible for strongly decreasing the surface tension of the water, until the CMC is reached. Since the surface tension of an emulsion is defined as the surface tension of the continuous phase, in particular when designing an oil-in-water emulsion, it is extremely relevant to have available a model that can quantify this phenomenon, to be able to select/design a surfactant and be sure that its effect on the overall surface tension is well described.

Since the maximum effect on the overall surface tension is obtained once the CMC is reached, by considering constant temperature it is reasonable to imagine that a model based on the molecular structure of the surfactant can produce reliable results. The model developed by Wang *et al.* [85] is a good QSAR model for a first trial:

$$\gamma_0 = 11.98 + 0.4780 \times NO + 0.5845 \times KHO - 0.0007763 \times E_T$$
$$-0.01053 \times \Delta H_f + 0.09734D - 0.1345 \times NO \times KHO \tag{5.40}$$

where NO, KHO, ET, ΔH_f, and D are descriptors for the compound that must be estimated first before the property can be calculated. In this case, development of GC models like the M&G GC$^+$ method also appears to be feasible since sufficient data are available.

5.7
Software Tools

A wide variety of property prediction tools can be found in the commercial process simulators. In this chapter, however, two special tools (ThermoDB and ICAS-Property Package) that are particularly suitable for product design/analysis related calculations are mentioned. More details on the models, methods and tools for chemical product design can be found in Kontogeorgis and Gani [86].

5.7.1
ThermoData Engine (TDE)

The NIST ThermoData Engine (*TDE*) software represents a full-scale implementation of the concept of dynamic data evaluation for thermophysical properties. This concept requires large electronic databases capable of storing essentially all relevant experimental data known to date with metadata and uncertainties [87]. The combination of these databases with expert-system software, designed to automatically generate recommended property values based on available experimental and predicted data, leads to the ability to produce critically evaluated data dynamically or on demand. TDE provides many tools for process-product design, such as the thermophysical properties of pure compounds, experiment planning and product design tools, a tool for evaluation of properties of mixtures including consistency tests, evaluation of properties of material streams, and a tool for recommendation of solvents. TDE is used in various applications ranging from data quality assurance, to validation of new experimental data, to chemical process and product design. More information on TDE can be found in published papers such as Reference [87].

5.7.2
ICAS-Property Package

The ICAS-Property package is part of the Integrated Computer Aided System developed by the authors [25]. It contains several databases (of organic chemicals, solvents, ionic liquids, etc.), a large collection of pure compound property models (primary, secondary, and functional), and mixture properties (bulk and equilibrium related). In addition, it includes tools for model parameter estimation and calculation of phase diagrams. Figure 5.5 gives a schematic representation of the ICAS-property prediction package.

Based on the ICAS-property package, Conte *et al.* [88] developed the virtual Product-Process Design Laboratory software. This tool is ideally suited for model-based chemical product (formulation) design.

5.8
Conclusions

Through a collection of thermophysical properties and the models most commonly employed to estimate them, an overview of property prediction for liquid formulated

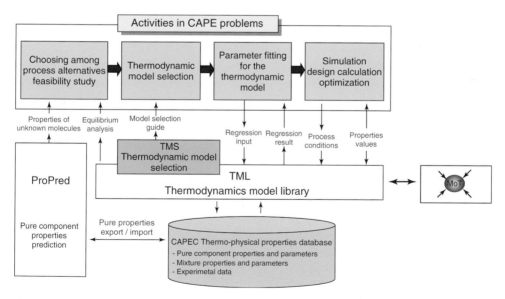

Figure 5.5 ICAS-property prediction package (pure compound property estimation; property model library; model parameter regression; applications in process-product design).

(homogeneous or emulsion) products is provided. It is shown that, even though the target properties of the liquid product refer to mixture properties, to estimate them the pure compound properties are also needed. In addition, in many cases, the mixture property could be modeled as a primary pure compound property. Taking advantage of this special feature, predictive GC based models have been developed. Furthermore, for mixture properties, predictive GC models have been developed and new properties are being added to the list. This is very promising from the point of view of computer-aided model-based formulation design as a first stage to identifying the promising candidates to be verified experimentally in the second stage. This allows an enlargement of the search space to identify truly novel alternatives very fast and without too much cost. On the other hand, the experimental effort is efficiently used and focused on a few selected and very promising candidates. The properties and their models, however, play very important roles. The search space depends on the application area of the property models. The reliability of the formulated product depends on the reliability of the property models, which, in turn, depends on the uncertainty of the experimentally measured data. In addition, as the liquid product design/analysis problems become more complex the need for data, models, and theory also increases.

Appendix 5.A: Overview of the M&G GC$^+$ Method

Property prediction models based on the MG (Marrero–Gani) method [5] are considered here since this method is based exclusively on the molecular structure of the pure component and exhibits a good accuracy and a wide range of applicability covering chemical, biochemical, and environment-related pure components. In the MG method, property estimation is performed at three levels. The first level has a large set of simple groups that allow for the representation of a wide variety of organic chemicals. However, these groups only partially capture the proximity effects and are unable to distinguish among isomers. The second level of estimation involves groups that provide a better description of proximity effects and can differentiate among isomers. Hence, a second level of estimation is intended to deal with polyfunctional, polar or non-polar, and cyclic chemicals. The third level of estimation includes groups that provide more structural information about molecular fragments of chemicals whose description is insufficient through the first- and second-order groups; hence, this level allows estimation of complex heterocyclic and polyfunctional acyclic chemicals. The MG method includes 220 first-order groups, 130 second-order groups, and 74 third-order groups to represent the molecular structure of the organic chemicals. The property prediction model used to estimate the properties of organic chemicals by employing the MG method has the form:

$$f(X) = \sum_i N_i C_i + w \sum_j M_j D_j + z \sum_k E_k O_k \tag{5.A1}$$

The function $f(X)$ is a function of property X and it may contain additional adjustable model parameters (universal constants) depending on the property involved. In Equation 5.A1, C_i is the contribution of the first-order group of type-i that occurs N_i times, D_j is the contribution of the second-order group of type-j that occurs M_j times, and E_k is the contribution of the third-order group of type-k that has O_k occurrences in the molecular structure of pure component. For determination of the contributions, C_i, D_j, and E_k, Marrero and Gani [5] suggested a multilevel estimation approach. Equation 5.A1 is a general model for all the properties and the definition of $f(X)$ is specific for each property X (model details together with detailed regression statistics for the 29 properties modeled are given by Hukkerikar et al. [41]).

Often, when applying GC-based models, the required GC parameter may be missing. The most common way to obtain these parameters is to collect new experimentally measured data and estimate the missing contributions (parameters). A more efficient, faster, and cheaper but also reliable option is to predict them with the recently developed atom connectivity method [23]. For modeling of pure compound properties based on the CI method, the following property model is employed:

$$f(X) = \sum_i a_i A_i + b(^v\chi^0) + 2c(^v\chi^1) + d \qquad (5.A2)$$

where a_i is the contribution of atom of type-i that occurs A_i times in the molecular structure, $^v\chi^0$ is the zeroth-order (atom) connectivity index, $^v\chi^1$ is the first-order (bond) connectivity index, b and c are adjustable parameters, and d is a universal parameter. Please note that $f(X)$ of the models in the MG method and in the CI method (i.e., left-hand side of Equations 5.A1 and 5.A2) has the same functional form for a particular pure compound property X and the values of universal constants for the CI models are the same as those for the GC models. A detailed procedure to calculate $^v\chi^0$ and $^v\chi^1$ for the groups as well as for the entire molecule is given by Gani et al. [23]. Once these indices are calculated, the following CI model equations are applied to the missing groups to compute $f(X_m)$ and $f(X^*)$:

$$f(X_m) = \sum_i a_{m,i} A_{m,i} + b(^v\chi^0)_m + 2c(^v\chi^1)_m \qquad (5.A3)$$

$$f(X^*) = \left(\sum_m n_m f(X_m) \right) + d \qquad (5.A4)$$

where m is the number of different missing groups and n_m indicates the number of times a missing group appears in the molecule. Finally, value of property X is estimated using the following model equation:

$$f(X) = \sum_i N_i C_i + f(X^*) + w \sum_j M_j D_j + z \sum_k E_k O_k \qquad (5.A5)$$

Again, a detailed model description and regression statistics for the 29 properties modeled are given in Hukkerikar et al. [41].

Appendix 5.B: Prediction of the UNIFAC Group Interaction Parameters

The generic form of the GC-based UNIFAC method [74] is written as:

$$\ln (\gamma)^{\text{UNIFAC}} = f(\underline{x}, T, \underline{a}, \underline{R}, \underline{Q})$$ (5.B1)

where $\underline{x}$ are the molar fractions of each compound, T is the temperature of the system, $\underline{a}$ are the group interaction parameters (*GIPs*) (obtained through regression with data), $\underline{R}$ and $\underline{Q}$ are the group van der Waals volumes and group the surface area, respectively. Often, the GIPs $\underline{a}$ are missing for the liquid mixture under study. As in the case of pure compound property prediction, these parameters may also be predicted through an atom-connectivity model based on the UNIFAC method. This combined model is called *UNIFAC-CI*.

Atom connectivity indices can also represent the groups used in the UNIFAC-method and the regressed atom connectivity index (CI)-interaction parameters can be used to predict the issuing group-interaction parameters [75, 76]. For application of the UNIFAC-CI approach, the atom interaction parameters (*AIPs*) a, b, c, and d are used to predict the missing GIPs, a_{kl}; the equations needed for their calculation are given below:

$$a_{kl} = \underbrace{b_{\text{C-C}}(A^{\text{CC}}_{kl}) + b_{\text{C-O}}(A^{\text{CO}}_{kl}) + b_{\text{C-N}}(A^{\text{CN}}_{kl}) + \cdots}_{\text{for 0 order interactions}}$$

$$\underbrace{+c_{\text{C-C}}(A^{\text{CC}}_{kl}) + c_{\text{C-O}}(A^{\text{CO}}_{kl}) + c_{\text{C-N}}(A^{\text{CN}}_{kl}) + \cdots}_{\text{for 1st order interactions}}$$

$$\underbrace{+d_{\text{C-C}}(A^{\text{CC}}_{kl}) + d_{\text{C-O}}(A^{\text{CO}}_{kl}) + d_{\text{C-N}}(A^{\text{CN}}_{kl}) + \cdots}_{\text{for 2nd order interactions}}$$

$$\underbrace{+e_{\text{C-C}}(A^{\text{CC}}_{kl}) + e_{\text{C-O}}(A^{\text{CO}}_{kl}) + e_{\text{C-N}}(A^{\text{CN}}_{kl}) + \cdots}_{\text{for 3rd order interactions}}$$ (5.B2)

with:

$$(A^{XY}_{kl})_0 = \frac{n^{(k)}_X {}^{\nu}X^0_{(l)} - n^{(l)}_Y {}^{\nu}X^0_{(k)}}{{}^{\nu}X^0_{(l)} {}^{\nu}X^0_{(k)}}$$ (5.B3)

$$(A^{XY}_{kl})_1 = \frac{n^{(k)}_X {}^{\nu}X^1_{(l)} - n^{(l)}_Y {}^{\nu}X^0_{(k)}}{{}^{\nu}X^1_{(l)} {}^{\nu}X^0_{(k)}}$$ (5.B4)

$$(A^{XY}_{kl})_2 = \frac{n^{(k)}_X {}^{\nu}X^1_{(l)} - n^{(l)}_Y {}^{\nu}X^1_{(k)}}{{}^{\nu}X^1_{(l)} {}^{\nu}X^1_{(k)}}$$ (5.B5)

$$(A^{XY}_{kl})_1 = \frac{n^{(k)}_X {}^{\nu}X^2_{(l)} - n^{(l)}_Y {}^{\nu}X^0_{(k)}}{{}^{\nu}X^2_{(l)} {}^{\nu}X^0_{(k)}}$$ (5.B6)

where $n^{(k)}_X$ is the number of atoms of type X in the group k, ${}^{\nu}X^m{}_{(k)}$ is the mth order valence connectivity index for the group k, A^{CC}_{kl} is an intermediate variable used to predict the GIP a_{kl} between the groups k and l, and the regressed coefficients a, b, c, d, and e represent the atomic interactions between the C, H, O, N, and Cl atoms.

Examples of the application of the UNIFAC-CI method can be found in References [75, 76].

References

1. Gani, R. (2004) *Comput. Chem. Eng.*, **28**, 2441.
2. Conte, E., Gani, R., and Ng, K.M. (2011) *AIChE J.*, **57**, 2431.
3. Conte, E., Gani, R., Cheng, Y.S., and Ng, K.M. (2012) *AIChE J.*, **58**, 173.
4. Pashley, R.M. and Karaman, M.E. (2004) *Applied Colloid and Surface Chemistry*, John Wiley & Sons, Ltd., Chichester.
5. Marrero, J. and Gani, R. (2001) *Fluid Phase Equilib.*, **183**, 183–184.
6. Fredenslund, A., Gmehling, J., Michelsen, M.L., Rasmussen, P., and Prausnitz, J.M. (1977) *Ind. Eng. Chem. Proc. Des. Dev.*, **16**, 450.
7. Slater, J.C. (1951) *Phys. Rev.*, **81**, 385.
8. Karayiannis, N.C., Mavrantzas, V.G., and Theodorou, D.N. (2004) *Macromolecules*, **37**, 2978.
9. Van Speybroeck, V., Gani, R., and Meier, R.J. (2010) *Chem. Soc. Rev.*, **39**, 1764.
10. Soave, G. (1972) *Chem. Eng. Sci.*, **27**, 1197.
11. Kontogeorgis, G.M., Voutsas, E.C., Yakoumis, I.V., and Tassios, D.P. (1996) *Ind. Eng. Chem.*, **35**, 4310.
12. Renon, H. and Prausnitz, J.M. (1968) *AIChE J.*, **14**, 135.
13. Klamt, A. (1995) *J. Phys. Chem.*, **99**, 2224.
14. Rackett, H.G. (1970) *J. Chem. Eng. Data*, **15**, 514.
15. Spencer, C.F. and Danner, R.P. (1972) *J. Chem. Eng. Data*, **17**, 236.
16. Yamada, T. and Gunn, R.D. (1973) *J. Chem. Eng. Data*, **18**, 234.
17. Elbro, H.S., Fredenslund, A., and Rasmussen, P. (1991) *Ind. Eng. Chem. Res.*, **30**, 2576.
18. Tsibanogiannis, I.N., Kalospiros, N.S., and Tassios, D.P. (1994) *Ind. Eng. Chem. Res.*, **33**, 1641.
19. Imhels, E.C. and Gmehling, J. (2003) *Ind. Eng. Chem. Res.*, **42**, 408.
20. Horvath, A.L. (1992) *Molecular Design*, Elsevier Science, Amsterdam.
21. Sheldon, T.J., Adjiman, C.S., and Cordiner, J.L. (2005) *Fluid Phase Equilib.*, **231**, 27.
22. Megnassan, E., Legoff, D., and Proutiere, A. (1994) *J. Mol. Liq.*, **59**, 37.
23. Gani, R., Harper, P.M., and Hostrup, M. (2005) *Ind. Eng. Chem. Res.*, **44**, 7262.
24. Constantinou, L. and Gani, R. (1994) *AIChE J.*, **40**, 1697.
25. Gani, R., Hytoft, G., Jaksland, C., and Jensen, A.K. (1997) *Comput. Chem. Eng.*, **21**, 1135.
26. Sastri, S.R.S. and Rao, K.K. (1992) *Chem. Eng. J.*, **50**, 9.
27. Conte, E., Martinho, A., Matos, H.A., and Gani, R. (2008) *Ind. Eng. Chem. Res.*, **47**, 7940.
28. Van Wesenbeeck, I., Driver, J., and Ross, J. (2008) *Bull. Environ. Contam. Toxicol.*, **80**, 315.
29. Klein, J.A., Wu, D.T., and Gani, R. (1992) *Comput. Chem. Eng.*, **16**, 229.
30. Conte, E. (2010) Innovation in integrated chemical product-process design: development through a model-based system approach, PhD thesis, Technical University of Denmark, Lyngby.
31. Nielsen, T.L., Abildskov, J., Harper, P.M., Papaeconomou, I., and Gani, R. (2001) *J. Chem. Eng. Data*, **46**, 1041.
32. Kroenlein, K., Muzny, C.D., Diky, V., Kazakov, A.F., Chirico, R.D., Magee, J.W., Abdulagatov, I., and Frenkel, M. (2011) *J. Chem. Inf. Model.*, **51**, 1506.
33. American Institute of Chemical Engineers (2003) DIPPR (Design Institute for Physical Property Data) files Version 17.0.
34. Gross, J. and Sadowski, G. (2001) *Ind. Eng. Chem. Res.*, **40**, 1244.
35. Ceriani, R., Gani, R., and Meirelles, A.J.A. (2010) *J. Food Process Eng.*, **33**, 208.

36. Karunanithi, A.T., Achenie, L., and Gani, R. (2007) *Chem. Eng. Sci.*, **12**, 3276.

37. Harper, P.M. and Gani, R. (2000) *Comput. Chem. Eng.*, **24**, 677.

38. Abildskov, J. and O'Connel, J.P. (2003) *Ind. Eng. Chem. Res.*, **42**, 5622.

39. Gani, R., Jimenez-Gonzalez, C., and Constable, D.J.C. (2005) *Comput. Chem. Eng.*, **29**, 1661.

40. Modarresi, H., Conte, E., Abildskov, G.R., and Crafts, P. (2008) *Ind. Eng. Chem. Res.*, **42**, 5234.

41. Hukkerikar, A., Sarup, B., Ten Kate, A., Abildskov, J., Sin, G., and Gani, R. (2012) *Fluid Phase Equilib.*, **321**, 25–43.

42. Hildebrand, J.H. (1936) *The Solubility of Non-Electrolytes*, Reinhold Publishing Corporation, New York.

43. Delgado, E.J. and Diaz, G.A. (2006) *SAR QSAR Environ. Res.*, **17**, 483.

44. Hukkerikar, A., Kalakul, S., Sarup, B., Young, D.M., Sin, G., and Gani, R. (2012) *J. Chem. Inf. Model.*, **52** (11), 2823.

45. Gao, D.W., Wang, P., Yang, L., Peng, Y.Z., and Liang, H. (2001) *J. Environ. Sci. Health*, **37**, 601.

46. Porter, M.R. (1994) *The Handbook of Surfactants*, Chapman & Hall, New York.

47. Jaycock, M.J. and Parfitt, G.D. (1981) *Chemistry of Interfaces*, Ellis Horwood, Chichester.

48. Yuan, S.L., Cai, Z.T., Xu, G.Y., and Wang, W. (2003) *Acta Phys. Chim. Sin.*, **19**, 334.

49. Ren, Y., Liu, H.X., Yao, X.J., Liu, M.C., Hu, Z.D., and Fan, B.T. (2006) *J. Colloid Interface Sci.*, **302**, 669.

50. Ren, Y., Zhao, B.W., Chang, Q., and Yao, X.J. (2011) *J. Colloid Interface Sci.*, **358**, 202.

51. Tadros, T.F. (2005) *Applied Surfactants: Principles and Applications*, Wiley-VCH Verlag GmbH, Weinheim.

52. Ruckenstein, N. and Nagarajan, R. (1980) *J. Phys. Chem.*, **84**, 1349.

53. Li, X.F., Zhang, G.Y., Dong, J.F., Zhou, X.H., Yan, X.C., and Luo, M.D. (2004) *J. Mol. Struct.*, **710**, 119.

54. Huibers, P.D.T., Lobanov, V.S., Katritzky, A.R., Shah, D.O., and Karelson, M. (1996) *Langmuir*, **12**, 1462.

55. Griffin, W.C. (1949) *J. Soc. Cosmet. Chem.*, **1**, 311.

56. Griffin, W.C. (1954) *J. Soc. Cosmet. Chem.*, **5**, 249.

57. Atlas Powder Company (1948) *Atlas Surface Active Agents*, Industrial Chemicals Department, Atlas Powder Company, Wilmington.

58. Delgado Charro, M.B., Otero Espinar, F.J., and Blanco Méndez, J. (1997) *Technol. Farm.*, **1**, 207.

59. Griffin, W.C. (1955) *Am. Perf. Essent. Oil. Rev.*, **65** 26.

60. Pasquali, R.C., Taurozzi, M.P., and Bregni, C. (2008) *Int. J. Pharm.*, **356**, 44.

61. Beerbower, A. and Hill, M.W. (1971) *McCutcheon's Detergents and Emulsifiers Annual*, Allured Publishing Company, Ridgewood, NJ.

62. Israelachvili, J.N., Mitchell, D.J., and Ninham, B.W. (1976) *J. Chem. Soc., Faraday Trans.*, **72**, 1525.

63. Salager, J.L. (1996) *Prog. Colloid Polym. Sci.*, **100**, 137.

64. Schott, H. (1996) *Tens. Surf. Det.*, **6**, 457.

65. Shaw, D.J. (1983) *Introduction to Colloid and Surface Chemistry*, Butterworth-Heinemann, Oxford.

66. Li, Y., Xu, G., Luan, Y., Yuan, S., and Xin, X. (2007) *J. Disp. Sci. Technol.*, **26**, 799.

67. Broze, G. (1999) *Surfactant Science Series*, Vol. 82, Marcel Dekker, New York.

68. Smith, J.M., Van Ness, H.C., and Abbott, M.M. (2001) *Introduction to Chemical Engineering Thermodynamics*, McGraw-Hill, New York.

69. Suarez, J.T., Torres-Marchal, C., and Rasmussen, P. (1989) *Chem. Eng. Sci.*, **44**, 782.

70. Cao, W., Knudsen, K., Fredenslund, A.A., and Rasmussen, P. (1993) *Ind. Eng. Chem. Res.*, **32**, 2077.

71. Van der Waals, J.D. (1873) Over de continuiteit van den gas- en vloeistoftoestand, PhD thesis, Leiden University.

72. Peng, D.Y. and Robinson, D.B. (1976) *Ind. Eng. Chem.*, **15**, 59.

73. Wilson, G.M. (1964) *J. Am. Chem. Soc.*, **86**, 127.

74. Abrams, D.S. and Prausnitz, J.M. (1975) *AIChE J.*, **21**, 116.

75. Gonzalez, H.E., Abildskov, J., and Gani, R. (2007) *Fluid Phase Equilib.*, **261**, 199.

76. Mustaffa, A.A., Kontogeorgis, G.M., and Gani, R. (2011) *Fluid Phase Equilib.*, **302**, 274.

77. Bogdanic, G. and Fredenslund, A. (1994) *Ind. Eng. Chem. Res.*, **33**, 1331.

78. Bogdanic, G. and Vidal, J. (2000) *Fluid Phase Equilib.*, **173**, 241.

79. Le Chatelier, H. (1891) *Ann. Mines*, **19**, 388.

80. Affens, W.A. and McLaren, G.W. (1972) *J. Chem. Eng. Data*, **17**, 482.

81. Krieger, I.M. and Dougherty, T.J. (1959) *Trans. Soc. Rheol.*, **3**, 137.

82. Pal, R. (1995) *AIChE J.*, **41**, 783.

83. Princen, H.M. and Kiss, A.D. (1989) *J. Colloid Interface Sci.*, **112**, 427.

84. Pilemand, C. (2002) Surfactants–Their Abilities and Important Physico-Chemical Properties, DTKommunikation A/S, Copenhagen.

85. Wang, Z.W., Feng, J.L., Wang, H.J., Cui, Z.G., and Li, G.Z. (2006) *J. Dispersion. Sci. Technol.*, **26**, 441.

86. Kontogeorgis, G.M. and Gani, R. (2004) *Computer Aided Property Estimation for Process and Product Design* (eds G.M. Kontogeorgis and R. Gani), Elsevier B.V., Computer Aided Chemical Engineering, vol. 19, pp. 1–425.

87. Frenkel, M. (2011) *Comput. Chem. Eng.*, **35**, 393.

88. Conte, E., Gani, R., and Malik, T.I. (2011) *Fluid Phase Equilib.*, **302**, 294.

6
Sources of Thermophysical Properties for Efficient Use in Product Design

Richard Sass

6.1
Introduction

Process synthesis, design, and optimization and also detailed engineering for chemical plants and equipment depend heavily on the availability and reliability of thermophysical property data of the pure components and mixtures involved. To illustrate this fact we can analyze the needs for one of the essential engineering processes, the separation of fluid mixtures. For the design of such a typical separation process, for example, distillation, we require the thermodynamic properties of a mixture, in particular for a system that has two or more phases at a certain temperature or pressure. Missing property data, for example, binary parameters for the phase equilibrium, can be calculated by group contribution or by quantum mechanical methods. These calculation methods are mainly used in the first step of process studies. Later in the process life cycle, when higher accuracy is needed, they are increasingly replaced by experimental values.

Nevertheless, modeling results have a great impact on the design and construction of single chemical apparatus as well as whole plants or production lines. Inaccurate data may lead to very expensive misjudgments over whether to proceed with a new process, or modification of it, or not to go ahead. Inadequate or unavailable data may result in a promising and profitable process being delayed or in the worst case rejected, solely because it was not properly modeled in a simulation [1]. This can be the case if azeotropic compositions are wrong or if the miscibility gaps are not recognized. Another potential danger is that the credibility of the results of a thermophysical model calculation generated by computer software is very high, even when the result is wrong. Consequently, the expert has a duty to prove that the most sophisticated software will not lead automatically to the most cost-effective solution, in order to save energy effectively, without a background that includes a correspondingly accurate database of physical and thermodynamic data. In all cases, it is the responsibility of the user to verify the results of the calculation and to consult, if necessary, an expert in thermodynamic calculations.

Product Design and Engineering: Formulation of Gels and Pastes, First Edition.
Edited by Ulrich Bröckel, Willi Meier, and Gerhard Wagner.
© 2013 Wiley-VCH Verlag GmbH & Co. KGaA. Published 2013 by Wiley-VCH Verlag GmbH & Co. KGaA.

6.2
Overview of the Important Thermophysical Properties for Phase Equilibria Calculations

The data that are needed and which types of data are available are described below. Table 6.1 gives a short, general overview of the data needed in the simulation and design of processes and products.

6.3
Reliable Sources of Thermophysical Data

For many years, the most popular way to find thermophysical property data was to look them up in classical papers, books, or favorite book collections, starting with the *Handbook of Chemistry and Physics* [2] up to data collection handbooks issued by data producers, such as the *Landolt–Börnstein* sets [3], and the DECHEMA *Chemistry Data Series* [4].

The preferred route to reliable data sources today is to search within an electronic database for the components, mixtures, and properties one needs. A numerical database or collection typically contains the literature references as well as measurement data. Databases are typically compiled and/or maintained by individuals or groups having a well-known reputation in that field. Therefore, they have an overview of the primary literature publishing physical property data and are able to continuously add new data to their collections. In many cases these groups are able to check the consistency of data published in the literature. Efforts are even made today by publishers to avoid the publication of data over which there are doubts about the repeatability and uncertainty during the referee process (e.g., NIST data

Table 6.1 Important categories of property data.

Property type	Specific properties
Phase equilibrium	Boiling and melting points, vapor pressure, fugacity and activity coefficients, solubility (Ostwald or Bunsen coefficients), distribution coefficient
P-V-T behavior	Density, volume, compressibility, critical constants
Caloric properties	Specific heat, enthalpy, entropy
Transport properties	Viscosity, thermal conductivity, ionic conductivity, diffusion coefficients
Boundary properties	Surface tension
Chemical equilibrium	Equilibrium constants, association/dissociation constants, enthalpies of formation, heat of reaction, Gibbs energy of formation, reaction rates
Acoustic	Velocity of sound
Optical	Refractive index, polarization
Molecular properties	Virial coefficients, binary interaction parameters, ion radius, and volume

tools [5, 6]). In most cases these groups also use their own collections and methods for model development [7, 8].

In the following a survey is given of existing and still maintained collections and databases on solutions.

6.4
Examples of Databases for Thermophysical Properties

Because dozens of sources of thermodynamic data are available on the World Wide Web, only a few major providers will be mentioned in this chapter. A wider overview of what is available on the Web is given, for example, on the web pages of the University of Illinois [9]. Table 6.2 gives a few examples of larger databases.

Four examples of databases are now described in the following pages.

Table 6.2 Provider list for thermophysical data.

Producer	Database name	URL (status: 8 August 2012)
DECHEMA	DETHERM	*http://www.dechema.de/detherm-lang-en.html*
DDBST	Dortmund Data Bank DDB	*http://www.ddbst.de/ddb.html*
DTU Lyngby	CERE	*http://www.cere.dtu.dk/Expertise/Data_Bank.aspx*
NIST	Chemistry WebBook	*http://webbook.nist.gov/chemistry/*
IUPAC-NIST	Solubility database	*http://srdata.nist.gov/solubility/*
K&K Associates	Thermal Resource Center	*http://www.tak2000.com/*
FIZ CHEMIE	Infotherm	*http://www.fiz-chemie.de/en/home/products-services/chemical-data/thermophysics.html*
Reed Elsevier	Reaxis	*https://www.reaxys.com/info/about-overview*
AIChE/DIPPR	DIPPR	*http://www.aiche.org/dippr*
G&P Engineering Software	MIXPROPS	*http://www.gpengineeringsoft.com/pages/pdtmixprops.html*
G&P Engineering Software	PHYPROPS	*http://www.gpengineeringsoft.com/pages/pdtphysprops.html*
Ecole Polytechnique de Montreal	FACT	*http://www.crct.polymtl.ca/fact/index.php*
JESS	JESS	*http://jess.murdoch.edu.au/jess_home.htm*
S. Ohe	Fundamental Physical Properties	*http://data-books.com/bussei_e/bs_index.html*
Prode	Prode Properties	*http://www.prode.com/en/ppp.htm*
TUV SUD NEL	PPDS	*http://www.tuvnel.com/tuvnel/rd_physical_properties_of_fluids/*
THERMODATA	THERMODATA	*http://thermodata.online.fr*

6.4.1
DETHERM

The DETHERM database [10] provides thermophysical property data for about 36 500 pure compounds and 124 000 mixtures. DETHERM contains literature values, together with bibliographical information, descriptors, and abstracts. At the time of publication, some 7.6 million data sets are stored. DETHERM is a collection of data packages produced by well-known providers of thermophysical packages, unified in under a common graphical user interface. The data packages shown in Table 6.3 are part of DETHERM.

Figure 6.1 shows an example of the possibilities for the presentation of modeling results.

6.4.2
Dortmund Database (DDB)

With a view to the synthesis and design of separation processes, fitting, and critical examination of model parameters used for process simulation and the development of group contribution methods, in 1973 a computerized data bank for phase equilibrium data was started by J. Gmehling and U. Onken at the University of Dortmund [8].

While initially mainly VLE (vapor–liquid equilibria) data for non-electrolyte mixtures ($T_b > 0\,°C$) were considered, also stored later on were VLE including compounds with $T_b < 0\,°C$, LLE (liquid–liquid equilibria), h^E, γ^∞, azeotropic, $C_p{}^E$, SLE (solid–liquid equilibria), V^E, adsorption equilibrium, polymer data for non-electrolyte and electrolyte systems, as well as pure component properties. The DDB currently contains nearly 5.23 million data points for 43 000 components from 54 930 references. The DDB contains the properties shown in Table 6.4.

The Dortmund database is distributed by DDBST GmbH, Oldenburg and as part of DETHERM by DECHEMA e.V. (Section 6.4.1).

6.4.3
DIPPR Database

The major content of the database of the Design Institute for Physical Property Data (DIPPR) a subsidiary of the American Institute of Chemical Engineers (AIChE) (*https://www.aiche.org/resources/databases*) is mainly data collections of pure component properties, but also included are data for selected properties of mixtures and the results of a project related to environmental, safety, and health data. In total, data for 2040 compounds in the database cover mainly the components of primary interest to the process industries. The special focus of the DIPPR database is to provide reliable data of thermophysical properties, including the temperature dependency of the properties, that are approved by technical committees, where industrial experts are involved in the design of the database and in the evaluation of the data.

Table 6.3 Content of DETHERM.

Dortmund Data Bank (DDB) (Professor Gmehling, University of Oldenburg and DDBST GmbH)	**Phase equilibrium data** Vapor–liquid equilibria Liquid–liquid equilibria Vapor–liquid equilibria of low boiling substances Activity coefficients at infinite dilution Gas solubilities Solid–liquid equilibria Azeotropic data **Excess properties** Excess enthalpies Excess heat capacities Excess volume **Pure component properties** Transport properties Vapor pressures Critical data Melting points Densities Caloric properties Others
Electrolyte data collection ELDAR (Professor Barthel, Professor Kunz, University of Regensburg)	Caloric data Electrochemical properties Phase equilibrium data PVT properties Transport properties
Thermophysical database Infotherm (FIZ CHEMIE)	PVT-data Transport properties Surface properties Caloric properties Phase equilibrium data: vapor–liquid equilibria gas–liquid equilibria liquid–liquid equilibria solid–liquid equilibria Pure component basic data Thermophysical parameter (COMDOR)
Data collection **C-DATA** (Institute for Chemical Technology, Prague) Additional collections (DECHEMA e.V.)	20 Physicochemical properties for 593 pure components Vapor pressures Transport properties: thermal conductivities viscosities Caloric properties PVT-data: PVT-data critical data Eutectic data Solubilities Diffusion coefficients

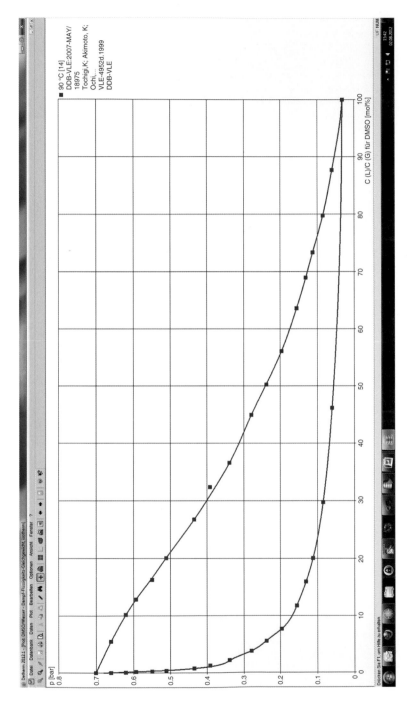

Figure 6.1 Graphical display of a VLE data set in DETHERM for the system dimethyl sulfoxide/water.

Table 6.4   Content of DDB.

Property package	Number of data sets (status: December 2011)
Vapor–liquid equilibria of normal boiling substances	31 640
Vapor–liquid equilibria of low boiling substances	31 480
Vapor–liquid equilibria of electrolyte systems	8 450
Liquid–liquid equilibria	22 575
Activity coefficients at infinite dilution (pure solvents)	60 750[a]
Activity coefficients at infinite dilution (mixtures)	1 425
Gas solubilities (non-electrolyte systems)	19 710
Gas solubilities (electrolyte systems)	1 650
Critical data of mixtures (critical lines)	2 450
Solid–liquid equilibria (mainly organic compounds)	38 000
Salt solubilities (mainly in water)	30 950
Azeotropic data	52 750[a]
Partition coefficients (octanol–water)	10 730[a]
Polymer mixtures	18 150
Excess enthalpies	20 525
Heat capacities of mixtures (including excess heat capacities)	3 700
Mixture densities (including excess volumes)	47 550
Mixture viscosities (including also viscosity deviations)	10 700
Thermal conductivities of mixtures	1 100
Surface tensions of mixtures	900
Speed of sound of mixtures	3 250
Pure component data	206 500

[a] Data tuples.

Table 6.5 gives an overview of the content of the DIPPR database.

6.4.4
NIST Chemistry WebBook

The NIST Chemistry WebBook provides access to data compiled and distributed by NIST under the Standard Reference Data Program.

The NIST Chemistry WebBook contains:

- thermochemical data for over 7000 organic and small inorganic compounds:
  - enthalpy of formation,
  - enthalpy of combustion,
  - heat capacity,
  - entropy,
  - phase transition enthalpies and temperatures,
  - vapor pressure;

Table 6.5 Properties in the DIPPR® 801 Database (DIPPR, Design Institute for Physical Property).

Constant properties: property	Units
Acentric factor	—
Auto-ignition temperature	K
Dielectric constant	—
Dipole moment	C·m
Absolute entropy of ideal gas at 298.15 K and 1 bar	J kmol^{-1} K^{-1}
Lower flammability limit temperature	K
Upper flammability limit temperature	K
Lower flammability limit percent	Vol.% in air
Upper flammability limit percent	Vol.% in air
Flash point	K
Gibbs energy of formation for ideal gas at 298.15 K and 1 bar	J kmol^{-1}
Standard state Gibbs energy of formation at 298.15 K and 1 bar	J kmol^{-1}
Net standard state enthalpy of combustion at 298.15 K	J kmol^{-1}
Enthalpy of formation for ideal gas at 298.15 K	J kmol^{-1}
Enthalpy of fusion at melting point	J kmol^{-1}
Standard state enthalpy of formation at 298.15 K and 1 bar	J kmol^{-1}
Heat of sublimation	J kmol^{-1}
Liquid molar volume at 298.15 K	m^3 kmol^{-1}
Melting point at 1 atm	K
Molecular weight	kg kmol^{-1}
Normal boiling point	K
Parachor	—
Critical pressure	Pa
Radius of gyration	m
Refractive index	—
Solubility parameter at 298.15 K	(J m^{-3})$^{\frac{1}{2}}$
Standard state absolute entropy at 298.15 K and 1 bar	J kmol^{-1} · K^{-1}
Critical temperature	K
Triple point pressure	Pa
Triple point temperature	K
Critical volume	m^3 kmol^{-1}
van der Waals area	m^2 kmol^{-1}
van der Waals reduced volume	m^3 kmol^{-1}
Critical compressibility factor	—
Temperature-dependent properties: property	
Heat capacity of ideal gas	J kmol^{-1} · K^{-1}
Heat capacity of liquid	J kmol^{-1} · K^{-1}
Heat capacity of solid	J kmol^{-1} · K^{-1}
Heat of vaporization	J kmol^{-1}
Liquid density	kmol m^{-3}
Second virial coefficient	m^3 kmol^{-1}
Solid density	kmol m^{-3}

Table 6.5 (Continued)

Constant properties: property	Units
Surface tension	$N\ m^{-1}$
Thermal conductivity of liquid	$W\ m^{-1}\ K^{-1}$
Thermal conductivity of solid	$W\ m^{-1}\ K^{-1}$
Thermal conductivity of vapor	$W\ m^{-1}\ K^{-1}$
Vapor pressure of liquid	Pa
Vapor pressure of solid or sublimation pressure	Pa
Viscosity of liquid	Pa s
Viscosity of vapor	Pa s

- reaction thermochemistry data for over 8000 reactions:
  - enthalpy of reaction,
  - free energy of reaction;
- IR spectra for over 16 000 compounds;
- mass spectra for over 15 000 compounds;
- UV/Vis spectra for over 1600 compounds;
- electronic and vibrational spectra for over 4500 compounds;
- constants of diatomic molecules (spectroscopic data) for over 600 compounds;
- ion energetics data for over 16 000 compounds:
  - ionization energy,
  - appearance energy,
  - electron affinity,
  - proton affinity,
  - gas basicity,
  - cluster ion binding energies;
- thermophysical property data for 34 fluids:
  - density, specific volume,
  - heat capacity at constant pressure (C_p),
  - heat capacity at constant volume (C_v),
  - enthalpy,
  - internal energy,
  - entropy,
  - viscosity,
  - thermal conductivity,
  - Joule–Thomson coefficient,
  - surface tension (saturation curve only),
  - sound speed.

Data on specific compounds in the Chemistry WebBook can be searched based on name, chemical formula, CAS registry number, molecular weight, chemical structure, or selected ion energetics, and spectral properties.

6.5
Special Case and Challenge: Data of Complex Solutions

A much bigger challenge than normal solutions, with an impact in many applications in the product design, is the modeling of complex solutions. Chemical and process engineers, for example, are nowadays able to model or even predict a vapor–liquid-equilibrium, the density, or viscosity of a multicomponent mixture containing numerous different species with sufficient reliability. But if only traces of salt are contained in the mixture, nearly all models tend to fail. This is especially true when the components of a mixture have an increasing complexity concerning the molecular structure and the intermolecular interaction, for example, for polar components such as fatty acids, for salts, polymers, biological material, or food components. An area influenced greatly by electrolyte modeling is biochemical engineering. Until now it was nearly impossible to predict quantitatively the salting-out effect of proteins, crystallization processes of biomolecules, the influence of ions on nanoparticle formation, their size morphology and crystal structure, zeolite synthesis, and so on. However, the development of new production processes in this intensively growing area requires accurate macroscopic physical property models capturing accurately the underlying physics. In some cases there is a limited understanding of these mechanisms, but no real predictability. Process as well as model development – either predicting or even only interpolating – requires large amounts of reliable thermophysical property data for electrolytes and electrolyte solutions. Among the most important property types are:

- vapor–liquid equilibrium data,
- activity coefficients,
- osmotic coefficients,
- electrolyte and ionic conductivities,
- transference numbers,
- viscosities,
- densities,
- frequency-dependent permittivity data.

How can such data be obtained?

The following pages give a survey of actually maintained databases for electrolyte properties [11].

6.6
Examples of Databases with Properties of Electrolyte Solutions

6.6.1
DETHERM/ELDAR Database

The *Electrolyte Database Regensburg* (ELDAR) (*www.uni-regensburg.de/ Fakultaeten/nat_Fak_IV/Physikalische_Chemie/Kunz/eldar/eldhp.html*) is a numerical property database for electrolytes and electrolyte solutions from the Institute of Physical and Theoretical Chemistry of the University of Regensburg. It started in 1976 and contains data on pure substances and aqueous as well as organic solutions. The database was designed to provide a literature reference, numerical data, and also as a model database for fundamental electrochemical research, applied research, and also the design of production processes.

ELDAR contains data of more than 2000 electrolytes in more than 750 different solvents with in total 56 000 chemical systems, 15 000 literature references, 45 730 data tables, and 595 000 data points. ELDAR contains data on *physical properties* like densities, dielectric coefficients, thermal expansion, compressibility, *P-V-T* data, state diagrams, critical data, *thermodynamic properties* like solvation and dilution heats, phase transition values (enthalpies, entropies, Gibbs' free energies), phase equilibrium data, solubilities, vapor pressures, solvation data, standard and reference values, activities and activity coefficients, excess values, osmotic coefficients, specific heats, partial molar values, apparent partial molar values, and *transport properties* like electrical conductivities, transference numbers, single ion conductivities, viscosities, thermal conductivities, and diffusion coefficients.

ELDAR is distributed as part of DECHEMA's numerical database for thermophysical property data DETHERM (Section 6.4.1). To access ELDAR one can therefore has a couple of options:

- in-house client-server installation as part of the DETHERM database;
- Internet access using DETHERM ... on the World Wide Web.

To obtain an overview of the data available, the Internet access option is recommended, because the existence of data for a specific problem can be checked free of charge and even without registration.

Parts of the ELDAR database are published inside DECHEMA's *Chemistry Data Series* as the *Electrolyte Data Collection* [12]. The printed collection and the database ELDAR have complementary functions. The data books give a clear arrangement of selected recommended data for each property of an electrolyte solution. The electrolyte solutions are classified according to their solvents and solvent mixtures. All solution properties have been recalculated from the original measured data with the help of compatible property equations. A typical page of the books contains for the described system:

- general solute and solvent parameters,
- fitted model parameter values,
- measured data together with deviations against the fit,
- a plot,
- the literature references.

The *Electrolyte Data Collection* has 20 volumes and consists of 7400 printed pages. Covered properties are:

- specific conductivities,
- transference numbers,
- limiting ionic conductivities,

- dielectric properties of water, aqueous, and non-aqueous electrolyte solutions,
- viscosities of aqueous and non-aqueous electrolyte solutions.

6.6.2
The Dortmund Database (DDB)

The Dortmund Database published by DDBST GmbH, Professor J. Gmehling from the University of Oldenburg, is well known for its data collections in the areas of vapor–liquid equilibria and related properties (Section 6.4.2). While the major part of the data collections deals with non-electrolyte systems, two collections contain exclusively electrolyte data. They are focused on:

- vapor–liquid equilibria,
- gas solubility.

The two collections together currently contain around 11 000 data sets. Access to these collections is possible either online using the DETHERM on the Web service or in-house using special software from DDBST or DECHEMA.

6.6.3
Data Bank for Electrolyte Solutions at CERE DTU Chemical Engineering

The Center for Energy Resources Engineering (CERE) of the Technical University of Denmark (DTU) operates a data bank for electrolyte solutions (*http://www.cere.dtu.dk/Expertise/Data_Bank.aspx*). It is a compilation of experimental data for (mainly) aqueous solutions of electrolytes and/or non-electrolytes. The database is a mixture of a literature reference database and a numerical database. Currently, references to more than 3000 papers are stored in the database together with around 150 000 experimental data. The main properties are activity and osmotic coefficients, enthalpies, heat capacities, gas solubilities, and phase equilibria like VLE, LLE, and SLE. Access to the literature reference database is free of charge. The numerical values must be ordered at CERE.

6.6.4
JESS

A very useful tool is provided by the Joint Expert Speciation System (JESS), a joint project between two groups in Australia and South Africa (*http://jess.murdoch.edu.au/jess_onjess.htm*) [13]. JESS has a large database for the physicochemical properties of electrolytes in aqueous solution. This consists of about 300 000 property values for over 100 electrolytes. Data from the literature are available for activity coefficients, osmotic coefficients, heat capacities, and densities/volumes.

6.6.5
Springer Materials – The Landolt-Börnstein Database

Based on the *Landolt–Börnstein* (New Series) book collection (*http://www.springer.com/springermaterials*) the Springer Materials database also provides data on electrochemical systems, for example, on electrochemical processes at the boundaries of electrodes and electrolytes or electrical conductivities and equilibrium data [14].

6.6.6
Closed Collections

In addition to the above-described publicly available and still maintained databases old electrolyte data collections were built up in the past. Among them is, for example, the ELYS database, which was compiled by Professor Victor M.M. Lobo, Department of Chemistry, University of Coimbra, Portugal, with thermodynamic and transport property data, such as density, viscosity, diffusion, and coefficients, or the DIPPR 811 and 861 Electrolyte Database Projects [15]. However, these closed collections are, typically, no longer maintained and also not publicly available. It is likely that the references and/or data published in these collections could be also found within the above-mentioned, ongoing collections.

6.7
Properties of New Component Classes: Ionic Liquids and Hyperbranched Polymers

The operating and investment costs of many thermal separation processes are significantly influenced by the properties of additives. High selectivity and capacity, good chemical and thermal stability combined with low volatility, good separability, and low price are requirements that the entrainer or adsorbent cannot often meet simultaneously. Considering this background, the potential of two substance classes, ionic liquids and hyperbranched polymers, have been discussed for multiple applications in thermal separation processes.

Large collections of these properties have been realized inside two IUPAC projects in cooperation with NIST [16]:

- thermodynamics of ionic liquids, ionic liquid mixtures, and the development of standardized systems;
- ionic liquids database.

Meanwhile, other database producers have enriched their collection with this type of data (DETHERM or DDB).

The properties of hyperbranched polymers can be found in numerous papers in the literature [17–23]. Compared to linear polymers, the mainly amorphous hyperbranched polymers show low melt and solution viscosities, high thermal stability, and remarkably high selectivity and capacity.

Overall, it can be stated that the configurable properties of these two classes of compounds allow a wide variety of applications in process engineering and product design.

References

1. Carlson, E.C. (1996) Don't gamble with physical properties for simulations. *Chem. Eng. Progress*, **92** (10), 35–46.
2. Haynes, W.M. (ed.) (2012) *CRC Handbook of Chemistry and Physics*, CRC Press LLC, Boca Raton, FL.
3. Cibulka, I., Sosnkowska-Kehiaian, K., Fontaine, J.-C., Vill, V., Schopper, H., Guelachvili, G., Hüttner, W., Gupta, A., Kaul, A., Predel, F., Franke, P., Seifert, H.J., Villars, P., Cenzual, K., and Rössler, U. (eds) (2013) *Landolt–Börnstein: Numerical Data and Functional Relationships in Science and Technology – New Series, Set 2012*, Berlin, Heidelberg, Springer-Verlag.
4. Behrens, D. *et al.* (eds) (1977–2011) *Chemistry Data Series*, DECHEMA e.V., Frankfurt.
5. Frenkel, M., Chirico, R.D., Diky, V., Yan, X., Dong, Q., and Muzny, C. (2005) Thermo Data Engine (TDE): software implementation of the dynamic data evaluation concept. *J. Chem. Inf. Model*, **45** (4), 816–838.
6. Frenkel, M., Chirico, R.D., Diky, V., Muzny, C., Dong, Q., Marsh, K.N., Dymond, J.H., Wakeham, W.A., Stein, S.E., Königsberger, E., Goodwin, A.R.H., Magee, J.W., Thijssen, M., Haynes, W.M., Watanasiri, S., Satyro, M., Schmidt, M., Johns, A.I., and Hardin, G.R. (2006) New global communication process in thermodynamics: impact on quality of published experimental data. *J. Chem. Inf. Model*, **46**, (6), 2487–2493.
7. Barthel, J. and Popp, H. (1992) Methods of the knowledge based system ELDAR for the simulation of electrolyte solution properties. *Anal. Chim. Acta*, **265**, 259–266.
8. Onken, U., Rarey-Nies, J., and Gmehling, J. (1989) The Dortmund data bank: a computerized system for retrieval, correlation, and prediction of thermodynamic properties of mixtures. *Int. J. Thermophys.*, **10** (3), 739–747.
9. Mansoori, G.A. (2012) Data & Property Calculation Websites. *http://tigger.uic.edu/~mansoori/Thermodynamic.Data.and.Property_html* (accessed 9 March 2013).
10. Westhaus, U., Droege, T., and Sass, R. (1999) DETHERM – a thermophysical property database. *Fluid Phase Equilib.*, **158–160**, 429–435.
11. Westhaus, U. and Sass, R. (2004) Reliable thermodynamic properties of electrolyte solutions – a survey of existing data sources, *Z. Phys. Chem.*, **218**, 1–8.
12. Barthel, J. *et al.* (1992–2010) *Electrolyte Data Collection*, Chemistry Data Series, Vol. XII, Parts 1–5, DECHEMA e.V., Frankfurt.
13. May, P.M., Rowland, D., Königsberger, E., and Hefter, G. (2010) JESS, a joint expert speciation system – IV: a large database of aqueous solution physico-chemical properties with an automatic means of achieving thermodynamic consistency, *Talanta*, **81** (1-2), 142–148.
14. Holze, R. and Lechner, M.D. (2014) *Electrochemistry, Subvolume B: Electrical Conductivities and Equilibria of Electrochemical Systems*, Springer-Verlag, Berlin, Heidelberg, to be published in.
15. Thomson, G.H. and Larsen, A.H. (1996) DIPPR: satisfying industry data needs. *J. Chem. Eng. Data*, **41**, 930–934.
16. Meyers, F. (ed.) (2005) Ionic liquids database. *Chem. Int.*, **27** (5), 23.
17. Fréchet, J.M.J. and Tomalia, D.A. (2001) *Dendrimers and Other Dendritic Polymers*, John Wiley & Sons, Ltd, Chichester.
18. Froehling, P.E. (2004) Development of DSM's Hybrane® hyperbranched polyester amides. *J. Polym. Sci., Part A: Polym. Chem.*, **42**, 3110–3115.

19. Hult, A., Johansson, M., and Malmström, E. (1999) Hyperbranched polymers. *Adv. Polym. Sci.*, **143**, 1–34.

20. Seiler, M., Jork, C., Kavarnou, A., Hirsch, R., and Arlt, W. (2004) Separation of azeotropic mixtures using hyperbranched polymers or ionic liquids. *AIChE J.*, **50**, 2439–2454.

21. Seiler, M. (2004) Phase behavior and new applications of hyperbranched polymers in the field of chemical engineering, Dissertation, Fortschritt-Berichte VDI, Reihe 3, No. 820, Universität Erlangen-Nürnberg.

22. Sunder, A., Heinemann, J., and Frey, H. (2000) Controlling the growth of polymer trees: concepts and perspectives for hyperbranched polymers. *Chem. Eur. J.*, **6** (14), 2499–2506.

23. Rolker, J. (2009) Verzweigte polymere: neue absorbentien in der thermischen verfahrenstechnik, Dissertation, Universität Erlangen-Nürnberg.

7
Current Trends in Ionic Liquid Research

Annegret Stark, Martin Wild, Muhammad Ramzan, Muhammad Mohsin Azim, and Anne Schmidt

7.1
Introduction

Much has been said and written about areas of applications and advantages of ionic liquids (ILs), in particular as solvents in chemical processes [1–7], which arise from their unique properties. As many authoritative reviews and books, for example, References [7–9], have been written about properties and applications, this chapter aims to point out four cases of *emerging ionic liquid* R&D fields:

1) ionic liquids as acido-basic media;
2) binary mixtures of ionic liquids: properties and applications;
3) ionothermal synthesis of nanoporous materials;
4) hydrogenation reactions in and with ionic liquids.

Before doing so, the most important features of ionic liquids are summarized, as these are the basis of often surprising effects. Ionic liquids are salts composed of organic cations and organic or inorganic anions (details of the nomenclature used are given in the subsection below). This gives rise to a tremendous number of potential salts [10]. Hence, in a feeble attempt to cut down on the number of potential ionic liquid candidates, 100 °C was suggested as the maximum melting point limit above which salts would no longer qualify as ionic liquids [11]. In the authors' eyes, however, no good reason exists for this definition. More importantly, ionic liquids are tunable materials that exhibit beneficial properties for a given application [11–15], even, or maybe especially, if this is carried out above 100 °C.

In reactions, this tunability relates mostly to the unusual solvation environments that can be provided by the choice and nature of the ions. Therefore, not only have tremendous effects on selectivity and reactivity been observed, but also the engineering can profit, for example, by allowing for the design of (thermomorphic) multiphasic reaction systems, facile regeneration, and improved stability of catalysts, as will be demonstrated in the case study on hydrogenation reactions in and with ionic liquids [16–20]. These advantages in combination with benefits arising from the negligibly low vapor pressure may accumulate to a reduced energy requirement

Product Design and Engineering: Formulation of Gels and Pastes, First Edition.
Edited by Ulrich Bröckel, Willi Meier, and Gerhard Wagner.
© 2013 Wiley-VCH Verlag GmbH & Co. KGaA. Published 2013 by Wiley-VCH Verlag GmbH & Co. KGaA.

and improved life-cycle assessment of a process [21–23]. Other beneficial properties may (but do not have to) include a wide liquid range, high thermal stability, low melting point, adaptable solvating capabilities, and high conductivities [1, 24].

To be able to determine structure-dependent effects on these and other properties, we have adopted a research approach in which we systematically investigate a series of ionic liquid homologues, which are built by variation of any of three structural moieties: the cation, the cation's substituent, or the anion. It is, however, this abundance of different anion–cation combinations that makes the search for an optimal ionic liquid as difficult as finding the proverbial needle in a haystack. Presently, with a sufficient database of physicochemical properties being described in the literature for selected ionic liquids, knowledge-based approaches such as group-contribution methods [25–32] or other computer-based predictions are being developed. If the multifunctionality and the number of different interactions possible in ionic liquids will allow for the precise prediction of properties, the use of such approaches will be seen in the future.

In the sections below, we summarize the state of the art of the four emerging areas of ionic liquid research listed above, for which the number of publications is still relatively low, but exciting application areas have been postulated.

7.1.1
Ionic Liquid Abbreviations

The ionic liquid shorthand in this chapter follows the accepted format of the community: The most frequently used 1-alkyl-3-methylimidazolium cation is abbreviated $[C_nMIM]^+$, where n indicates the number of carbon atoms on a linear alkyl group. If the C2-position is methylated, for example, to inhibit carbene formation in basic environment, the abbreviation is $[C_ndMIM]^+$ for 1-alkyl-2,3-dimethylimidazolium. Atom labeling in the text is made according to the following:

$$R-N \underset{2}{\overset{5 \quad 4}{\overbrace{}}} \overset{3}{N^+}-CH_3$$

Other cations are abbreviated as follows: pyridinium-based cations: $[C_nPY]^+$, pyrrolidinium-based: $[C_nPYR]^+$, ammonium-based: $[C_wC_xC_yC_zN]^+$. For acido-basic media, the same nomenclature is used. The anion is abbreviated in general according to their conventional shorthand. Exceptions are in following:

$[CH_3\text{-}(CH_2)_7\text{-}O\text{-}SO_3]^-$	octyl sulfate
$[NTf_2]^-$	bis(trifluoromethanesulfonyl)amide
$[p\text{-}OTs]^-$	p-toluenesulfonate
$[NPf_2]^-$	bis(pentafluoroethanesulfonyl)amide

7.2
Ionic Liquids as Acido-basic Media

A special subgroup of ionic liquids is the group of acido-basic ionic liquids, which shows distinctively different properties and also different interactions with reactants of a reaction under investigation. The name of these special ionic liquids is related to their chemical structure, which contains an acidic and a basic structure moiety. This can be explained with an imaginary acido-basic ionic liquid with the structure $[C_1NH_3][X]$. For a superacid (such as trifluoromethanesulfonic acid), proton transfer from the acid to the base methylamine can be presumed to be quantitative and the protonated methylamine (methylammonium) may, hence, act as an acid, while the anion acts as a base. With a weak acid (such as phenol), proton transfer is limited and hence methylamine remains as the base and the still protonated phenol can act as acid. Because of this equilibrium between amine + acid and the resulting acido-basic ionic liquid under non-aqueous conditions, it is difficult to judge, at this point, which amine + acid pairs lead to quantitative proton transfer and hence adhere truly to the definition of ionic liquids (i.e., consist solely of ions).

This section introduces the state of the art of this relatively new subgroup of ionic liquids.

7.2.1
Synthesis

Conventional ionic liquids are in most instances prepared by alkylation of an amine followed by ion exchange, if the application requires a less nucleophilic anion (Scheme 7.1). This synthetic strategy can result in ionic liquids with many advantages, such as low melting point and viscosity, and high thermal stability. However, the preparation is time and solvent consuming to obtain products with high purity.

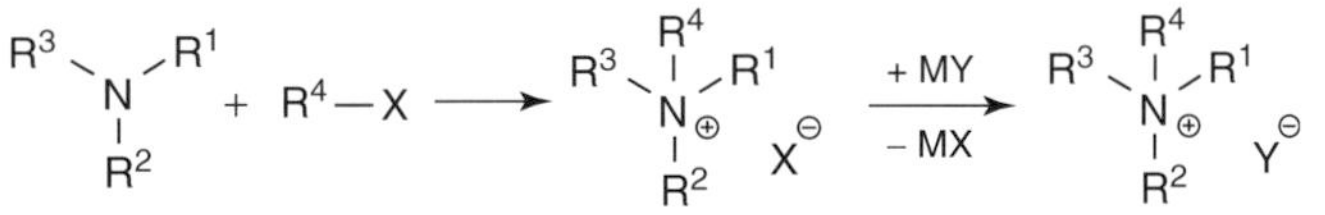

Scheme 7.1 Preparation of conventional ionic liquids by alkylation followed by an ion exchange [R^1, R^2, R^3: H, alkyl, or aryl; heterocycles (e.g., pyridine, imidazole); and R^4: alkyl chain].

Acido-basic ionic liquids are much easier to prepare, as only equimolar amounts of the acid are added to the base (in most cases, an amine), which is dissolved in water to prevent strong exotherms (Scheme 7.2). The reaction has several advantages, such as being a single-step-reaction and, if the acid addition step and the successive water removal are carried out at low temperatures, there are no byproducts, leading to products of high purity [33, 34]. Acido-basic ionic liquids with a melting point below room-temperature have been prepared from ammonia [35, 36], pyridine [37], pyrazole, and imidazole (IM) [33, 38].

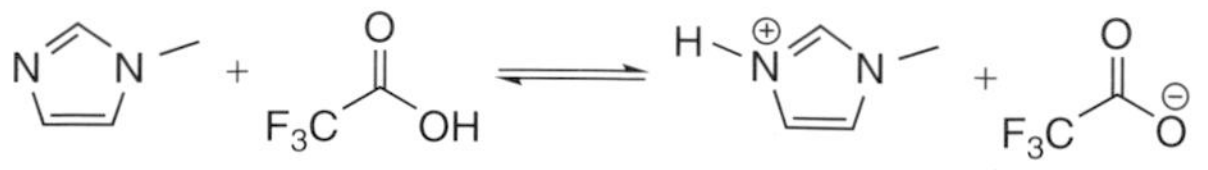

Scheme 7.2 Preparation of the acido-basic ionic liquid 1-methylimidazolium trifluoroacetate by mixing equimolar amounts of 1-methylimidazole with trifluoroacetic acid.

The position of the equilibrium shown (Scheme 7.2) is influenced by many different factors and the product, the acido-basic ionic liquid, is in any case in equilibrium with the used starting materials (amine and acid). Factors that have an influence on the equilibrium are the pK_a of the acid and the pK_b of the amine. Furthermore, the temperature and the concentration of the compounds have an effect on the equilibrium. The lower the pK_a of the acid, the more the equilibrium is shifted to the product side (right-hand side). The same applies for the pK_b of the base. An increasing temperature promotes at first the dissociation of the acid, which results in a shift of the equilibrium to the product side (right). If the temperature is further increased, cleavage of the N-H-bond is promoted and the equilibrium is shifted backwards to the amine and acid (left). In addition, the equilibrium is shifted to the product side (right) if an excess of either the acid or the base is added.

To make things even more complex, the equilibrium is shifted by adding one or more reactants of a reaction that is catalyzed by the ionic liquid. In this case, the reactant could interact either with the starting materials (amine and acid) or with the acido-basic ionic liquid. In the first case, one of the compounds is "removed" from the equilibrium and the result is a shift to the left-hand side (Scheme 7.2). In the second case, a strong interaction with the acido-basic ionic liquid "removes" the product from the equilibrium and this results in a shift to the right-hand side (Scheme 7.2). The equilibrium simply follows the principle of Le Chatelier.

7.2.2
Structure

From the discussion in the previous section, it is obvious that the microscopic organization of an acido-basic ionic liquid must be very complex, and the degree of organization influences its properties. At present, relatively little is known about the structure in the liquid state, and what types of interactions prevail.

In principle, however, it is possible that the ions formed exist either as ion pairs or as larger macromolecular aggregates, or, if the proton transfer is not quantitative, the ionic liquid also contains neutral species [39]. To clarify the degree of organization, different methods have been used: X-ray and neutron diffraction [40], Raman [41] and light [42] scattering, nuclear magnetic resonance (*NMR*) spectroscopy [43], as well as molecular modeling [41, 44, 45].

Judeinstein *et al.* have investigated several acido-basic ionic liquids by NMR spectroscopy to clarify the organization of the individual molecules in dependence of the pK_a of the acid. This study was designed to show if there are ion pairs, dissociated ions, or associated ion pairs of the ammonium cation and the anion

[46]. The acido-basic ionic liquids investigated were derived from triethylamine as the base and either of three different acids, hence varying the pK_a of the acids: the superacid bis(trifluoromethanesulfonyl)amide (pK_a: -12.20 [47]), the strong acid trifluoroacetic acid (pK_a: 0.23), and the weak acid acetic acid (pK_a: 4.76 [48]). The self-diffusion coefficients of the ions give an idea of the prevailing structural motif in the ionic liquid.

For the strong acid, trifluoroacetic acid, the self-diffusion coefficients of the acidic proton, the nitrogen-containing moiety, and the fluorine-containing moiety are the same, indicating the presence of associated ion pairs. For the system made up of the superacid, bis(trifluoromethanesulfonyl)amide, the self-diffusion coefficients of the acidic proton and the nitrogen-containing moiety are the same and are somewhat larger than that of the fluorine-containing moiety, indicating the presence of dissociated ions. The weak acid, acetic acid, shows different self-diffusion coefficients for all three moieties. In this case, the ionic liquid is thought to consist of triethylamine, the corresponding anion of the added acid, and a dissociated proton. The results of these heteronuclear NMR measurements are summarized in Table 7.1 [46].

Table 7.1 NMR Results and proposed structure of the acido-basic ionic liquids triethylammonium trifluoroacetate, triethylammonium bis(trifluoromethanesulfonyl)amide, and triethylammonium acetate [46].

	$[C_2C_2C_2NH][CF_3CO_2]$	$[C_2C_2C_2NH][NTf_2]$	$[C_2C_2C_2NH][OAc]$
Diffusion coefficient at 67 °C $(10^6\ cm^2\ s^{-1})^a$	$D_{H+} = 1.90$ $D_N = 1.97$ $D_{A-} = 1.94$ $D_{H+} = D_N = D_{A-}$	$D_{H+} = 1.29$ $D_N = 1.27$ $D_{A-} = 0.97$ $D_{H+} = D_N > D_{A-}$	$D_{H+} = 3.54$ $D_N = 2.51$ $D_{A-} = 2.95$ $D_{H+} > D_{A-} > D_N$
^{15}N NMR N–H coupling	Yes	Yes	No
HOESY (^{15}N...^{1}H)	N–H interaction	N–H interaction	N–H interaction
HOESY (^{13}C...^{1}H)	Interaction between anion, amine, and H$^+$	Intermolecular interaction	Interaction of H$^+$ with acetate and amine
Proposed structure	Mostly associated ion pairs	Dissociated ammonium ion and $[NTf_2]^-$	Proton fully dissociated

^{a} D_{H+} = diffusion coefficient of the acidic proton, D_N = diffusion coefficient of the amine, D_{A-} = diffusion coefficient of the anion.

To obtain further information, ^{1}H decoupled ^{15}N NMR measurements were carried out, and the results received compared to NMR data of triethylamine and tri-ethylammonium chloride. For $[C_2C_2C_2NH][NTf_2]$, the nitrogen resonance is close to the nitrogen resonance of triethylammonium chloride, indicating quantitative proton transfer. Conversely, for both $[C_2C_2C_2NH][CF_3CO_2]$ and $[C_2C_2C_2NH][OAc]$ the nitrogen resonance is near to the nitrogen resonance of the triethylamine, indicating incomplete proton transfer, which confirms the interpretations made above [46]. However, the relatively long relaxation times of the nuclei in NMR present a problem, which may be overcome by FTIR (Fourier-transformed infrared) measurements (Scheme 7.3). In this study, 1-methylimidazole was combined with either of four different acids (trifluoroacetic acid, dichloroacetic acid, phenylacetic acid, or acetic acid). The received data should verify the above-described classification and should show if there are some other complexes that cannot be identified by NMR spectroscopy. For this purpose, the asymmetric carbonyl stretching vibration was investigated, and the data of the acido-basic ionic liquids compared to the resonances of the methyl esters, the corresponding tetrabutylammonium salts of the acids, and the free acids. In this way, three different structural motives of acido-basic ionic liquids were identified [49, 50]:

I)   neutral complexes with weak acids ($pK_a \geq 2.2$), where the frequency of the asymmetric carbonyl stretching vibration is less than the frequency of the free acid and the methyl ester of the acid (i.e., structural motif similar as the one of $[C_2C_2C_2NH][OAc]$, Table 7.1);

II) ionic complexes with strong acids ($pK_a \leq 2.0$), where the frequency of the asymmetric carbonyl stretching vibration is less than the frequency of the tetrabutylammonium salt of the acid (i.e., structural motif similar to that of $[C_2C_2C_2NH][NTf_2]$, Table 7.1);

III) depolarized partially ionic complexes, which are in equilibrium with II and where the asymmetric carbonyl stretching vibration is between the frequency of the tetrabutylammonium salt (bond order: 1.5) and the free acid (bond order: 2.0); this corresponds to the structural motif presented for $[C_2C_2C_2NH][CF_3CO_2]$ in Table 7.1.

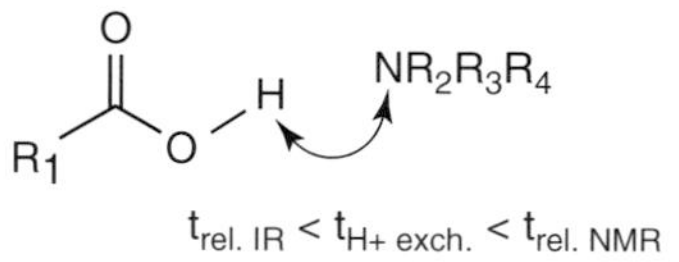

$$t_{rel.\ IR} < t_{H+\ exch.} < t_{rel.\ NMR}$$

Scheme 7.3 Proton exchange between a carboxylic acid (R_1COOH) and a nitrogen base ($NR_2R_3R_4$); and comparison of the relaxation times of the proton exchange ($t_{H+\ exch.}$) and the relaxation times of NMR spectroscopy ($t_{rel.\ NMR}$) and IR-spectroscopy ($t_{rel.\ IR}$).

By substitution of the methyl group in 1-methylimidazole with a tertiary butyl group, the equilibrium of II and III is shifted to III. The reason for this lies in the increased space requirements of the tertiary butyl group of the 1-alkylimidazole, and a face-to interaction of the ion pairs was postulated [50].

7.2.3
Physicochemical Properties

As with the structure, little is known about the properties of acido-basic ionic liquids, and making predictions of the resulting properties for new (base + acid) combinations is still more a question of gut-feeling rather than knowledge-based.

7.2.3.1 Thermal Properties

The thermal properties (i.e., melting and decomposition point) of acido-basic ionic liquids depend on many different factors: for example, the structure of both cation and anion and the degree of their intermolecular interaction. It was long believed that the melting point mostly depends on van der Waals forces and not on electrostatic Coulomb- and hydrogen-bond interactions. This impression arose because the melting point of imidazolium-based salts was higher if the C2-proton of 1-alkylimidazolium was substituted by a C2-alkyl, which intuitively should reduce anion–cation interactions (Scheme 7.4). This trend holds true for both conventional and acido-basic ionic liquids. Furthermore, increasing the length of the alkyl chain on N1 decreases the melting point [34]. In pyridinium-based salts, the position of alkyl substitution also affects the thermal properties. For example, in case of the lutidine cation (lutidine: dimethylpyridine) the melting or glass transition points increase with increasing symmetry [33]. However, no correlation was found between the melting point of the parent amine and the resulting acido-basic ionic liquid [33].

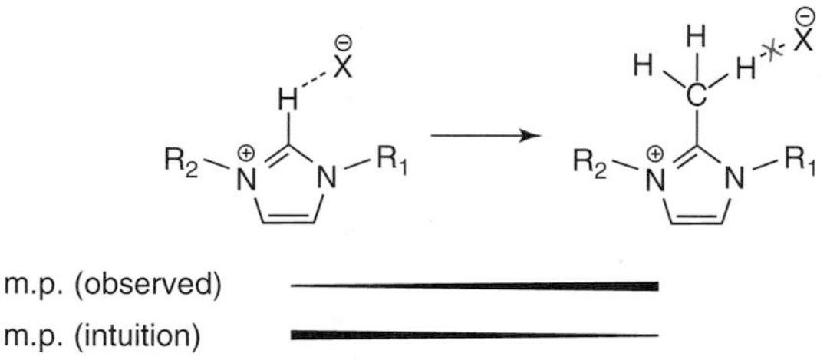

m.p. (observed)

m.p. (intuition)

Scheme 7.4 Trend for the melting point of imidazolium-based ionic liquids when exchanging the proton at C2 with a methyl group.

Beside the cation, the anion also has a strong influence on the thermal properties. In general, the addition of inorganic acids yields salts with high melting points (above 37 °C) while organic acids (except methanesulfonic acid) yield low melting acido-basic ionic liquids (below 11 °C) [34]. In most cases, the melting point decreases with increasing ion radius [51]. In general, fluorinated anions such as tetrafluoroborate and bis(trifluoromethanesulfonyl)amide show low melting points [38, 52, 53]. But there are exceptions!

Regarding the thermal stability, it can be said that acido-basic ionic liquids based on fluorinated anions possess a high thermal stability. For example, acido-basic ionic liquids with the bis(trifluoromethanesulfonyl)amide anion show decomposition temperatures from 190 to over 400 °C! This is a property such acido-basic ionic liquids have in common with conventional ionic liquids based on fluorinated anions [34]. Overall, it has been stated that acido-basic ionic liquids based on a superacid and a strong base possess very similar properties to conventional ionic liquids [54].

7.2.3.2 Conductivity

Just like the thermal properties, the conductivity depends on the cation and the anion. Additionally, one could expect a sudden increase in conductivity when the temperature is raised over the melting point. However, for acido-basic ionic liquids, this was not observed, indicating that the ionic liquids adopt a supercooled state at low temperatures [33].

To determine the dependence of conductivity on the cation, many systematic studies were performed in recent years. For a bis(trifluoromethanesulfonyl)amide, it was found that the conductivity increases somewhat with increasing molecular weight of the cation, but the effect is small [34]. Additionally, measurements with tetrafluoroborate-based acido-basic ionic liquids show that the conductivity decreases when the symmetry of the cation decreases [33]. It is also dependent on the site of substitution (e.g., on an indolium cation): For example, 1,2-substituted indolium cations show relatively high conductivities of about 10^{-2} S cm^{-1} (at 50 °C). In contrast, other substituted indolium cations (1-alkyl, 2-alkyl, 2,3-dialkyl) have conductivities of about 10^{-9} S cm^{-1} (at 50 °C) [33].

The effect of the anion was investigated with 1-ethylimidazolium-based acido-basic ionic liquids. It was found that nitrate, tetrafluoroborate, trifluoromethanesulfonate, and bis(trifluoromethanesulfonyl)amide show high conductivities of about 10^{-2} S cm^{-1}, while halide anions yield salts with low conductivities of around 10^{-4} S cm^{-1} at room temperature [34].

An interesting experiment was carried out to clarify if proton conductivity also occurs in acido-basic ionic liquids. For this purpose, imidazolium bis(trifluoromethanesulfonyl)amide ([HIM][NTf$_2$]) was used. The anode compartment was set either under a dinitrogen or dihydrogen atmosphere. In the former case, the detected current was very low. However, when switching to dihydrogen the current increases and dihydrogen evolved at the cathode. The following equations explain the processes lying behind this:

anode: $H_2 + 2\,IM \rightarrow 2\,[HIM]^+ + 2\,e^-$
electrolyte: proton conductivity
cathode: $2\,[HIM]^+ + 2\,e^- \rightarrow H_2 + 2\,IM$.

The same experiment carried out with conventional ionic liquids found no increase in current on switching the atmosphere [55].

7.2.3.3 Dynamic Viscosity

If the viscosity of acido-basic ionic liquids is depicted as function of the temperature, slightly curved trends are observed (decreasing viscosity with increasing temperature), similar to conventional ionic liquids. One could argue that if a correlation between the dynamic viscosity and the electrostatic interactions exists the acido-basic ionic liquid with the largest anion should show the lowest viscosity. To verify this hypothesis, the dynamic viscosities of different acido-basic ionic liquids were determined. Table 7.2 shows that the lowest viscosities were obtained with 1-ethylimidazolium-based acido-basic ionic liquids. In this case, the hexafluorophosphate gave the highest viscosity, while the lowest was obtained with tetrafluoroborate. This clearly demonstrates that it is not only electrostatic interactions and ion radii that play a role, although the large bis(fluoroalkanesulfonyl)amides do in fact possess mostly low viscosities. Table 7.2 shows, however, that small cations yield in general in lower viscosities [34].

7.2.4
Applications

Just like the much more investigated conventional ionic liquids, acido-basic ionic liquids have great potential in several applications. For example, they can serve as two-in-one solvent-and-catalyst in organic synthesis or as electrolytes, for example, in fuel cells or photovoltaic systems [39, 56–59].

7.2.4.1 Organic Synthesis

With regard to sustainability, one of the most important goals when optimizing an organic reaction is to increase the yield and the selectivity for the required product. The latent acidity and relatively low viscosity (in particular when using liquid starting materials) of acido-basic ionic liquids make them prime candidates as solvent-cum-catalyst for acid-catalyzed reactions.

For example, cross coupling reactions, such as Suzuki coupling, are important in many organic synthesis protocols. However, the catalysts (e.g., palladium or nickel species) are frequently deactivated under the harsh reaction conditions and by various chemical functionalities, making the use of protecting groups indispensable. Hence, alcohols and thiols are frequently protected with diphenylmethanol via ether formation, which has been carried out with catalysts such as $PdCl_2$, $NbCl_5$, Brønsted acids, Lewis acids, and carbenes [60–63]. Recently, acido-basic ionic

Table 7.2 Viscosity $(cP)^a$ of different acido-basic ionic liquids at 25 °C.

Alkyl imidazole	Acid					
	$H[BF_4]$	$H[ClO_4]$	$H[PF_6]$	$H[CF_3SO_3]$	$H[NTf_2]$	$H[NPf_2]$
1-Methylimidazole	+	+	+	+	81	218
1-Ethylimidazole	41	112	550	58	54	133
1,2-Dimethylimidazole	100	+	+	+	100	+
1-Ethyl-2-methylimidazole	67	+	+	+	69	186
1-Benzyl-2-ethylimidazole	+	+	+	+	252	552

^{a} + = >1000 cP [34].

Table 7.3 Ether formation of diphenylmethanol at 80 °C (microwave), after 10 min reaction time, with the acido-basic ionic liquid triethylammonium methanesulfonate [63].

Alcohol	Yield (%)	Alcohol	Yield (%)
Methanol	93	Benzyl alcohol	91
Ethanol	95	Propargyl alcohol	97
Propanol	82	2-Methylpropan-2-ol	19
Carboxybenzyl-NH-$(CH_2)_2$-O-$(CH_2)_2$-OH	46	Phenol	48

liquids have gained significance due to their tunable proton activity [64–66]. For example, the versatility of this reaction was demonstrated using triethylammonium hydrogen sulfate or methanesulfonate, giving good to excellent yields (Table 7.3) in short reaction times (<30 min) and under moderate reaction conditions (microwave, 80 °C). An exception is 2-methylpropan-2-ol (i.e., *tert*-butanol), which is due to the facile formation of the *tert*-butyl cation under the reaction conditions, leading to several side-reactions [63].

In comparison, phenol forms the corresponding diphenylmethyl phenyl ether with a yield of only 4% (10 mol.% $PdCl_2$, 80 °C, 96 h, in dichloromethane), while the reaction in acido-basic ionic liquids yields 48% of the desired product [60, 62].

Another important condensation reaction is the formation of substituted 1,2,4-triazoles. This reaction is very sensitive to electronic effects and steric hindrance, but for organometallic catalysis the 3-phenyl-4-mesityl-1,2,4-triazole is required as ligand. By mineral acid catalysis, the reaction did not proceed when starting from the corresponding oxazoline and mesitylamine [67–71], while with the acido-basic ionic liquid [HPY][CF_3CO_2] (PY = pyridine) the reaction gave 55% yield. To verify what moiety of the ionic liquid catalyzes the reaction, the product was obtained in a conventional ionic liquid ([C_4MIM]Br (MIM = 1-methylimidazole) or [C_4PY]Br, which are not catalytically active themselves) by adding an organic acid. In this case, the desired product was observed in low yields when increasing the temperature and reaction time. This results in the assumption that both ions of the acido-basic ionic liquid act in the catalytic cycle. Scheme 7.5 shows a possible reaction mechanism.

Furthermore, esterification reactions have also been carried out with acido-basic ionic liquids. The products of this reaction type have high industrial importance. Esters are used as plasticizers, in perfumes, as solvents, flavors, and precursors for pharmaceuticals, agro-, and fine chemicals [72, 73]. In general, acids such as sulfuric acid or *p*-toluenesulfonic acid are used as catalysts. These catalysts have several disadvantages, such as the generation of different side-products and their corrosivity. Furthermore, the neutralization of the acids after the reaction results in a large amount of salts, which have to be disposed of [74, 75]. To overcome these disadvantages, different ionic liquids were tested. Conventional ionic liquids do not contain an acidic moiety, and the resulting yields are lower than in the

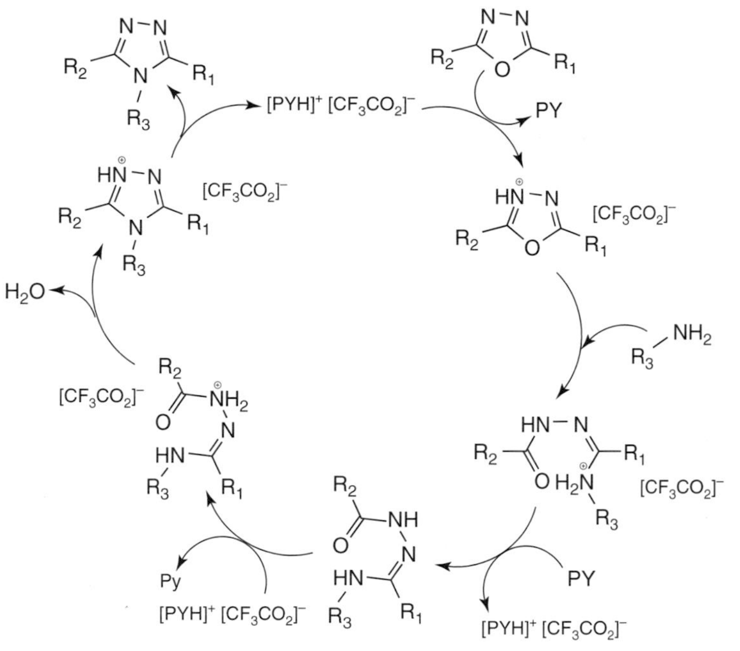

Scheme 7.5 Possible reaction mechanism of the formation of substituted 1,2,4-triazoles catalyzed by [HPY][CF$_3$CO$_2$] [71].

absence of ionic liquid. However, when acido-basic ionic liquids were used good yields resulted while allowing for the recyclability (due to facile phase separation) of the acido-basic ionic liquids without loss of reactivity [76, 77]. For example, 91% benzyl acetate was obtained with a selectivity of 100% with 1-methylimidazolium tetrafluoroborate [76].

To clarify the effect of the anion and the cation, some systematic studies of the esterification of 1-octanol and acetic acid were performed [78]. The results are summarized in Table 7.4. For the anions, the following order of decreasing yield was found: [HSO$_4$]$^-$ > [p-OTs]$^-$ > [H$_2$PO$_4$]$^-$ > [BF$_4$]$^-$, while the effect of the cation was negligible (due to their fairly similar basicity: [C$_2$C$_2$NH$_2$] 3.16 < [C$_2$C$_2$C$_2$NH] 3.25 < [C$_2$NH$_3$] 3.35) [78, 79]. However, using more bulky cations also reduced the yield of the ester. In addition to these cations and anions, pyridinium methanesulfonate and pyridinium p-toluenesulfonate were used, affording n-butyl acetate in high yields (>70%) and excellent selectivities of over 98% [77].

At present, few industrial applications of acido-basic ionic liquids are publicly known, but a very prominent example is the BASIL process of BASF SE, where 1-methylimidazole scavenges the hydrochloric acid formed in a condensation reaction that produces alkoxydiphenyl- and dialkoxyphenyl phosphines, leading to formation of the acido-basic ionic liquid 1-methylimidazolium chloride (Scheme 7.6) [80].

Table 7.4 Yields of *n*-octyl acetate in the esterification of 1-octanol and acetic acid catalyzed by different acido-basic ionic liquids[a] [78].

Acido-basic ionic liquid	Yield (%)	Acido-basic ionic liquid	Yield (%)
$[C_2C_2C_2NH][HSO_4]$	81	$[C_2C_2C_2NH][H_2PO_4]$	26
$[C_2C_2NH_2][HSO_4]$	88	$[C_2C_2NH_2][H_2PO_4]$	25
$[C_2NH_3][HSO_4]$	94	$[C_2NH_3][H_2PO_4]$	28
$[C_2C_2(PhCH_2)NH][HSO_4]$	86	$[C_2C_2C_2NH][BF_4]$	18
$[C_4C_4C_4NH][HSO_4]$	70	$[C_2C_2NH_2][BF_4]$	50
$[C_8C_8C_8NH][HSO_4]$	70	$[C_2NH_3][BF_4]$	25
$[i\text{-}C_3\text{-}i\text{-}C_3NH_2][HSO_4]$	72	$[C_2C_2C_2NH][p\text{-}OTs]$	42

[a] Reaction conditions: acetic acid: 20 mmol, 1-octanol: 24 mmol, acido-basic ionic liquid: 1 g, temperature: 90 °C, reaction time: 4 h.

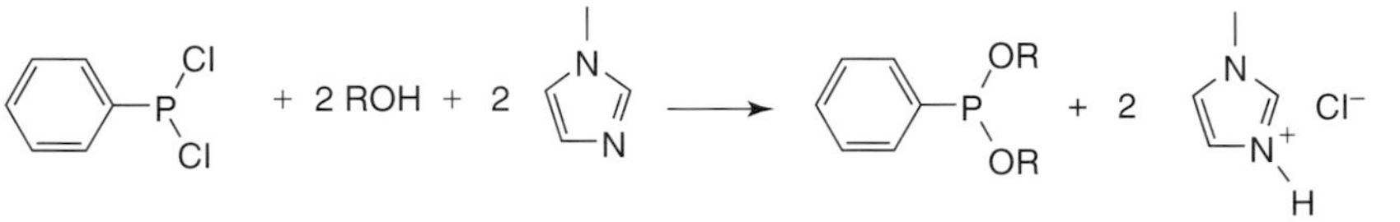

Scheme 7.6 BASIL process by BASF SE using an acido-basic ionic liquid.

In contrast to the former process (which used triethylamine), in the BASIL process a liquid–liquid biphasic system is obtained, which can be easily separated. In addition, 1-methylimidazole acts as nucleophilic catalyst, which increases the rate of reaction dramatically, leading to residence times below 1 s. This, in turn, permitted the application of a new reactor concept (jet reactor), leading to an 80 000-fold increase of the space–time yield!

The pharmaceutical company Eli Lilly and Company demonstrated on a pilot-plant scale that pyridinium chloride acts as demethylating agent in the reaction of 4-methoxyphenylbutanoic acid to give 4-hydroxyphenylbutanoic acid, a precursor in the manufacture of a preclinical candidate [81].

7.2.4.2 Fuel Cells

Beside the use as solvent-cum-catalyst material, acido-basic ionic liquids have been described as electrolytes in fuel cell applications.

Fuel cells are special types of galvanic cells in which dihydrogen and dioxygen react to produce water and electric energy. For this application, it is important to employ a proton-conducting electrolyte that is stable under the fuel cell conditions. The first step is an oxidation of dihydrogen to protons, which migrate through the electrolyte to the other electrode (cathode) where the reduction of dioxygen takes place. For optimal cell performance, cell conditions above 100 °C would be beneficial due to a resulting higher activity of the electrode material, hence leading to a reduced fuel cell volume and lower electrode material consumption. However, this excludes the use of a water-based electrolyte [82]. Electrolyte requirements are thus: a low vapor pressure, high thermal stability and ionic conductivity, and a wide electrochemical window.

In the past, many different water-free and proton-conducting electrolytes were investigated in fuel cells. For example, phosphoric acid, showing a self-diffusion of about 15%, was tested under fuel-cell conditions and some companies in fact generate their electricity with this type of fuel cell [83–85]. However, there are some problems associated with the use of phosphoric acid, such as absorption of the acid into the silica carbide membrane and the volatility of phosphoric acid, which decreases cell life-time [86–91]. Other investigations focused on the use of solid electrolytes, such as $CsHSO_4$ and CsH_2PO_4. These proton-conducting salts have a limited temperature window as they possess a solid–plastic crystalline phase transition at 140 and 170 °C, respectively. In addition, dehydration of the salts takes place in the plastic crystalline phase [92].

Ionic liquids have the required properties for many electrolyte-based applications, and they have indeed been investigated, for example, for lithium batteries [93, 94], electric double-layer capacitors [95, 96], actuators [97, 98], and dye-sensitized solar cells [99–102].

With regards to fuel cells, conventional ionic liquids, such as 1-ethyl-3-methylimidazolium fluorohydrogenate $[F(HF)_x^-\ x = 1.3$ or $2.3]$ has been tested. This ionic liquid enables fuel cell conditions of 120 °C [103, 104]. Lee *et al.* used diethylmethylammonium trifluoromethanesulfonate and bis(trifluoromethane-sulfonyl)amide. Both acido-basic ionic liquids show high ion conductivities of about $50\,mS\,cm^{-1}$ at 150 °C and proton transfer occurs by the Grotthuß mechanism [82]. Interestingly, the voltammogram of diethylmethylammonium trifluoromethanesulfonate shows different behavior for the first cycle than for the following cycles. In the first cycle, the dihydrogen oxidation reaction occurs at about 0.4 V, while in subsequent cycles it occurs at 0.0 V. The reason for this phenomenon is the availability of a proton acceptor: in acido-basic ionic liquids, the acceptor for the proton can in principle be either the amine (higher basicity) or the anion (generally lower basicity). In the first cycle, the proton acceptor must be the anion as there is no free amine available, owing to the almost quantitative proton transfer of the trifluoromethanesulfonic acid to the amine. Since the anion is a relatively weak base, an overpotential is observed. During the first cycle, amine is generated by reaction of the protons with dioxygen at the cathode, hence liberating free amine, which is then available for subsequent cycles, leading to a lower overpotential. If dioxygen is bubbled through the ionic liquid the reductive potential was observed at 1.0 V [82]. In the case of diethylmethylammonium bis(trifluoromethanesulfonyl)amide the open circuit voltage was determined as 0.7 V, in contrast to about 1.0 V for diethylmethylammonium trifluoromethanesulfonate [82]. This shows that acido-basic ionic liquids with similar bulk properties may in fact possess different electrochemical properties [82].

7.2.5
Conclusions: Ionic Liquids as Acido-basic Media

Acido-basic ionic liquids are a new subgroup of ionic liquids that can be prepared in a very simple, cost-effective way. The microscopic organization of these materials is very complex and depends on the pK_a of the parent acid and the pK_b of the parent base (e.g., the amine). This structure has to be investigated in more detail in future to determine the limits of the three groups of identified substructures in dependence of the pK_a of the corresponding acid.

The properties of the acido-basic ionic liquids depend on the structure of the cation and the anion and the degree of their intermolecular interaction. The precise position and the size of substituents on the cation affect the melting point, conductivity, and dynamic viscosity. In particular if strong acid–base pairs are chosen, the resulting acido-basic ionic liquids show the same trends as conventional ionic liquids. However, currently there is no general relation between the structure and the properties of acido-basic ionic liquids.

These new materials can be used in different reactions as solvents and catalysts, that is, as two-in-one materials. They show very good reaction performances and in certain cases they enable the preparation of the target molecules in acceptable to excellent yields. Sometimes, acido-basic ionic liquids show better performances than the pure acid- or the pure base-catalyst. Furthermore, acido-basic ionic liquids have been shown to be recyclable by removal of the reactants and products by extraction or simple phase separation. In addition, quite unusual properties, for example, a low vapor pressure, high proton conductivity, and wide electrochemical window, have led to the applicability of acido-basic ionic liquids as electrolytes in fuel cells.

7.3
Binary Mixtures of Ionic Liquids: Properties and Applications

As pointed out in the introduction, ionic liquids are referred to as *designer solvents* due to the tunability of their physicochemical properties to suit a given task or application. In general, there are three strategies by which to fine-tune the physical properties of a (in principle) suitable ionic liquid (while maintaining its chemical properties) to achieve improved physical properties:

1) As in most instances, the prevailing chemical properties are defined by the choice of the anion, the cation and/or the cations substitution pattern can be changed to fine-tune properties such as viscosity, density, surface tension, conductivity, solubility, and so on.

2) The physicochemical properties can be modified by the addition of a co-solvent. This co-solvent may, for example, decrease the viscosity and hence improve issues such as mass transfer. However, this strategy also lessens the advantages of ionic liquids, in particular those relating to the low vapor pressure (volatility and flammability) [105, 106].

3) Alternatively, the properties can be fine-tuned by employing a system composed of two (or more) ionic liquids (binary mixtures of conventional ionic liquids).

This third strategy has, however, not been extensively applied yet, and there are only a few reports that discuss the behavior of ionic liquid binary mixtures from a physicochemical point of view. Combined analysis of literature data of the very limited set of ionic liquids has shown that such ionic liquid binary mixtures seem to behave as ideal mixtures, that is, their bulk properties (such as densities [107–109], isobaric thermal expansivities [107], excess enthalpies [107], viscosities [107, 109, 110], conductivities [110], effective dipole moment [107], refractive index [108, 109], and NMR chemical shift) are a linear function of the molar composition [111]. This is surprising, because one could have expected a very complex mixing behavior affecting the properties, considering that binary ionic liquid mixtures consist of three chemical moieties (i.e., a number of ion types).

However, pronounced deviations in the excess properties point to reorientation processes occurring in the mixture. Hence, to understand underlying structural changes and the resulting effect on their properties, detailed investigations of physicochemical properties are necessary. The following subsections summarize the state of the art of binary ionic liquid mixtures regarding physicochemical properties, spectroscopic investigations, and potential applications.

7.3.1
Physicochemical Properties

In a very detailed study, Fox *et al.* investigated the density, viscosity, conductivity, and thermal phase behavior of six binary mixtures of N-alkyl-N-methylpyrrolidinium bis(trifluoromethanesulfonyl)amide with various alkyl chain lengths (CH_3 to C_5H_{11}). The results show that the addition of an ionic liquid with a smaller cation, such as $[C_1C_1PYR]^+$, $[C_2C_1PYR]^+$, or $[C_3C_1PYR]^+$, to an ionic liquid with a longer alkyl chain on the cation, such as $[C_3C_1PYR]^+$, $[C_4C_1PYR]^+$, or $[C_5C_1PYR]^+$, results in a mixture with a higher *density* than that of the longer alkyl chain substituted ionic liquid (Figure 7.1). This means that the more the mixture contains of the ionic liquid of lower molecular weight, the more potential charge carriers are present in a given volume.

The *viscosity* of the binary mixtures ($[C_4C_1PYR][NTf_2] + [C_1C_1PYR][NTf_2]$), ($[C_5C_1PYR][NTf_2] + [C_1C_1PYR][NTf_2]$), and ($[C_3C_1PYR][NTf_2] + [C_2C_1PYR][NTf_2]$) increases upon addition of the smaller ionic liquid. This is because both $[C_1C_1PYR][NTf_2]$ ($T_m = 133\,^\circ C$) and $[C_2C_1PYR][NTf_2]$ ($T_m = 90\,^\circ C$) are solids at $80\,^\circ C$. However, as seen for the example of ($[C_5C_1PYR][NTf_2] + [C_3C_1PYR][NTf_2]$), it is possible to obtain lower viscosities when adding as a liquid a second ionic liquid, which possesses a higher ion mobility. In this instance, the *conductivity* increases quite dramatically, which is due to the higher charge concentration and lower viscosity [110]. In several cases, this has even led to non-ideal behavior of the conductivity, meaning that the mixture possesses a distinctively higher conductivity [107, 112].

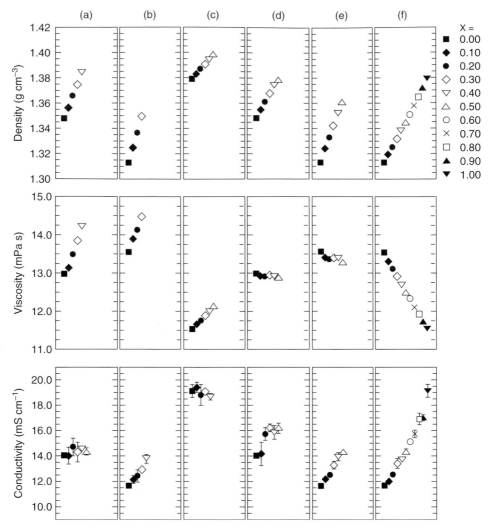

Figure 7.1 Density, viscosity, and ionic conductivity of IL$_1$+IL$_2$ mixtures at 80 °C. From left to right: (a) (1−X) [C$_4$C$_1$PYR][NTf$_2$] + (X) [C$_1$C$_1$PYR][NTf$_2$], (b) (1−X) [C$_5$C$_1$PYR][NTf$_2$] + (X) [C$_1$C$_1$PYR][NTf$_2$], (c) (1−X) [C$_3$C$_1$PYR][NTf$_2$] + (X) [C$_2$C$_1$PYR][NTf$_2$], (d) (1−X) [C$_4$C$_1$PYR][NTf$_2$] + (X) [C$_2$C$_1$PYR][NTf$_2$], (e) (1−X) [C$_5$C$_1$PYR][NTf$_2$] + (X) [C$_2$C$_1$PYR][NTf$_2$] and (f) (1−X) [C$_5$C$_1$PYR][NTf$_2$] + (X) [C$_3$C$_1$PYR][NTf$_2$]. Reprinted with permission from [110]. Copyright 2012 American Chemical Society.

Studies on 1,3-dialkylimidazolium- [107], pyridinium- [108, 109], and pyrrolidinium-based [110] ionic liquids, in which either the cation was varied, for example ([C$_6$MIM]$^+$ + [C$_2$MIM]$^+$), while keeping the anion constant or the anion was varied, for example ([BF$_4$]$^-$ + [CH$_3$SO$_4$]$^-$), while keeping the cation constant, showed that although ideal mixing behavior was found for all the mixtures, a closer look at the *excess molar volume* can give some insights into underlying structural effects.

The *excess molar volume* is defined as the difference between the experimental and the ideal molar volume. Hence, a positive value indicates that the mixture requires more space than the ideal mixture, that is, it is less dense, indicating possibly less effective packing. Figure 7.2 demonstrates, with an imaginary example of a binary ionic liquid mixture, the relations between experimental and theoretical (i.e., ideal) density, and the resulting (in this case positive) excess molar volume.

The excess molar volume is positive for many binary mixtures investigated: ($[C_4MIM][BF_4]$ + $[C_6MIM][BF_4]$), ($[C_4MIM][NTf_2]$ + $[C_4MIM][BF_4]$), ($[C_4MIM][NTf_2]$ + $[C_4MIM][PF_6]$), ($[C_2MIM][N(CN)_2]$ + $[C_2MIM][BF_4]$), ($[C_4PY][BF_4]$ + $[C_4PY][NTf_2]$), and ($[C_4C_1PY][BF_4]$ + $[C_8C_1PY][BF_4]$) [107–109, 113]. On the other hand, the excess molar volumes obtained from ($[C_4MIM][BF_4]$ + $[C_4MIM][PF_6]$), ($[C_2MIM][BF_4]$ + $[C_6MIM][BF_4]$), ($[C_4MIM][BF_4]$ + $[C_4MIM][CH_3SO_4]$), and ($[C_4C_1PY][BF_4]$ + $[C_4C_1PY][N(CN)_2]$) [107, 108, 113] mixtures are slightly negative. They deviate in the relative degree of deviation, and also the maxima or minima shift between $0.3 < X < 0.9$. In rare cases, a non-symmetrical variation and even trends with two maxima/minima are observed.

The interpretation of these results is difficult, and hampered somewhat by the fact that strongly coordinating anions (acetate or chloride) have not yet been investigated, which should indicate the limits of deviations. In a first approximation, negative values indicate more efficient packing, for example, due to the utilization of larger voids [108].

The *excess enthalpy* is a reflection of the overall destruction or formation of interactions in the mixing process. Positive excess enthalpy values indicate that the dissolution of the second ionic liquid requires more energy and is less voluntary than theoretically presumed, that is, interactions are destroyed. Negative excess

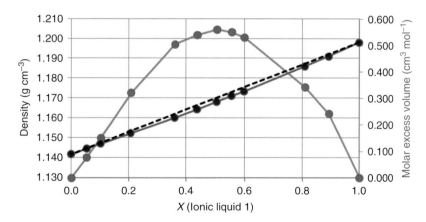

Figure 7.2 Density and excess molar volume (V^E) for imaginary ionic liquid mixtures. Solid lines are added to the experimental data to guide the eye; the dashed line shows the ideal density.

enthalpy values, on the other hand, indicate the formation of stronger interactions. Navia *et al.* [113] argued that if an ionic liquid that deviates only in the length of the alkyl chain of the cation is added to another ionic liquid, only dispersive interactions should be affected. The accommodation of this second cation is energy consumptive, especially if the difference in alkyl chain length is larger. If the cation is kept constant and the anions are varied, the formation of overall stronger interactions occurs. This is more the case if a more basic anion, such as methanesulfonate, is added than a less basic one, such as hexafluorophosphate (Figure 7.3) [113]. However, this interpretation does not correlate directly with the conclusions drawn from the excess molar volume behavior, where a negative deviation from ideality was found for the ($[C_6MIM][BF_4] + [C_2MIM][BF_4]$) mixture.

Similar to conductivity, the thermal phase behavior, determined by differential scanning calorimetry (*DSC*), also exhibits non-ideal behavior, and the presence of a mixture of cations decreases the *melting/phase transition point* in comparison with both pure ionic liquids [110].

Overall, this study clearly demonstrates the potential of tuning the physical properties of ionic liquids by using binary mixtures.

7.3.2
Structural Investigations

The strength of the aggregation between anions ($[PF_6]^-$, $[CF_3CO_2]^-$, $[InCl_4]^-$, $[BPh_4]^-$, and $[BF_4]^-$) and cations ($[C_4MIM]^+$) has been investigated in detail using electrospray ionization mass spectrometry (*ESI-MS*). In positive and negative ion

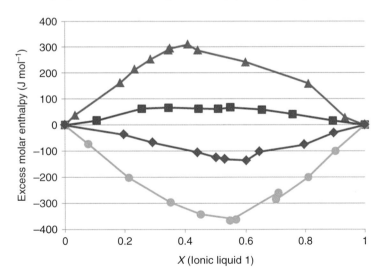

Figure 7.3 Excess molar enthalpies as a function of composition at $T = 30\,°C$: (▲) $X[C_6MIM][BF_4] + (1 - X)[C_2MIM][BF_4]$; (■) $X[C_4MIM][BF_4] + (1 - X) [C_6MIM][BF_4]$; (♦) $X[C_4MIM][PF_6] + (1 - X)[C_4MIM][BF_4]$; (●) $X[C_4MIM][BF_4] + (1 - X)[C_4MIM][MeSO_4]$. Redrawn from Reference [113].

mode, aggregates detected contained as most favorable assemblies $[(X_1)_{m+1}(A)_m]^+$ and $[(X_1)_m(A)_{m+1}]^-$, respectively, but also some doubly and triply charged species (Figure 7.4). A tandem mass spectrometric strategy was employed via low energy collision activation to further fragment loosely bonded ions from mixed $[A^- \ldots X^+ \ldots B^-]$ and higher aggregates to evaluate the solvent-free hydrogen bond strength. For example, when the supramolecule $[CF_3CO_2^- \ldots X^+ \ldots PF_6^-]$ was selected, isolated, and dissociated, $[PF_6]$ was formed exclusively, indicating that $[PF_6]^-$ has weaker interactions than $[CF_3CO_2]^-$ with the cation. The relative hydrogen bond strength was measured by using Cooks' kinetic method and gave the relative order (for $[X]^+ = $ 1-butyl-3-methylimidazolium): $[CF_3CO_2]^- > [BF_4]^- > [PF_6]^- > [InCl_4]^- > [BPh_4]^-$ [114].

Andanson *et al.* investigated ionic liquid mixtures involving a common cation (1-butyl-3-methylimidazolium) with several anions [chloride, tetrafluoroborate, hexafluorophosphate, and bis(trifluoromethanesulfonyl)amide] by means of IR spectroscopy. In mixtures of $[C_4MIM][BF_4]$ and $[C_4MIM][PF_6]$ significant changes occurred in the anion's IR bands, indicating that highly symmetric anions mix at the molecular level. According to these findings, nano-segregation, that is, the molecular alignment to yield well-defined polar and non-polar domains in the liquid phase, does not occur in these ionic liquid mixtures [115].

Aparicio and Atilhan worked on a combined experimental (refractive index, density, IR spectroscopy) and computational study on the molecular level of a binary ionic liquid mixture containing N-alkyl-3-methylpyridinium pyridinium cations. The effect of either the anion ($[C_4C_1PY][BF_4]$ + $[C_4C_1PY][N(CN)_2]$) or the cation ($[C_4C_1PY][BF_4]$ + $[C_8C_1PY][BF_4]$) on liquid structure was analyzed over the whole molar fraction range. ATR-FTIR (attenuated total reflection Fourier-transform infrared) spectroscopy and molecular dynamics results of ($[C_4C_1PY][BF_4]$ + $[C_4C_1PY][N(CN)_2]$) mixtures indicate ideal mixing behavior over the whole composition range, with areas where either cation–$[BF_4]$ interactions prevail with

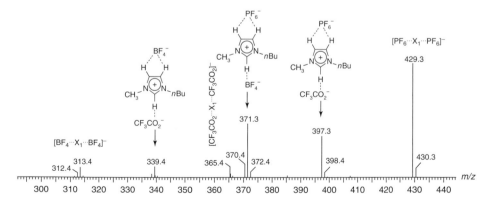

Figure 7.4 ESI-MS mass spectrum in negative ion mode of an equimolar mixture of $[X_1][BF_4]$, $[X_1][PF_6]$, and $[X_1][CF_3CO_2]$ ($X_1 = $ 1-butyl-3-methylimidazolium). Adapted with permission from [114]. Copyright 2004 Wiley-VCH.

$[N(CN)_2]^-$ being dispersed or vice versa, when $[N(CN)_2]^-$ is in excess over $[BF_4]^-$. Negative excess molar volumes are obtained, indicating improved packing in the mixed system. However, for the $([C_4C_1PY][BF_4] + [C_8C_1PY][BF_4])$ system, ATR results point to the existence of microheterogeneities with non-polar domains where dispersive interactions of the alkyl chains are strong. These domains are in particular favored in mixtures with high mole fraction of $[C_8C_1PY][BF_4]$. Adding $[C_8C_1PY][BF_4]$ to the mixture leads to a disturbance of the microstructure, which requires more space and leads hence to positive excess molar volumes [108].

Using UV/Vis spectroscopy and solvatochromic reporter probes to investigate the polarity, dielectric, hydrogen bonding interactions in binary ionic liquid mixtures, Fletcher *et al.* reported already in 2003 that hydrogen bonding is a very important mode of interaction between anions and cations. Unfortunately, only mixtures of ionic liquids with relatively low coordinating ability were investigated, that is, $([C_4MIM][NTf_2] + [C_4MIM][PF_6])$, $([C_2MIM][NTf_2] + [C_4MIM][PF_6])$, and $([C_2MIM][NTf_2] + [C_4MIM][NTf_2])$, and it would be interesting to see results for anions that tend to strong hydrogen bond formation [116].

Brüssel *et al.* used *ab initio* molecular dynamics simulations to gain insight into the structural changes of a $[C_2MIM][SCN] + [C_2MIM]Cl$ (1:1) mixture compared with the pure compounds. They revealed that thiocyanate coordinates with its nitrogen atom at C2-H (the most acidic site) and preferentially with its sulfur atom to C4-H and C5-H of the imidazolium cation. Addition of the more basic chloride partially suppresses the C2-H coordination of thiocyanate via the nitrogen atom. Instead, the coordination via the sulfur atom with the C4-H and C5-H is displaced by a specific interaction with the nitrogen atom of the thiocyanate. The structural arrangement is, hence, mostly governed by competition of the anions for hydrogen bond interactions with the cation. The cations in the system are $\pi-\pi$-stacked, and the chloride-based ionic liquid features a higher degree of order than the thiocyanate-based one, as also witnessed by the higher melting point of the former. This study is a good example of the delicate structural changes that occur upon mixing, and which may affect the degree of deviation from ideal mixing behavior [111, 117].

7.3.3
Potential Applications

Recently, the binary mixture $([C_4PY][BF_4] + [C_4PY][NTf_2])$ was proposed as an alternative for sulfolane when used in the extraction of toluene from alkanes. One criterion for potentially improved performance was a higher density than that of sulfolane to positively affect phase separation, which is fulfilled at molar fractions of $[C_4PY][BF_4] < 0.8$. Another criterion was the separation factor, which is higher than that of sulfolane at a molar fraction of $[C_4PY][BF_4] = 0.7$. At present, one criterion affecting pumping and mixing costs is not yet fulfilled, that is, the viscosity, with all molar fractions exhibiting a higher viscosity than sulfolane. Further investigations will need to show if this disadvantage can be overcome [109].

In organic chemistry, a binary (ionic liquid + ionic liquid) mixture has shown surprising efficiency in affecting the selectivity of reactions by tuning anion nucleophilicity. Bini and Chiappe showed that in nucleophilic dediazoniation the apparent nucleophilicity of an anion such as $[NTf_2]^-$ is higher than that of bromide or chloride. When $[PhN_2][BF_4]$ with a mixture of $([C_4MIM]Br + [C_4MIM][NTf_2])$ was used as solvent, not the expected bromobenzene but two unusual $[NTf_2]$-derivatives were formed (Scheme 7.7). This phenomenon can be explained by the strong and preferential hydrogen bond interaction between the cation C2-H and bromide [118], which decreases its activity for nucleophilic attack. The less coordinated $[NTf_2]^-$ anion is then able to substitute the diazo-group. The same was found when bromide was replaced by chloride, but in the presence of an iodide-based ionic liquid the expected iodobenzene was formed [119].

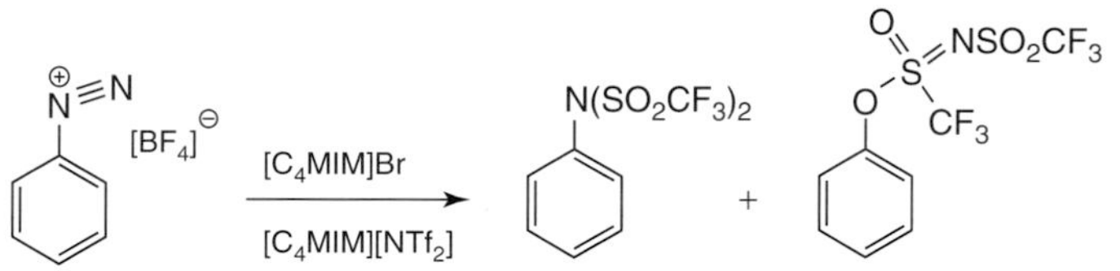

Scheme 7.7 Proposed reaction scheme of dediazoniation of $[PhN_2][BF_4]$ in $([C_4MIM]Br + [C_4MIM][NTf_2])$ [119].

Similarly, our group was able to show that such a type of preferential interactions can also be used in the dissolution of cellulose. In general, to achieve the dissolution of cellulose, the strong intra- and intermolecular hydrogen bond network of the polymer chains must be broken by providing alternative hydrogen bond acceptor sites, for example, basic anions of ionic liquids such as acetate or chloride (but not bromide) [120–122].

^{1}H NMR studies of pure ionic liquids with ethanol (as cellulose model) had shown that a higher downfield chemical shift of the C2-H was observed for halide-based ionic liquids than for all other anions (Figure 7.5), indicating preferential interaction with the cation rather than with the cellulose model. It was hence argued that when using a binary ionic liquid mixture consisting of a dissolving ionic liquid (i.e., $[C_2MIM][OAc]$) and a non-dissolving ionic liquid (i.e., $[C_2MIM]Br$) only so much dissolving ionic liquid is required to interact with all cellulose-OH groups just to achieve dissolution, and the non-dissolving ionic liquid can be used to provide a fluid medium. In an extensive study, this hypothesis was proven to be indeed correct, and a cellulose solution resulted at a ratio of $1:1$ (acetate : OH-groups), even if the non-dissolving ionic liquid was in tenfold excess (at $100\,^\circ C$) [118, 122].

7.3.4
Conclusion: Binary Mixtures of Ionic Liquids

In conclusion, it appears as if, to a first approximation, many of the bulk physical properties can be predicted with good accuracy by presuming ideal mixing behavior.

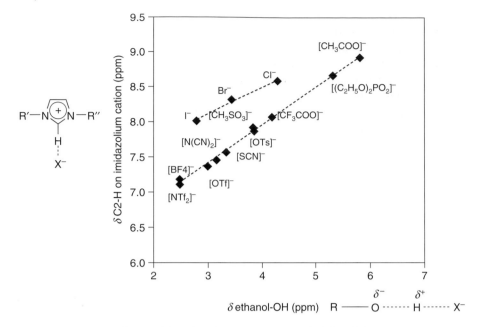

Figure 7.5 Correlation of the ethanol-OH proton chemical shift and the cationic C2-H chemical shift in ¹H NMR in equimolar ethanol and $[C_4MIM]^+$-based ionic liquid mixtures. Reprinted with permission from [118]. Copyright 2012, Springer.

Hence, only the data of both pure components need to be known. This is certainly surprising considering the fact that binary ionic liquid mixtures constitute in fact ternary, that is, ([cation 1][anion 1] + [cation 1][anion 2]) or ([cation 1][anion 1] + [cation 2][anion 1]), mixtures of ions, while quaternary ([cation 1][anion 1] + [cation 2][anion 2] have not been investigated at all yet. However, a literature survey has shown that, on a molecular level, very subtle changes in interactions between the ions can occur in binary ionic liquid mixtures, and it is reasonable to presume that non-ideal behavior will be found for some higher cation-functionalized ionic liquids and possibly those with more basic and coordinating anions.

The combined physicochemical data clearly show that binary ionic liquids can be used to tune the bulk physical properties of an ionic liquid. Because the conductivity is strongly dependent on the viscosity and the number of charge carriers in a given volume, non-ideal behavior can be expected as a combined result of deviations from ideality of these two properties. It could be of interest for electrochemical applications to determine such mixtures with especially large deviations from ideal mixing behavior.

On a molecular level it can be stated that, just as for pure ionic liquids, hydrogen bond interaction between the cations and anions is the most important feature (beside Coulomb interactions) affecting chemical and physical properties [123]. In particular in the mostly used 1,3-dialkylimidazolium salts, the C2-H is the most acidic and hence it is the main interaction site for such specific interactions. This is also the case in binary ionic liquid mixtures, and the series of hydrogen bond acceptor strength (determined by ESI-MS) is in agreement with that of pure ionic liquids [124, 125]. Delicate changes in the interaction pattern can occur upon adding an ionic liquid featuring a second and competing anion, and further investigations, in particular spectroscopic and molecular dynamic studies, should prove fruitful in elucidating the underlying structural arrangements.

Some examples can be found in the literature of beneficial applications of binary ionic liquid mixtures. Preferential interactions may tune the nucleophilicity of one of the anions, which can be exploited in the organic chemistry and catalysis, hinting at new research avenues to exploit in the future.

7.4
Nanoporous Materials from Ionothermal Synthesis

Nanoporous solids (Figure 7.6) are an important class of materials due to their defined pore topologies. Because of their large number of applications [126, 127] in catalysis, adsorption, and ion exchange, research interest focuses on both preparing new structures and scouting innovative areas of application. Conventionally, these materials are prepared by so-called *hydrothermal* synthetic routes. In a general hydrothermal protocol, the reagents [i.e., the tetrahedral atom source(s), a mineralizer, and a structure-directing agent (*SDA*)] are mixed with water as solvent and sealed in a polytetrafluoroethylene-lined steel autoclave. The mixture is then heated to 150–200 °C under high autogenous pressure for a certain reaction time. The frameworks and preparation methods of these materials are well documented and are reproducible [128].

An alternative route to hydrothermal preparation is the so-called *solvothermal* method. This strategy employs the use of non-aqueous organic solvents, such as hexanol, propanol, glycol, glycerol, and pyridine [129].

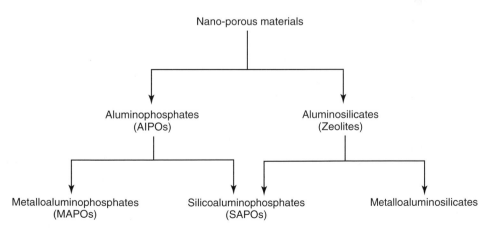

Figure 7.6 Classification of zeolites and zeotypes (nanoporous materials).

However, these methods have several drawbacks, in particular relating to the high autogenous pressure developing during the reaction, which necessitates the use of autoclaves. This implicates a batch-wise manufacture in sealed autoclaves of limited size, and safety concerns (high pressure, use of hydrofluoric acid as mineralizer, etc.). One approach to solve these problems can be the use of solvents with low vapor pressure and low toxicity: ionic liquids. The physicochemical properties, such as the solubility for zeolite precursors, can be tuned selectively by choosing an appropriate cation and anion combination. They may serve as built-in SDAs in ionothermal synthesis due to their pre-organized structures [130].

To exploit this concept, a new strategy, the so-called *ionothermal synthesis*, was introduced by Morris and coworkers [131], who used an ionic liquid both as solvent and SDA in the synthesis of porous materials. This method involves the crystallization of porous materials from a tetrahedral atom source in ionic liquids, in the presence of a mineralizer (e.g., fluoride), at atmospheric pressure (Figure 7.7). The most important feature of this method is the dual function of the ionic liquid, that is, as solvent as well as SDA, due to which the competition between solvent and SDA to interact with the growing framework is eliminated. Ionic liquids possess very low vapor pressures and, hence, they do not contribute to the autogenous pressure when heated, and reactions can be carried out at higher temperatures. This removes the safety concerns associated with the high pressure of hydrothermal methods and, in a laboratory fume-cupboard, reactions can be performed in beakers and flasks rather than sealed autoclaves.

Recent publications using ionothermal synthesis have reported the formation of aluminophosphates (*AlPOs*), metalloaluminophosphates (MAPOs), and silicoaluminophosphates (*SAPOs*). The state of the art of these is discussed in the following subsection.

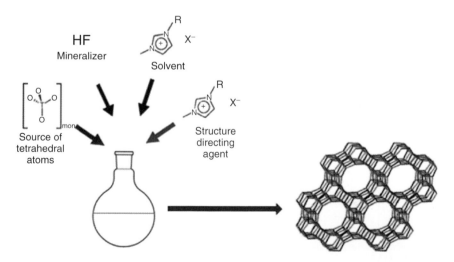

Figure 7.7 Schematic representation of an ionothermal synthesis. Adapted from [132] by permission of The Royal Society of Chemistry.

7.4.1
Aluminophosphates

AlPOs were the first materials prepared by ionothermal synthesis [131]. Since the first breakthrough, there has been extensive success in this area. [C$_2$MIM]Br was used as the first ionic liquid solvent and Table 7.5 shows several of the frameworks that can be synthesized from one particular ionic liquid as well as by different ionic liquids. An interrupted novel structure type, *SIZ-1* (St. Andrews ionic liquid zeolite-1), was formed in the absence of HF (Table 7.5) while adding HF resulted in the *AEL* (AlPO4-11) framework type [131]. On the other hand, if water is quantitatively removed from the reaction mixture, the product formed has a chabazite (*CHA*) framework type. The reaction, after evaporation of water, in the absence of HF and at higher temperatures (200 °C) yielded SIZ-6 [133], a layered AlPO structure. Thus, the reaction temperature, the presence of fluoride, and water affect the framework type of the product.

AEL [134], *SOD* [135] (sodalite), and *LTA* (Linde type A) [136] framework types can also be prepared in the same ionic liquid, [C$_2$MIM]Br, but under different reaction conditions. This shows the versatility of ionothermal synthesis in obtaining various porous materials. The effect of the synthesis conditions [136], such as P/Al and F/Al ratios, on the LTA framework was investigated. It was observed that phase transformation between quartz-type material and the LTA framework is sensitive to P/Al ratios. An optimum value of F/Al is also necessary for crystallization of the framework. An AlPO coating [141] with an AEL framework was prepared using [C$_2$MIM]Br for corrosion resistant layers. The effect of different

Table 7.5 Reaction conditions (mole equivalents, temperature, reaction time) for AlPO framework types prepared in different ionic liquids.

Ionic liquid	Reference	Framework type	IL	Al^a	P^b	HF	H$_2$O	Temperature (°C)	Time (h)
[C$_2$MIM]Br	[131]	Interrupted (novel)	43.0	1	3.0	0.0	2.9	150	66
[C$_2$MIM]Br	[131]	AEL	43.0	1	3.0	0.7	3.8	150	68
[C$_2$MIM]Br	[131]	CHA	39.0	1	3.0	0.7	0.0	150	68
[C$_2$MIM]Br	[133]	Chain structure	39.0	1	3.0	0.0	0.0	200	96
[C$_2$MIM]Br	[134]	AEL	82.0	1	4.0	0.4	c	150	68
[C$_2$MIM]Br	[135]	SOD	40.0	1	3.0	3.6	c	190	72
[C$_2$MIM]Br	[136]	LTA	40.0	1	2.1	3.2	c	180	120
[C$_2$MIM][NTf$_2$]	[137]	Chain	27.0	1	2.9	0.69	3.6	170	96
[BenzMIM]Cl	[138]	LTA	40.0	1	3.0	0.7	c	160	10
[C$_2$PY]Br	[139]	CHA	52	1	2.9	0.69	c,d	180	1
[C$_2$PY]Br	[140]	AFI	7.4	1	1.02	0.75	1.9	160	4

^{a}Al(O*i*Pr)$_3$.
^{b}H$_3$PO$_4$.
^{c}Not specified.
^{d}Microwave heating.

aluminum precursors [142] showed that $Al(OiPr)_3$, $Al_2O(CH_3CO_2)_4 \cdot 4H_2O$, and $Al(NO_3)_3 \cdot 9H_2O$ can be used for the preparation of the AEL framework type.

Ionic liquids have proven to be good media for absorbing microwaves [143, 144]. This property of ionic liquids was used to synthesize AlPOs with the AEL type framework by employing [C_2MIM]Br as solvent [145]. Microwave heating was compared to conventional oven heating. It was observed that the crystallization time was reduced to minutes as compared to several hours by convective heating.

Another ionic liquid, [C_4MIM]Br, has also been investigated as reaction medium in ionothermal synthesis. This has only been used for the synthesis of the AEL-type framework of AlPOs. A systematic study [140] was carried out by our group, using an initial molar composition of [C_4MIM]Br : $Al(OiPr)_3$: H_3PO_4 : HF : H_2O of 7.37 : 1 : 1.02 : 0.75 : 1.86, at 160 °C [146] to investigate the effect of reaction conditions on the framework of the product. It was shown that the initially formed mixture of the *AFI* (AlPO4-5) and the AEL frameworks is converted into a pure AEL phase (Figure 7.8) at longer reaction times. The AEL framework, with ten-membered ring pores, is more thermodynamically stable than the AFI framework, which contains 12-membered ring pores. This shows that a thermally stable framework (AEL) forms, which is in accordance with earlier observations [145, 147] but under different reaction conditions. SEM (scanning electron microscope) images of these materials (Figure 7.9) show that the materials are crystalline and possess rod-like crystals.

The use of HF is vital for the crystallization of porous material as fluoride is supposed to help the crystallization process by increasing the solubility of

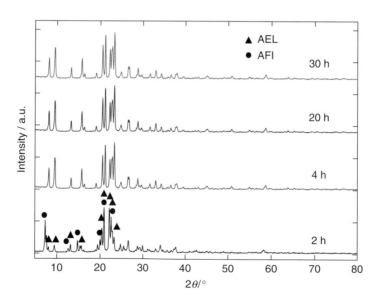

Figure 7.8 XRD (X-ray diffraction) patterns of AlPOs synthesized with different reaction times in [C_4MIM]Br [140].

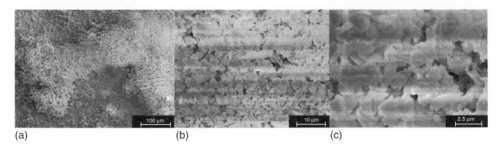

(a)   (b)   (c)

Figure 7.9 SEM images of AlPOs with an AEL framework type synthesized in [C$_4$MIM]Br: (a) 100, (b) 10, and (c) 2.5 μm [140].

starting materials [148]. Indeed, no framework was formed in the absence of HF in [C$_4$MIM]Br, though previously an interrupted framework was reported in the absence of HF [131]. An optimum F/Al ratio is required for crystallization of the framework [136, 140]. Likewise, the P/Al ratio has a strong impact on the crystallization of porous frameworks as the acid concentration controls the hydrolyzing rate of Al(OiPr)$_3$. Therefore, there is a certain P/Al ratio range [136, 140] under reaction conditions where crystallization of porous materials occurs. As HF and H$_3$PO$_4$ are employed in general as aqueous solutions, increasing the amounts of these reagents also increases the water content in the reaction, which may also influence the ionothermal conditions.

As mentioned above, HF is very important for crystallization but it is very toxic and difficult to handle. Therefore, an attempt was made to use [NH$_4$]F as mineralizer under above-mentioned reaction conditions [140]. [NH$_4$]F is solid and requires higher amounts and higher temperature (Figure 7.10) to crystallize

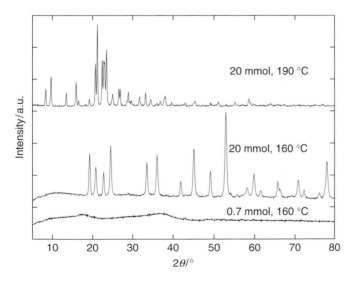

Figure 7.10 XRD patterns of AlPOs synthesized with different [NH$_4$]F/Al ratios and variations in temperature in [C$_4$MIM]Br [140].

the porous (AEL) framework. Research is in progress to optimize the reaction conditions for the use of [NH$_4$]F as mineralizer. [NH$_4$]F has been successfully used under anhydrous conditions (AlO(OH) and [NH$_4$][H$_2$PO$_4$]) [149] to prepare porous materials.

As insinuated above, water plays a very important role in ionothermal synthesis. To elucidate the detailed effect on crystallization time, water was added incrementally to anhydrous reagents. It was demonstrated that a small quantity of water is beneficial for the ionothermal synthesis in terms of reducing the crystallization time [149]. Likewise, when using aqueous reagents [140], no porous framework but a dense phase is formed when water is evaporated prior to crystallization (Figure 7.11).

As stated above, the ionic liquid does not contribute to the autogenous pressure when using closed reaction vessels. However, some pressure will develop nevertheless during the reaction due the aqueous reagent solutions and alcohol produced during reaction of phosphoric acid and Al(OiPr)$_3$. Interestingly, in closed vessels, and at somewhat elevated pressures, a quartz-type product (Figure 7.11) forms, again demonstrating the benefit of ambient condition reactions.

Furthermore, the ionic liquid/Al ratio has no particular effect on the framework type of product, as shown by experiments with ionic liquid/Al ratios between 4.90 and 14.72 [140]. As the ionic liquid acts as both solvent and SDA, under the reaction conditions employed in Reference [140], an ionic liquid/Al ratio of 7.37 was used to perform this bimodal function and obtain a crystalline product.

The effect of systematically increasing the alkyl chain length from two to five carbons in 1-alkyl-3-imidazolium bromide ionic liquids was investigated in the

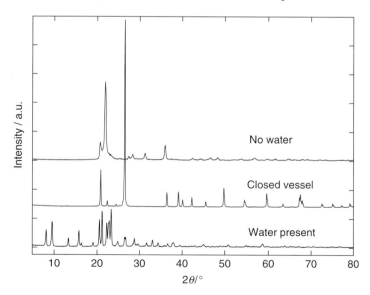

Figure 7.11 XRD patterns of AlPOs demonstrating the effect of the water content in [C$_4$MIM]Br [140].

ionothermal synthesis of AlPOs. In the whole series of ionic liquids, the same framework topology (CHA) was obtained. Solid-state ^{13}C NMR data suggest that the SDA occluding in the materials is not the original cation from the ionic liquid but the 1,3-dimethylimidazolium cation (except when alkyl chain length is two), owing to its decomposition in a closed reaction system, in the presence of HF [150].

As ionic liquids act as SDAs, different frameworks should theoretically be possible by changing the ionic liquid. However, one has to consider the properties of both cation and anion. Table 7.5 shows the frameworks obtained using different ionic liquids in ionothermal synthesis [137–140, 151].

[C_2PY]Br can also be used to synthesize different framework types (Table 7.5) depending on the reaction conditions, as is the case with [C_2MIM]Br. It was used to synthesize the CHA [138] and AFI [140] framework types. This is the first example of the synthesis of a pure AFI framework (Figure 7.12) in pyridinium-based ionic liquids without using co-SDAs. The framework is stable over 24 h of reaction time and is homogeneously crystallized with rod-like morphology (Figure 7.13). Although the AFI framework is thermally stable over a long time (>24 h), a mixture of AFI and AEL framework types resulted if the reaction temperature was increased from 160 to 175 °C. With a further increase in temperature to 210 °C, an AlPO having an AEL (major) framework was found with a very small fraction of AFI type framework (Figure 7.14). Therefore, it can be stated that denser frameworks form at higher temperatures [140].

There is a link between solvothermal and ionothermal synthesis when materials are synthesized in corresponding amine and amine-based ionic liquids. Wragg *et al.* [139] demonstrated this relation and obtained materials with the same framework

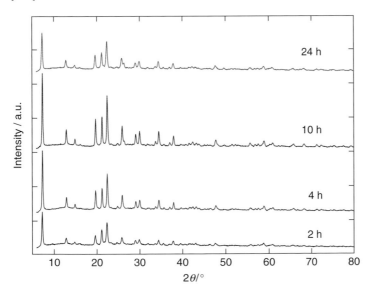

Figure 7.12 XRD patterns of AlPOs synthesized with different reaction times in [C_2PY]Br [140].

(a)   (b)   (c)

Figure 7.13 SEM images of AlPOs with an AFI framework type synthesized in [C$_2$PY]Br: (a) 100, (b) 10, and (c) 2.5 μm [140].

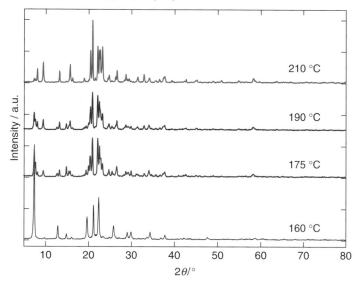

Figure 7.14 XRD patterns of AlPOs synthesized at different reaction temperatures in [C$_2$PY]Br [140].

types (e.g., CHA framework in both ionothermal and solvothermal synthesis using 1-alkyl-3-methylimidazolium bromide and 1-methylimidazole, respectively).

7.4.2
Co-structure Directing Agents and Aluminophosphates

Co-structure-directing agents (*co-SDAs*) can alter the crystallization process of AlPOs and are used to obtain a higher phase selectivity. This strategy was used to prepare the AFI framework type, using aliphatic and aromatic amines as co-SDAs with ionic liquids [146, 147]. A detailed study was conducted to determine the role of the co-SDA morpholine in the ionothermal synthesis with [C$_4$MIM]Br, by using both *in situ* NMR spectroscopy and density functional theoretical studies [152]. Morpholine forms a hydrogen-bonded complex with the imidazolium cation

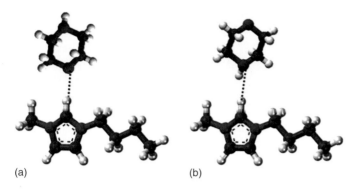

(a)   (b)

Figure 7.15 Complexes of the $[C_4MIM]^+$ cation with (a) morpholine-*O* and (b) morpholine-*N*. Redrawn, from [152].

(Figure 7.15) during the crystallization of AlPOs, which acts in the gel stage as a co-SDA for the AFI framework.

Less thermodynamically favorable frameworks (larger) can be stabilized in two ways: by increasing the molar ratios of co-SDA [146, 153] or using co-SDA of different sizes [146, 154–156]. This approach can also be used to selectively tune the framework between the AEL, AFI, LTA, and CHA [146, 155, 156] types through control of the size of the SDA.

7.4.3
Silicoaluminophosphates

At present, little evidence exists showing the feasibility of ionothermal synthesis of SAPOs. The first synthesis [134] of SAPOs was claimed using $[C_2MIM]Br$ in ionothermal synthesis by introducing tetraethyl orthosilicate (*TEOS*) as silicon source in AlPO preparation. $[C_2MIM]Br$ is the preferred SDA for the AEL framework of AlPOs, and the same framework was found for the resulting SAPO material. The AEL type framework was also obtained in the same ionic liquid with microwave heating for 20 min [145] rather than hours with convective heating. AEL framework material of higher crystallinity was obtained by increasing the amount of HF (i.e., HF/Al ratios from 0.8 to 1.6). However, the yield was decreased due to higher digestion of the Si- and Al-reagents by the higher concentration of fluoride ions [145]. Using microwave heating, SAPOs coatings (AEL) were reported [141] for corrosion resistant applications.

7.4.4
Metalloaluminophosphates

Introducing heteroatoms to substitute for Al^{3+} or P^{5+} in the AlPO framework will impart catalytic activity, and this can be interesting in heterogeneous catalysis. Few efforts have been made to synthesize metal substituted AlPOs in ionic liquids.

Cobalt aluminophosphate (*CoAlPO*) was prepared using the above-mentioned method [131] by adding $Co(OH)_2$ to the reaction mixture [157]. The crystals of three different framework types [*AEI* (AlPO4-18), SOD, and a yet unidentified type] were formed but could not be isolated. Later on, a pure SOD type CoAlPO-framework was synthesized using $CoCl_2 \cdot 6H_2O$ [135] as Co source; 27% of Al sites were replaced by Co. The same material can be synthesized in [C_2MIM]Br, [C_4MIM]Br, and [C_4MIM][BF_4]. The relative crystallinity of the product increases when the temperature is increased and reaches an optimum value at 180 °C. Iron aluminophosphate (*FeAlPO*) [135] can also be prepared using the same molar composition.

Ionic liquids [C_2MIM]Br and [C_4MIM]Br were used as reaction media in the ionothermal synthesis of magnesium AlPO. The AEL type framework was formed in [C_2MIM]Br. A mixture of AEL and AFI type frameworks were formed in [C_4MIM]Br. It was found that the Mg/Al ratio affects the framework type and that the heating method (convective and microwave) has a strong influence on the crystallization time [158]. A manganese AlPO with an AFI framework [159] can be prepared in [C_2dMIM]Br.

7.4.5
Zeolites (Aluminosilicates)

The ionothermal synthesis of AlPOs is relatively straightforward. The synthesis of zeolites is, however, a much greater challenge for ionothermal synthesis. There has been only one report for ionothermal synthesis of siliceous zeolites. The reaction systems in conventional zeolite synthesis use hydroxide or fluoride ions as mineralizing agents [129] to aid dissolution of the silicon precursor. For the ionothermal protocol, the properties of the ionic liquids where tailored by using a mixture of [C_4MIM]Br and [C_4MIM][OH], resulting in the solubility of the silicon precursor TEOS in the presence of HF [160]. In this way, the first siliceous zeolite was synthesized with the MFI (silicate-1) type framework.

Presently, there is no report for the ionothermal synthesis of zeolites (aluminosilicates), which may be due to difficulties in dissolution of the starting materials. However, as ionic liquids are known as designer solvents, tailoring the properties of ionic liquids to dissolve the starting materials appears possible. Further fundamental studies are required to fully understand the chemistry of the precursors in ionic liquids.

7.4.6
Conclusions: Nanoporous Materials from Ionothermal Synthesis

Ionothermal synthesis is a promising method for the synthesis of nanoporous materials. The same type of AlPO frameworks can be synthesized using a wide range of reaction conditions. The crystallization of a porous framework is affected by the reaction parameters, that is, reaction time, temperature, P/Al ratio, F/Al ratio, and the content of water. The framework type can be selected by changing the constitution of the ionic liquid, but more research is required to fully understand the effect of the structure of the ionic liquid. Considering the number of accessible ionic liquids, it is not unlikely that novel nanoporous structures may be achieved, and some yet to be identified structures have been indeed obtained. The thermodynamically stable frameworks form preferably with increasing crystallization time and temperature. In addition to the type of ionic liquid chosen, the phase stability and selectivity can be tuned by using co-SDAs. Silicon and metal atoms can be incorporated into the AlPO framework to extend the dimensions of possible applications. Despite the success in ionothermal synthesis of AlPOs, the field of true zeolites is yet to be explored. There is room for future research to find alternative mineralizers and tune the properties of ionic liquids for the synthesis of zeolites.

7.5
Catalytic Hydrogenation Reactions in Ionic Liquids

This section aims to demonstrate the state of the art of hydrogenation reactions in ionic liquids. This particular reaction has been selected as it belongs to those that have been investigated from the mid-1990s [161, 162], starting as liquid phase catalysis [albeit often (solid–liquid–liquid–gaseous) multiphasic], and which have experienced a notable transition due to realization of the design opportunities offered by the exceptional structures of ionic liquids.

The catalyst metals are introduced into the system either as homogeneously dissolved precursors or as metal (nano)particles. As discussed by Geldbach and Dyson, discerning the true structure of the active catalyst in hydrogenation reactions (i.e., if it is a molecular complex, a soluble metal cluster, or a soluble colloid or if nanoparticles are involved), and which mechanism prevails, is a difficult task [163]. In this context, we do not aim to judge the nature of the active catalyst but rather focus on the advantages offered with respect to the chemistry and engineering of hydrogenation reactions carried out in ionic liquids. In any case, the ionic liquid selected may act as an inert solvent (i.e., exhibiting only interactions that bring about dissolution of the starting materials) or may feature specific interactions that can cause co-catalytic effects. Examples are ionic liquids with latent Lewis acidity or basicity, which interact with the catalytically active site [164]. Again, the many possible structural variations described above can be advantageously applied to tune the solubility of the starting material, catalyst, and/or product, hence affecting the chemical engineering of the process. From a chemical viewpoint, the structure can be also tuned to affect the rate of reaction and selectivity, as will be shown below [1]. In general, the solubility of dihydrogen is low in ionic liquids due to their relatively high polarity, but it is sufficient to allow for high rate of reactions. In addition, the solubility of alkenes bearing no further functionalities is relatively low, but it can be altered by choosing either longer alkyl chains on the cation or relatively large anions that allow for charge delocalization. As an example, Table 7.6 shows the solubility of 1-octene and 1-hexene [165–167].

Table 7.6 Solubility of 1-hexene and 1-octene in various ionic liquids (20 °C) [165, 166].

Ionic liquid	Solubility of 1-hexene (mol. %)	Ionic liquid	Solubility of 1-octene (mol. %)
$[C_4MIM][BF_4]$	2.0	$[C_4MIM][BF_4]$	1.5
$[C_4MIM][PF_6]$	5.0	$[C_6MIM][BF_4]$	5.0
$[C_4C_1PYR][CF_3SO_3]$	8.0	$[C_8MIM][BF_4]$	17.0
$[C_4MIM][CF_3CO_2]$	9.0	$[C_6PY][BF_4]$	5.5
$[C_6MIM][CF_3CO_2]$	17.5	$[C_4MIM][PF_6]$	2.5
$[C_4DMIM][NTf_2]$	11.0	$[C_4MIM][CF_3SO_3]$	3.5
$[C_4C_1PYR][NTf_2]$	15.5	$[C_4MIM][NTf_2]$	8.0
$[C_4MIM][NTf_2]$	16.0	$[C_6MIM][NTf_2]$	17.0
		$[C_8PY][NTf_2]$	48.0

7.5.1
Early Developments of Ionic Liquids in Hydrogenation Reactions

The first successful hydrogenation reactions were carried out by Chauvin *et al.* [161] and Suarez and coworkers [162, 168], using rhodium- or ruthenium-based catalysts. Already in 1995, Chauvin *et al.* [161] gave a prime example of ionic liquid technology at its best: They used a (nowadays quite unusual) ionic liquid, that is, $[C_4MIM][SbF_6]$, with $[Rh(NBD)(PPh_3)_2][PF_6]$ (*NBD*, norbornadiene) as catalyst precursor to demonstrate that a five-times higher reaction rate results in comparison to when acetone is used, even though the starting material solubility is very low. Furthermore, recycling of the catalyst-containing ionic liquid phase by simple phase separation was possible without loss of activity and at low leaching of the catalyst, below the detection limit (0.02%).

With regard to selectivity, a very interesting phenomenon was observed when the selective hydrogenation of cyclohexadiene was investigated. The selectivity of cyclohexene was found to be highly controllable, and at 96% conversion the selectivity for cyclohexene is as high as 98%. The high selectivity can be explained by the higher solubility of the diene, which after conversion into cyclohexene is expelled from the ionic liquid phase, hence avoiding over-hydrogenation to cyclohexane.

The last important point made in these early contributions is the purity of the ionic liquids used, which at the time were prepared in-house in small batches. The purity assessment of ionic liquids was still in its infancy, but since then several detailed analytical protocols have been published [169]. For the hydrogenation reaction, residual chloride, which can be present in ionic liquids if they are made by alkylation of, for example, 1-methylimidazole with an alkyl chloride, followed by ion exchange (Scheme 7.1), caused tremendous deactivation of the catalyst [161].

The formation of an explicitly heterogeneous hydrogenation reaction medium, that is, where the catalyst is present in the bulk ionic liquid in the form of multi-atomic (nano)particles, was observed for several transition metals stabilized by the ionic liquid, including iridium [3, 170], palladium, and platinum [163], which have been reported to show higher activities than molecular catalysts [171]. Hydrogenation reactions occur in general under very mild conditions.

For example, Huang *et al.* [172] demonstrated quantitative conversions and high selectivities in the hydrogenation of 1-hexene, cyclohexene, or 1,3-cyclohexadiene with phenanthroline-stabilized palladium nanoparticles (2–5 nm diameter) in [C_4MIM][PF_6]. The reaction was carried out under mild conditions (0.1 MPa dihydrogen pressure and 20–60 °C). The selectivity of the reaction of 1,3-cyclohexadiene is a function of time, and either cyclohexene or cyclohexane can be obtained in 100% yield after 2 and 7 h, respectively. The reaction mixture was recyclable at least ten times without a decrease in activity. In the absence of phenanthroline, the palladium particles agglomerate and, in the second cycle, the conversion decreases from 95% to 35%.

7.5.2
Stereoselective Hydrogenation Reactions in Ionic Liquids

The next major development in the area of hydrogenation in and with ionic liquids was the induction of stereoselectivity and the establishment of enantioselective synthesis protocols. Ionic liquids featuring a chiral center on either the cation [173–177] or on both moieties [176] have been prepared, and the concept of organocatalytic interaction to evoke enantiomeric excesses has been proven to work. One example is the enantioselective hydrogenation of a keto function of an ionic liquid's cation in the presence of a (*R*)-camphorsulfonate anion, resulting in up to 80% ee. It was postulated that a strong interaction of the enantiomerically pure anion by ion-pairing was responsible for this induction [14].

Of course, ionic liquids can also be used as innocuous solvents in enantioselective reactions. For example, very complex polar Ru(BINAP)(DPEN)Cl_2 (*BINAP*, 2,2'-bis(diphenylphosphino)-1,1'-binaphthyl; *DPEN*, (*R*,*R*)-1,2-diphenylethylenediamine) pre-catalysts have been investigated in the enantioselective hydrogenation of various aromatic ketones [178]. While in 1,2,3-trialkyl-substituted ionic liquids the conversion was quantitative, in 1,3-dialkylimidazolium-based ionic liquids, for example, [C_4MIM][NTf_2], intermediate carbene formation leads to catalyst poisoning. The stereoselectivity was between 90% and 99% ee, depending on the aromatic ketones used as starting material.

Several authors have investigated enantioselective hydrogenations using ruthenium complexes, and the development of suitable, albeit very complex, ligands has been in focus for about a decade now [179–183]. Recently, Zhao *et al.* [184] presented the latest generation of ligands (Scheme 7.8). When using a biphasic mixture of toluene and [C_4MIM][BF_4], the hydrogenation of an enamide was possible with high conversions (up to 100%) and enantioselectivity (>99% ee) in the first run. The recycling performance depends on the substitution pattern of the ligand, with best results obtained for R^1 = H, R^2 = CH_3, and n = 6 (Scheme 7.8).

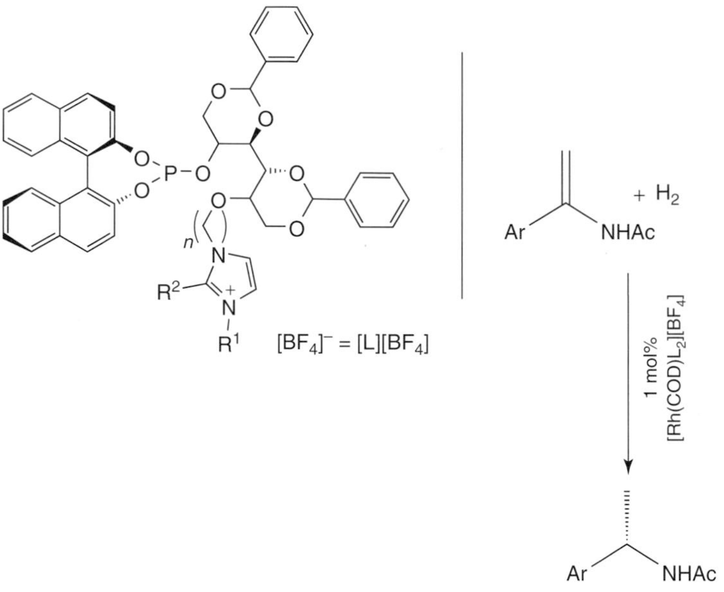

Scheme 7.8 Hydrogenation reaction employing a monodentate phosphite ligand in the catalyst; $n = 4, 6, 12$; $R^1 = CH_3, C_4H_9$; $R^2 = H, CH_3$; Ac = acyl, Ar = aromatic substituent. Adapted from [183].

In general, reports on enantioselective hydrogenations with heterogeneous catalysts in ionic liquid are rare [14]. Sano *et al.* [185] recently presented the hydrogenation of methyl benzoylformate with platinum on alumina in the presence of a chiral modifier. Several organic solvents and ionic liquids were explored. Quantitative yields and high enantiomeric excesses (80–90% ee) were found for ionic liquids with anions of low coordinating ability (5% cat., 24 h, 20 °C, 1 MPa) [185].

7.5.3
Ionic Liquids as Thermomorphic Phases

An elegant way to solve problems relating to multiphase reaction systems, in particular mass transfer limitations, is to exploit the thermomorphic properties of some ionic liquid–substrate mixtures: At a lower temperature, two phases (ionic liquid containing catalyst, starting material and/or product) co-exist. Increasing the temperature leads to mutual solubility of the two liquid phases, so that mass transfer is facilitated during the reaction. Decreasing the temperature induces phase separation. The reuse of the catalyst-containing ionic liquid phase has been demonstrated without a decrease in activity [186]. This principle is shown in Figure 7.16.

Daguenet *et al.* [187] demonstrated this principle by the hydrogenation of an aqueous solution of 2-butyn-1,4-diol with the homogeneous catalyst precursor $[Rh(\eta^4\text{-}C_7H_8)(PPh_3)_2][BF_4]$ dissolved in $[C_8MIM][BF_4]$. Under the reaction conditions (80 °C and 6 MPa dihydrogen pressure), a single phase is obtained.

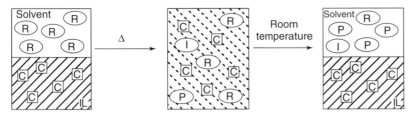

Figure 7.16 Catalysis in thermomorphic ionic liquid mixtures: C = catalyst (homogeneous or heterogeneous), IL = ionic liquid, R = reactant, I = intermediate, and P = product.

Although the ionic liquid chosen is not optimal and was not further optimized (it is nowadays well known that tetrafluoroborate-based ionic liquids tend to hydrolyze, and the relatively high water solubility of this particular ionic liquid does not allow for quantitative product phase separation), the proof-of-principle is clearly demonstrated. Later research using the precursor $RhCl_3 \cdot 3H_2O$ dissolved in $[CH_3(OCH_2CH_2)_{16}N(CH_2CH_3)_3][CH_3SO_3]$ for the hydrogenation of various alkenes showed quantitative conversions (60 °C, 1 MPa, 1 h), and a recyclability of the ionic liquid and catalyst of at least eight cycles [188].

7.5.4
SILCA-Type Materials

Finally, the general concept of SILCA (Supported Ionic Liquid Catalysts), has been developed and applied to gas-phase reactions, in particular hydrogenation reactions. SILCA-type materials are porous materials impregnated with a layer of an ionic liquid, where the ionic liquid is either chemisorbed or physisorbed on the surface. Furthermore, they contain a catalytically active component [20], which may be a homogeneously dissolved molecular transition metal complex (even as part of the anion of the ionic liquid) or (nano)particles stabilized by the ionic liquid ions. Even biocatalysts may be incorporated. In any case, the surface of the ionic liquid is much expanded when compared to bulk liquid–gas or liquid–liquid–gas reactions, which again facilitates mass transport. Additionally, much less ionic liquid is required, and the inventory ionic liquid costs of a process are therefore reduced by several orders of magnitude. The concept of SILCA combines the positive properties of homogeneous catalysis, for example, high activity and selectivity, and heterogeneous catalysis, for example, facile product–catalyst separation.

SILCA-type materials can be divided into three subgroups, that is, *SILP* (Supported Ionic Liquid Phase), *SILC* (Supported Ionic Liquid Catalyst), and *SCILL* (solid catalyst with ionic liquid layer) catalyst (Figure 7.17). Note that a clear definition of each sub-concept is difficult, and the one we present here is based on the original literature descriptions offered by the respective authors.

From a macroscopic viewpoint, SILCAs are heterogeneous catalysts and the (chemi- or physisorbed) ionic liquid serves as an immobilization phase for the catalytically active component. If, on a microscopic level, the metal catalyst precursor

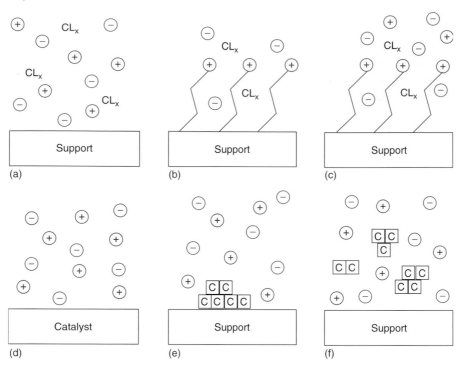

Figure 7.17 Schematic representations of different possible SILCA systems: (a) and (b) SILP; (c) SILC; and (d)–(f) SCILL. CL_x = organometallic catalyst and C = heterogeneous catalyst.

is present as a homogeneous catalyst, the material belongs to the subgroup called SILP [189].

The second type is called *SILC*. In this system, a monolayer of an ionic liquid is covalently attached to the surface of the support material and, furthermore, the surface is impregnated with a second physisorbed ionic liquid (i.e. chemi- and physisorbed ionic liquid). Again, the metal catalyst precursor is present as a homogeneous catalyst [16].

The third type, SCILL, features a heterogeneous solid metal catalyst, which may constitute the support material itself, be precipitated on the surface of an inert support, or be present as dispersed (nano)particles in the (chemi- or physisorbed) ionic liquid phase [19]. For hydrogenation reactions, most catalytically active SILCA-materials belong to the sub-class SCILL.

For example, the positive results for the stereoselective hydrogenation reported by Ngo *et al.* [178] (see above) triggered the development of a SILP catalyst by Lou *et al.* [190]. On the surface of MCM-48, MCM-41, SBA-15, or amorphous silica, an ionic liquid was bonded covalently, and a solution of the catalyst precursor $Ru(PPh_3)_2(S,S\text{-}DPEN)$ in $[C_4MIM][BF_4]$ was used to impregnate the modified silica materials. Investigations of the reaction performance (at quantitative conversion) of the hydrogenation of acetophenone gave similar enantioselectivities, that is, up to 78% (*R*) for the amorphous SiO_2 and MCM-48 as in the absence of the support, while the selectivity of the other two systems was lower. In recycling experiments, the amorphous silica especially gave good performance with quantitative conversions and high enantioselectivities [190]. For the MCM-48-based SILC system, higher reaction times or dihydrogen pressures were required to achieve quantitative conversions, and the selectivity was not negatively affected, as long as the temperature was not increased. Optimum conditions were determined as 20 °C, 3 MPa dihydrogen pressure, and 6 h reaction time, giving 100% conversion and an enantiomeric excess of 78%, with a preference for the (*R*)-enantiomer [190].

Kernchen *et al.* [19] investigated the hydrogenation of cyclooctadiene (COD) to cyclooctene and cyclooctane using a commercial porous nickel catalyst on silica impregnated with the ionic liquid $[C_4MIM][C_8H_{17}OSO_3]$, that is, a SCILL-type catalyst. The authors report an almost quantitative coating of the internal surface at a pore-filling degree of about 10%, reducing the surface by around 50%. The hydrogenation reaction was carried out at 5 MPa dihydrogen pressure and 50 °C. Under these conditions, no leaching was observed. Compared to the nickel catalyst not coated with an ionic liquid layer, the reaction rate decreased by a factor of six. However, the intrinsic selectivity to the intermediate cyclooctene increased from 40% to 70% [19].

The hydrogenation of cinnamaldehyde was studied by Virtanen *et al.* [191], who also applied a SCILL-type catalyst. Scheme 7.9 shows possible products and pathways of the reaction.

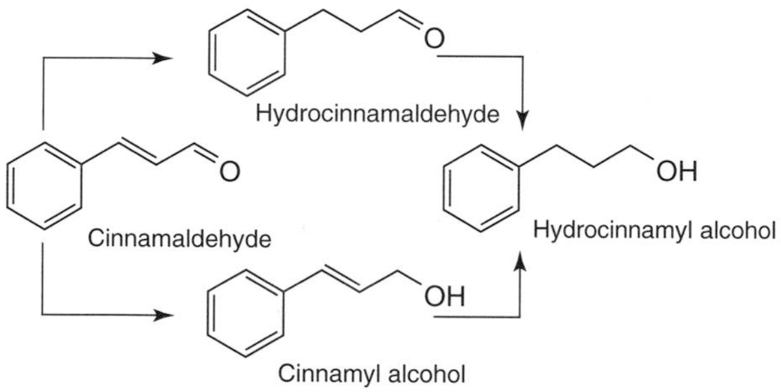

Hydrocinnamaldehyde

Cinnamaldehyde

Hydrocinnamyl alcohol

Cinnamyl alcohol

Scheme 7.9 Reaction pathways for the hydrogenation of cinnamaldehyde.

The catalyst material was prepared by immobilizing palladium particles in either $[C_4MIM][PF_6]$ or $[C_4C_1PY][BF_4]$ on activated carbon. At a dihydrogen pressure between 0.5 and 30 MPa, and temperatures between 60 and 150 °C, the selectivity of the reaction was the same as in the absence of ionic liquid, giving hydrocinnamaldehyde as the main product with a selectivity of around 80%. The conversion was somewhat higher with $[C_4C_1PY][BF_4]$, and lower with $[C_4MIM][PF_6]$, than without ionic liquid, which was explained by the low dihydrogen solubility in the latter ionic liquid [192].

In later studies [193], this system was modified and the hydrogenation of citral, which can give a large number of different products, was investigated (Scheme 7.10).

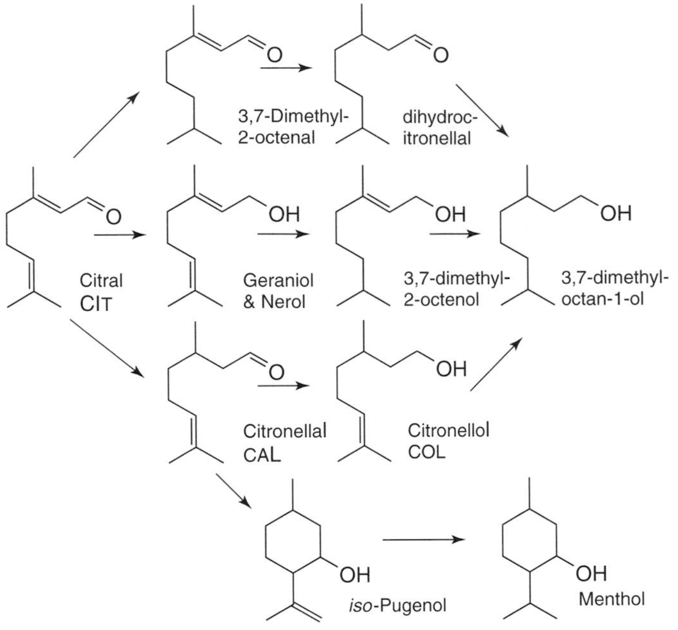

Scheme 7.10 Reaction pathways in the hydrogenation of citral.

Claus and his group extensively investigated the hydrogenation of citral with a SCILL-type catalyst [194–198]. Summarizing their results, a positive effect of the ionic liquid on the selectivity at decreased overall reaction rate was observed. As pointed out above, the effect on the selectivity is explained by a lower solubility of the intermediate products in the ionic liquids, which helps to expel the intermediate from the catalytically active site before further reaction takes place.

The transition metal used as catalyst can also have an effect on the reaction pathway, and hence on the selectivity. For example, in the hydrogenation of citral in ethanol at 70 °C the main product with palladium is citronellal while with nickel it is citronellol (in the absence of an ionic liquid) [199]. Furthermore, the choice of ionic liquid anion and cation were shown to affect the electronic environment. Hence, hydrogenation with a palladium-based SCILL catalyst showed no significant difference in performance in systems based on either $[PF_6]^-$- or $[NTf_2]^-$-based ionic liquids, while an $[N(CN)_2]^-$-based ionic liquid led to an exceptionally high selectivity of citronellal [196]. As for the cation, nitrile-functionalized imidazolium- and pyrrolidinium-based cations also improved the selectivity [197].

If as co-catalyst either the Lewis acid $ZnCl_2$ or the Brønsted acid HBF_4 was added, a co-immobilizing effect of the acids was observed. In the presence of strong acids, the formation of ring products is promoted, and menthol, which is only formed in traces in the absence of acids, becomes the main product at 100 °C and 0.5 MPa dihydrogen pressure [193].

7.5.5
Conclusions: Catalytic Hydrogenation Reactions in Ionic Liquids

In this section, we have summarized the state of the art regarding hydrogenation reactions making use of the unique properties of ionic liquids. Most of the very promising general concepts of catalyst and process engineering were demonstrated. These include:

- biphasic (liquid–liquid) mixtures (even thermomorphic) facilitating the separation of the product from the catalyst-containing ionic liquid phase;
- catalyst recycling and longevity of the active species;
- tunability of reactant solubility due to structural design of the ionic liquid moieties;
- tunability of the selectivity by choice of the ionic liquid (co-catalytic effect);
- increased stereoselectivity and organocatalytic effect of the ionic liquid structure on the product;
- heterogenization of catalytically active metals on a support.

7.6
Concluding Remarks

In this chapter, we have presented four cases of emerging areas in ionic liquid research and development:

1) ionic liquids as acido-basic media;
2) properties and applications of binary mixtures of ionic liquids;
3) the ionothermal synthesis of nanoporous materials;
4) catalytic hydrogenation reactions in and with ionic liquids.

Cases 1–3 have not nearly received enough attention to allow for a knowledge-based prediction of the properties of an imaginary ionic liquid, but gut-feeling and trial-and-error still prevail as strategies when selecting ionic liquids for a given application. Hence, for both conventional and acido-basic ionic liquids, predictive tools and comprehensive databases for bulk properties have just started to appear for selected ionic liquids. Understanding the interactions occurring on the molecular level, and exploiting these in chemical or engineering applications, is still in its infancy but some exciting examples have been presented in this chapter that allow a glimpse at what is still to come.

Case 4, the hydrogenation in and with ionic liquids, was chosen as a prime example of interdisciplinary research and development where ionic liquids have moved within barely 20 years from preliminary results to ready-to-use commercial ionic liquid technology. In this way, several overarching concepts have been identified and exploited, and the near future will show if industrial application will follow.

The enormous number of potential cation–anion combinations certainly is both boon and bane, depending on whether quick processing solutions have to be developed or whether a generous, inspiring research space allows a leisurely exploration.

Acknowledgements

The authors wish to thank the Deutsche Forschungsgemeinschaft (DFG) for funding within the priority programme Ionic Liquids (SPP 1191; STA1027/2-1 to /2-3) and Evonik Stiftung (PhD scholarship for M. M. Azim).

References

1. Wasserscheid, P. and Keim, W. (2000) Ionic liquids – new "solutions" for transition metal catalysis. *Angew. Chem., Int. Ed.*, **39**, 3772–3789.

2. Sheldon, R. (2001) Catalytic reactions in ionic liquids. *Chem. Commun.*, 2399–2407.

3. Dupont, J., de Souza, R.F., and Suarez, P.A.Z. (2002) Ionic liquid (molten salt) phase organometallic catalysis. *Chem. Rev.*, **102**, 3667–3691.

4. Parvulescu, V.I. and Hardacre, C. (2007) Catalysis in ionic liquids. *Chem. Rev.*, **107**, 2615–2665.

5. Beletskaya, I.P. and Kustov, L.M. (2010) Catalysis as an important tool of green chemistry. *Russ. Chem. Rev.*, **79**, 441–461.

6. Van Doorslaer, C., Wahlen, J., Mertens, P., Binnemans, K., and De Vos, D. (2010) Immobilization of molecular catalysts in supported ionic liquid phases. *Dalton Trans.*, **39**, 8377–8390.

7. Olivier-Bourbigou, H., Magna, L., and Morvan, D. (2010) Ionic liquids and catalysis: recent progress from knowledge to applications. *Appl. Catal. Gen.*, **373**, 1–56.

8. Welton, T. and Wasserscheid, P. (eds) (2007) *Ionic Liquids in Synthesis*, Wiley-VCH Verlag GmbH, Weinheim.

9. Kirchner, B. (ed.) (2009) *Ionic Liquids*, Topics in Current Chemistry, vol. 290, Springer, Berlin.

10. Seddon, K.R. (1999) in *The International George Papatheodorou Symposium* (eds S. Boghosian, V. Dracopoulos, C.G. Kontoyannis, and G.A. Voyiatzis), Institute of Chemical Engineering and High Temperature Processes, Patras, pp. 131–135.

11. Stark, A. and Seddon, K.R. (2007) in *Kirk-Othmer Encyclopedia of Chemical Technology*, 5th edn (ed. A. Seidel), John Wiley & Sons, Inc., Hoboken, NJ, pp. 836–920.

12. Welton, T. (2004) Ionic liquids in catalysis. *Coord. Chem. Rev.*, **248**, 2459–2477.

13. Arce, A., Earle, M.J., Katdare, S.P., Rodriguez, H., and Seddon, K.R. (2006) Mutually immiscible ionic liquids. *Chem. Commun.*, 2548–2550.

14. Schulz, P.S., Muller, N., Bosmann, A., and Wasserscheid, P. (2007) Effective chirality transfer in ionic liquids through ion-pairing effects. *Angew. Chem., Int. Ed.*, **46**, 1293–1295.

15. Stark, A. (2011) Ionic liquids in the biorefinery: a critical assessment of their potential. *Energy Environ. Sci.*, **4**, 19–32.

16. Mehnert, C.P., Cook, R.A., Dispenziere, N.C., and Afeworki, M. (2002) Supported ionic liquid catalysis – a new concept for homogeneous hydroformylation catalysis. *J. Am. Chem. Soc.*, **124**, 12932–12933.

17. Riisager, A., Fehrmann, R., Berg, R.W., van Hal, R., and Wasserscheid, P. (2005) Thermomorphic phase separation in ionic liquid-organic liquid systems – conductivity and spectroscopic characterization. *Phys. Chem. Chem. Phys.*, **7**, 3052–3058.

18. Riisager, A., Fehrmann, R., Flicker, S., van Hal, R., Haumann, M., and Wasserscheid, P. (2005) Very stable and highly regioselective supported

ionic-liquid-phase (SILP) catalysis: continuous flow fixed-bed hydroformylation of propene. *Angew. Chem., Int. Ed.*, **44**, 815–819.

19. Kernchen, U., Etzold, B., Korth, W., and Jess, A. (2007) Solid catalyst with ionic liquid layer (SCILL) – a new concept to improve selectivity illustrated by hydrogenation of cyclooctadiene. *Chem. Eng. Technol.*, **30**, 985–994.

20. Virtanen, P., Karhu, H., Kordas, K., and Mikkola, J.P. (2007) The effect of ionic liquid in supported ionic liquid catalysts (SILCA) in the hydrogenation of alpha,beta-unsaturated aldehydes. *Chem. Eng. Sci.*, **62**, 3660–3671.

21. Kralisch, D., Stark, A., Korsten, S., Kreisel, G., and Ondruschka, B. (2005) Energetic, environmental and economic balances: spice up your ionic liquid research efficiency. *Green Chem.*, **7**, 301–309.

22. Ott, D., Kralisch, D., and Stark, A. (2010) in *Handbook of Green Chemistry* (eds P. Wasserscheid and A. Stark), Wiley-VCH Verlag GmbH, Weinheim, pp. 309–334.

23. Deetlefs, M. and Seddon, K.R. (2010) in *Handbook of Green Chemistry* (eds P. Wasserscheid and A. Stark), Wiley-VCH Verlag GmbH, Weinheim, pp. 3–34.

24. Fuller, J., Breda, A.C., and Carlin, R.T. (1997) Ionic liquid-polymer gel electrolytes. *J. Electrochem. Soc.*, **144**, L67–L70.

25. Asakawa, T., Hisamatsu, H., and Miyagishi, S. (1995) Micellar pseudophase separation regions of 1H,1H,2H,2H-perfluoroalkylpyridinium chloride and hydrocarbon surfactants by group-contribution method. *Langmuir*, **11**, 478–482.

26. Domanska, U., Bogel-Lukasik, E., and Bogel-Lukasik, R. (2003) 1-octanol/water partition coefficients of 1-alkyl-3-methylimidazolium chloride. *Chem. – Eur. J.*, **9**, 3033–3041.

27. Belveze, L.S., Brennecke, J.F., and Stadtherr, M.A. (2004) Modeling of activity coefficients of aqueous solutions of quaternary ammonium salts with the electrolyte-NRTL equation. *Ind. Eng. Chem. Res.*, **43**, 815–825.

28. Kim, Y.S., Choi, W.Y., Jang, J.H., Yoo, K.P., and Lee, C.S. (2005) Solubility measurement and prediction of carbon dioxide in ionic liquids. *Fluid Phase Equilib.*, **228**, 439–445.

29. Waliszewski, D., Stepniak, I., Piekarski, H., and Lewandowski, A. (2005) Heat capacities of ionic liquids and their heats of solution in molecular liquids. *Thermochim. Acta*, **433**, 149–152.

30. Krossing, I. and Slattery, J.M. (2006) Semi-empirical methods to predict the physical properties of ionic liquids: an overview of recent developments. *Z. Phys. Chem.*, **220**, 1343–1359.

31. Preiss, U.P.R.M., Slattery, J.M., and Krossing, I. (2009) In silico prediction of molecular volumes, heat capacities, and temperature-dependent densities of ionic liquids. *Ind. Eng. Chem. Res.*, **48**, 2290–2296.

32. Preiss, U., Jungnickel, C., Thoming, J., Krossing, I., Luczak, J., Diedenhofen, M., and Klamt, A. (2009) Predicting the critical micelle concentrations of aqueous solutions of ionic liquids and other ionic surfactants. *Chem. – Eur. J.*, **15**, 8880–8885.

33. Hirao, M., Sugimoto, H., and Ohno, H. (2000) Preparation of novel room-temperature molten salts by neutralization of amines. *J. Electrochem. Soc.*, **147**, 4168–4172.

34. Ohno, H. and Yoshizawa, M. (2002) Ion conductive characteristics of ionic liquids prepared by neutralization of alkylimidazoles. *Solid State Ionics*, **154**, 303–309.

35. Sun, J., Forsyth, M., and MacFarlane, D.R. (1998) Room-temperature molten salts based on the quaternary ammonium ion. *J. Phys. Chem. B*, **102**, 8858–8864.

36. MacFarlane, D.R., Sun, J., Golding, J., Meakin, P., and Forsyth, M. (2000) High conductivity molten salts based on the imide ion. *Electrochim. Acta*, **45**, 1271–1278.

37. MacFarlane, D.R., Meakin, P., Sun, J., Amini, N., and Forsyth, M. (1999) Pyrrolidinium imides: a new family of molten salts and conductive plastic crystal phases. *J. Phys. Chem. B*, **103**, 4164–4170.

38. Bonhote, P., Dias, A.P., Papageorgiou, N., Kalyanasundaram, K., and Gratzel, M. (1996) Hydrophobic, highly conductive ambient-temperature molten salts. *Inorg. Chem.*, **35**, 1168–1178.

39. Ohno, H. (2006) Functional design of ionic liquids. *Bull. Chem. Soc. Jpn.*, **79**, 1665–1680.

40. Hardacre, C., Holbrey, J.D., Nieuwenhuyzen, M., and Youngs, T.G.A. (2007) Structure and solvation in ionic liquids. *Acc. Chem. Res.*, **40**, 1146–1155.

41. Fujii, K., Fujimori, T., Takamuku, T., Kanzaki, R., Umebayashi, Y., and Ishiguro, S.I. (2006) Conformational equilibrium of bis(trifluoromethanesulfonyl) imide anion of a room-temperature ionic liquid: Raman spectroscopic study and DFT calculations. *J. Phys. Chem. B*, **110**, 8179–8183.

42. Kuang, Q.L., Zhang, J., and Wang, Z.G. (2007) Revealing long-range density fluctuations in dialkylimidazolium chloride ionic liquids by dynamic light scattering. *J. Phys. Chem. B*, **111**, 9858–9863.

43. Bankmann, D. and Giernoth, R. (2007) Magnetic resonance spectroscopy in ionic liquids. *Prog. Nucl. Magn. Reson. Spectrosc.*, **51**, 63–90.

44. Del Popolo, M.G. and Voth, G.A. (2004) On the structure and dynamics of ionic liquids. *J. Phys. Chem. B*, **108**, 1744–1752.

45. Bagno, A., D'Amico, F., and Saielli, G. (2007) Computer simulation of diffusion coefficients of the room-temperature ionic liquid [bmim][BF_4]: problems with classical simulation techniques. *J. Mol. Liq.*, **131**, 17–23.

46. Judeinstein, P., Iojoiu, C., Sanchez, J.Y., and Ancian, B. (2008) Proton conducting ionic liquid organization as probed by NMR: self-diffusion coefficients and heteronuclear correlations. *J. Phys. Chem. B*, **112**, 3680–3683.

47. Gutowski, K.E. and Dixon, D.A. (2006) Ab initio prediction of the gas- and solution-phase acidities of strong bronsted acids: the calculation of pK(a) values less than −10. *J. Phys. Chem. A*, **110**, 12044–12054.

48. Lide, D.R. (ed.) (2002–2003) *Handbook of Chemistry and Physics*, Taylor & Francis, Boca Raton.

49. Tobin, J.B., Whitt, S.A., Cassidy, C.S., and Frey, P.A. (1995) Low-barrier hydrogen-bonding in molecular-complexes analogous to histidine and aspartate in the catalytic triad of serine proteases. *Biochemistry*, **34**, 6919–6924.

50. Cassidy, C.S., Reinhardt, L.A., Cleland, W.W., and Frey, P.A. (1999) Hydrogen bonding in complexes of carboxylic acids with 1-alkylimidazoles: steric and isotopic effects on low barrier hydrogen bonding. *J. Chem. Soc., Perkin Trans. 2*, 635–641.

51. Ue, M. (1994) Mobility and ionic association of lithium and quaternary ammonium-salts in propylene carbonate and gamma-butyrolactone. *J. Electrochem. Soc.*, **141**, 3336–3342.

52. Fuller, J., Carlin, R.T., and Osteryoung, R.A. (1997) The room temperature ionic liquid 1-ethyl-3-methylimidazolium tetrafluoroborate: electrochemical couples and physical properties. *J. Electrochem. Soc.*, **144**, 3881–3886.

53. McEwen, A.B., Ngo, H.L., LeCompte, K., and Goldman, J.L. (1999) Electrochemical properties of imidazolium salt electrolytes for electrochemical capacitor applications. *J. Electrochem. Soc.*, **146**, 1687–1695.

54. Belieres, J.P. and Angell, C.A. (2007) Protic ionic liquids: preparation, characterization, and proton free energy level representation. *J. Phys. Chem. B*, **111**, 4926–4937.

55. Susan, M.A.B.H., Noda, A., Mitsushima, S., and Watanabe, M. (2003) Bronsted acid–base ionic liquids and their use as new materials for anhydrous proton conductors. *Chem. Commun.*, 938–939.

56. Noda, A., Hayamizu, K., and Watanabe, M. (2001) Pulsed-gradient spin-echo H-1 and F-19 NMR ionic diffusion coefficient, viscosity, and ionic conductivity of non-chloroaluminate room-temperature ionic liquids. *J. Phys. Chem. B*, **105**, 4603–4610.

57. Wasserscheid, P. (2003) *Ionic Liquids in Synthesis*, John Wiley & Sons, Inc., Hoboken.

58. Tokuda, H., Hayamizu, K., Ishii, K., Susan, M.A.B.H., and Watanabe, M. (2004) Physicochemical properties and structures of room temperature ionic liquids. 1. Variation of anionic species. *J. Phys. Chem. B*, **108**, 16593–16600.

59. Ohno, H. (2005) *Electrochemical Aspects of Ionic Liquids*, John Wiley & Sons, Inc., Hoboken.

60. Bikard, Y., Weibel, J.M., Sirlin, C., Dupuis, L., Loeffler, J.P., and Pale, P. (2007) PdCl$_2$, a useful catalyst for protection of alcohols as diphenylmethyl (DPM) ethers. *Tetrahedron Lett.*, **48**, 8895–8899.

61. Yadav, J.S., Bhunia, D.C., Krishna, K.V., and Srihari, P. (2007) Niobium(V) pentachloride: an efficient catalyst for C-, N-, O-, and S-nucleophilic substitution reactions of benzylic alcohols. *Tetrahedron Lett.*, **48**, 8306–8310.

62. Mezaache, R., Dembele, Y.A., Bikard, Y., Weibel, J.M., Blanc, A., and Pale, P. (2009) Copper(II) bromide as an efficient catalyst for the selective protection and deprotection of alcohols as bis(4-methoxyphenyl)methyl ethers. *Tetrahedron Lett.*, **50**, 7322–7326.

63. Altimari, J.M., Delaney, J.P., Servinis, L., Squire, J.S., Thornton, M.T., Khosa, S.K., Long, B.M., Johnstone, M.D., Fleming, C.L., Pfeffer, F.M., Hickey, S.M., Wride, M.P., Ashton, T.D., Fox, B.L., Byrne, N., and Henderson, L.C. (2012) Rapid formation of diphenyl-methyl ethers and thioethers using microwave irradiation and protic ionic liquids. *Tetrahedron Lett.*, **53**, 2035–2039.

64. Gordon, C.P., Byrne, N., and McCluskey, A. (2010) A facile, protic ionic liquid route to N-substituted 5-hydroxy-4-methyl-3-oxoisoindoline-1-carboxamides and N-substituted 3-oxoisoindoline-4-carboxylic acids. *Green Chem.*, **12**, 1000–1006.

65. Henderson, L.C. and Byrne, N. (2011) Rapid and efficient protic ionic liquid-mediated pinacol rearrangements under microwave irradiation. *Green Chem.*, **13**, 813–816.

66. Thornton, M.T., Henderson, L.C., Byrne, N., and Pfeffer, F.M. (2012) Accessing highly-halogenated flavanones

using protic ionic liquids and microwave irradiation. *Curr. Org. Chem.*, **16**, 121–126.

67. Reitz, D.B. and Finkes, M.J. (1989) Reaction of 2,5-bis(trifluoromethyl)-1,3,4-oxadiazole with primary amines - synthesis of 4-substituted-3,5-bis(trifluoromethyl)-4H-1,2,4-triazoles. *J. Heterocycl. Chem.*, **26**, 225–230.

68. Kakefuda, A., Suzuki, T., Tobe, T., Tahara, A., Sakamoto, S., and Tsukamoto, S. (2002) Discovery of 4,5-diphenyl-1,2,4-triazole derivatives as a novel class of selective antagonists for the human V-1A receptor. *Bioorg. Med. Chem.*, **10**, 1905–1912.

69. Pekhtereva, T.M. and Shvaika, O.P. (2005) Recyclization reactions of 1,3,4-oxadiazoles and bis-1,3,4-oxadiazoles into 1,2,4-triazole derivatives. Synthesis of 5-unsubstituted 1,2,4-triazoles. *Khim. Geterotsikl. Soedin.*, 1026–1032.

70. Moulin, A., Bibian, M., Blayo, A.L., El Habnouni, S., Martinez, J., and Fehrentz, J.A. (2010) Synthesis of 3,4,5-trisubstituted-1,2,4-triazoles. *Chem. Rev.*, **110**, 1809–1827.

71. Chen, X.F., Liu, R., Xu, Y., and Zou, G. (2012) Tunable protic ionic liquids as solvent-catalysts for improved synthesis of multiply substituted 1,2,4-triazoles from oxadiazoles and organoamines. *Tetrahedron*, **68**, 4813–4819.

72. Larock, R.C. (ed.) (1999) *Comprehensive Organic Transformations*, Wiley-VCH Verlag GmbH, Weinheim.

73. Wuts, P.G.M. and Greene, T.W. (1999) *Green's Protective Groups in Organic Synthesis*, John Wiley & Sons, Inc., New York.

74. Hino, M. and Arata, K. (1985) Solid catalysts treated with anions. 13. Synthesis of esters from terephthalic and phthalic acids with N-octyl and 2-ethylhexyl alcohol, acrylic-acid with ethanol and salicylic-acid with methanol catalyzed by solid superacid. *Appl. Catal.*, **18**, 401–404.

75. Maki-Arvela, P., Salmi, T., Sundell, M., Ekman, K., Peltonen, R., and Lehtonen, J. (1999) Comparison of polyvinyl-benzene and polyolefin supported

sulphonic acid catalysts in the esterification of acetic acid. *Appl. Catal. Gen.*, **184**, 25–32.

76. Joseph, T., Sahoo, S., and Halligudi, S.B. (2005) Bronsted acidic ionic liquids: a green, efficient and reusable catalyst system and reaction medium for Fischer esterification. *J. Mol. Catal. Chem.*, **234**, 107–110.

77. Ganeshpure, P.A. and Das, J. (2007) Application of high-melting pyridinium salts as ionic liquid catalysts and media for Fischer esterification. *React. Kinet. Catal. Lett.*, **92**, 69–74.

78. Ganeshpure, P.A., George, G., and Das, J. (2008) Bronsted acidic ionic liquids derived from alkylamines as catalysts and mediums for Fischer esterification: study of structure-activity relationship. *J. Mol. Catal. A: Chem.*, **279**, 182–186.

79. Atkins, P.W. (2006) *Physikalische Chemie*, Wiley-VCH Verlag GmbH, Weinheim.

80. Saling, P., Maase, M., and Vagt, U. (2010) in *Handbook of Green Chemistry* (eds P. Wasserscheid and A. Stark), Wiley-VCH Verlag GmbH, Weinheim, pp. 291–308.

81. Schmid, C.R., Beck, C.A., Cronin, J.S., and Staszak, M.A. (2004) Demethylation of 4-methoxyphenylbutyric acid using molten pyridinium hydrochloride on multikilogram scale. *Org. Process Res. Dev.*, **8**, 670–673.

82. Lee, S.Y., Ogawa, A., Kanno, M., Nakamoto, H., Yasuda, T., and Watanabe, M. (2010) Nonhumidified intermediate temperature fuel cells using protic ionic liquids. *J. Am. Chem. Soc.*, **132**, 9764–9773.

83. Caires, M.I., Buzzo, M.L., Ticianelli, E.A., and Gonzalez, E.R. (1997) Preparation and characterization of matrices for phosphoric acid fuel cells. *J. Appl. Electrochem.*, **27**, 19–24.

84. Song, R.H., Dheenadayalan, S., and Shin, D.R. (2002) Effect of silicon carbide particle size in the electrolyte matrix on the performance of a phosphoric acid fuel cell. *J. Power Sources*, **106**, 167–172.

85. Yoon, K.H. and Yang, B.D. (2003) Preparation and characterization of matrix retaining electrolyte for a phosphoric acid fuel cell by non-volatile solvent, NMP. *J. Power. Sources*, **124**, 47–51.

86. Wainright, J.S., Wang, J.T., Weng, D., Savinell, R.F., and Litt, M. (1995) Acid-doped polybenzimidazoles – a new polymer electrolyte. *J. Electrochem. Soc.*, **142**, L121–L123.

87. Wang, J.T., Savinell, R.F., Wainright, J., Litt, M., and Yu, H. (1996) A H-2/O-2 fuel cell using acid doped polybenzimidazole as polymer electrolyte. *Electrochim. Acta*, **41**, 193–197.

88. Kreuer, K.D., Fuchs, A., Ise, M., Spaeth, M., and Maier, J. (1998) Imidazole and pyrazole-based proton conducting polymers and liquids. *Electrochim. Acta*, **43**, 1281–1288.

89. Glipa, X., Bonnet, B., Mula, B., Jones, D.J., and Roziere, J. (1999) Investigation of the conduction properties of phosphoric and sulfuric acid doped polybenzimidazole. *J. Mater. Chem.*, **9**, 3045–3049.

90. Bouchet, R. and Siebert, E. (1999) Proton conduction in acid doped polybenzimidazole. *Solid State Ion.*, **118**, 287–299.

91. Kawahara, M., Rikukawa, M., Sanui, K., and Ogata, N. (2000) Synthesis and proton conductivity of sulfopropylated poly(benzimidazole) films. *Solid State Ion.*, **136**, 1193–1196.

92. Jin, Y.C., Fujiwara, K., and Hibino, T. (2010) High temperature, low humidity proton exchange membrane based on an inorganic–organic hybrid structure. *Electrochem. Solid State Lett.*, **13**, B8–B10.

93. Matsumoto, H., Sakaebe, H., Tatsumi, K., Kikuta, M., Ishiko, E., and Kono, M. (2006) Fast cycling of Li/LiCoO$_2$ cell with low-viscosity ionic liquids based on bis(fluorosulfonyl)imide [FSI](−). *J. Power. Sources*, **160**, 1308–1313.

94. Seki, S., Kobayashi, Y., Miyashiro, H., Ohno, Y., Usami, A., Mita, Y., Watanabe, M., and Terada, N. (2006) Highly reversible lithium metal secondary battery using a room temperature ionic liquid/lithium salt mixture and a surface-coated cathode active material. *Chem. Commun.*, 544–545.

95. Nanjundiah, C., McDevitt, S.F., and Koch, V.R. (1997) Differential capacitance measurements in solvent-free ionic liquids at Hg and C interfaces. *J. Electrochem. Soc.*, **144**, 3392–3397.

96. Barisci, J.N., Wallace, G.G., MacFarlane, D.R., and Baughman, R.H. (2004) Investigation of ionic liquids as electrolytes for carbon nanotube electrodes. *Electrochem. Commun.*, **6**, 22–27.

97. Lu, W., Fadeev, A.G., Qi, B.H., Smela, E., Mattes, B.R., Ding, J., Spinks, G.M., Mazurkiewicz, J., Zhou, D.Z., Wallace, G.G., MacFarlane, D.R., Forsyth, S.A., and Forsyth, M. (2002) Use of ionic liquids for pi-conjugated polymer electrochemical devices. *Science*, **297**, 983–987.

98. Fukushima, T., Asaka, K., Kosaka, A., and Aida, T. (2005) Fully plastic actuator through layer-by-layer casting with ionic-liquid-based bucky gel. *Angew. Chem., Int. Ed.*, **44**, 2410–2413.

99. Oregan, B. and Gratzel, M. (1991) A low-cost, high-efficiency solar-cell based on dye-sensitized colloidal TiO_2 films. *Nature*, **353**, 737–740.

100. Papageorgiou, N., Athanassov, Y., Armand, M., Bonhote, P., Pettersson, H., Azam, A., and Gratzel, M. (1996) The performance and stability of ambient temperature molten salts for solar cell applications. *J. Electrochem. Soc.*, **143**, 3099–3108.

101. Kawano, R. and Watanabe, M. (2003) Equilibrium potentials and charge transport of an I-/I-3(−) redox couple in an ionic liquid. *Chem. Commun.*, 330–331.

102. Matsui, H., Okada, K., Kawashima, T., Ezure, T., Tanabe, N., Kawano, R., and Watanabe, M. (2004) Application of an ionic liquid-based electrolyte to a 100 mm x 100 mm sized dye-sensitized solar cell. *J. Photochem. Photobiol. Chem.*, **164**, 129–135.

103. Hagiwara, R., Nohira, T., Matsumoto, K., and Tamba, Y. (2005) A fluorohydrogenate ionic liquid fuel cell operating without humidification. *Electrochem. Solid State Lett.*, **8**, A231–A233.

104. Lee, J.S., Nohira, T., and Hagiwara, R. (2007) Novel composite electrolyte membranes consisting of fluorohydrogenate ionic liquid and polymers for the unhumidified intermediate temperature fuel cell. *J. Power. Sources*, **171**, 535–539.

105. Bayley, P.M., Lane, G.H., Nathalie, M.R., Bronya, R.C., Adam, S.B., Douglas, R.M., and Maria, F. (2009) Transport properties of ionic liquid electrolytes with organic diluents. *Phys. Chem. Chem. Phys.*, **11**, 7202–7208.

106. Lopes, J.N.C., Gomes, M.F.C., Husson, P., Padua, A.A.H., and Rebelo, L.P. (2011) Polarity, viscosity and ionic conductivity of liquid mixtures containing $[C_4C_1im][Ntf_2]$ and a molecular component. *Phys. Chem. B*, **115**, 6088–6099.

107. Stoppa, A., Buchner, R., and Hefter, G. (2010) How ideal are binary mixtures of room-temperature ionic liquids? *J. Mol. Liq.*, **153**, 46–51.

108. Aparicio, S. and Atilhan, M. (2012) Mixed ionic liquids: the case of pyridinium-based fluids. *J. Phys. Chem. B*, **116**, 2526–2537.

109. Larriba, M., Garcia, S., Navarro, P., Garcia, J., and Rodriguez, F. (2012) Physical properties of n-butylpyridinium tetrafluoroborate and n-butylpyridinium bis(trifluoromethylsulfonyl)imide binary ionic liquid mixtures. *J. Chem. Eng. Data*, **57**, 1318–1325.

110. Fox, T.F., Joshua, E.F., Weaver, J.E., and Henderson, W.A. (2012) Tuning binary ionic liquid mixtures: linking alkyl chain length to phase behavior and ionic conductivity. *J. Phys. Chem. C*, **116**, 5270–5274.

111. Brüssel, M., Brehm, M., Pensado, A.S., Malberg, F., Ramzan, M., Stark, A., and Kirchner, B. (2012) On the ideality of binary mixtures of ionic liquids. *Phys. Chem. Chem. Phys.*, **14**, 13204–13215.

112. Ning, H., Hou, M.Q., Mei, Q.Q., Liu, Y.H., Yang, D.Z., and Han, B.X. (2012) The physicochemical properties of some imidazolium-based ionic liquids and their binary mixtures. *Sci. Chin. Chem.*, **55**, 1509–1518.

113. Navia, P., Troncoso, J., and Romani, L. (2007) Excess magnitudes for ionic liquid binary mixtures with a common ion. *J. Chem. Eng. Data*, **52**, 1369–1374.

114. Gozzo, F.C., Santos, L.S., Augusti, R., Consorti, C.S., Dupont, J., and Eberlin, M.N. (2004) Gaseous supramolecules of imidazolium ionic liquids: "magic" numbers and intrinsic strengths of hydrogen bonds. *Chem. – Eur. J.*, **10**, 6187–6193.

115. Andanson, J.M., Beier, M.J., and Baiker, A. (2011) Binary ionic liquids with a common cation: insight into nanoscopic mixing by infrared spectroscopy. *J. Phys. Chem. Lett.*, **2**, 2959–2964.

116. Fletcher, K.A., Baker, S.N., Baker, G.A., and Pandey, S. (2003) Probing solute and solvent interactions within binary ionic liquid mixtures. *New J. Chem.*, **27**, 1706–1712.

117. Brüssel, M., Brehm, M., Voigt, T., and Kirchner, B. (2011) Ab initio molecular dynamics simulations of a binary system of ionic liquids. *Phys. Chem. Chem. Phys.*, **13**, 13617–13620.

118. Stark, A., Sellin, M., Ondruschka, B., and Massonne, K. (2012) The effect of hydrogen bond acceptor properties of ionic liquids on their cellulose solubility. *Sci. Chin. Chem.*, **55**, 1663–1670.

119. Bini, R., Chiappe, C., Marmugi, E., and Pieraccini, D. (2006) The "non-nucleophilic" anion [Tf$_2$N](−) competes with the nucleophilic Br-: an unexpected trapping in the dediazoniation reaction in ionic liquids. *Chem. Commun.*, 897–899.

120. Swatloski, R.P., Spear, S.K., Holbrey, J.D., and Rogers, R.D. (2002) Dissolution of cellulose with ionic liquids. *J. Am. Chem. Soc.*, **124**, 4974–4975.

121. Wasserscheid, P. and Stark, A. (2010) *Cellulose Dissolution and Processing with Ionic Liquids*, Wiley-VCH Verlag GmbH, Weinheim.

122. Sellin, M., Ondruschka, B., and Stark, A. (2010) Hydrogen bond acceptor properties of ionic liquids and their effect on cellulose solubility, in *Cellulose Solvents: For Analysis, Shaping and Chemical Modification*, ACS Symposium Series, Vol. **1033**, American Chemical Society, Washington, DC.

123. Weingärtner, H. (2008) Understanding ionic liquids at the molecular level: facts, problems, and controversies. *Angew. Chem., Int. Ed.*, **47**, 654–670.

124. Lungwitz, R. and Spange, S. (2008) A hydrogen bond accepting (HBA) scale for anions, including room temperature ionic liquids. *New J. Chem.*, **32**, 392–394.

125. Lungwitz, R., Friedrich, M., Linert, W., and Spange, S. (2008) New aspects on the hydrogen bond donor (HBD) strength of 1-butyl-3-methylimidazolium room temperature ionic liquids. *New J. Chem.*, **32**, 1493–1499.

126. Davis, M.E. (2002) Ordered porous materials for emerging applications. *Nature*, **417**, 813–821.

127. Corma, A. (1997) From microporous to mesoporous molecular sieve materials and their use in catalysis. *Chem. Rev.*, **97**, 2373–2419.

128. International Zeolite Association (IZA-SC) (2007) Database of Zeolite Structures. *http://www.iza-structure.org/databases/* (accessed 12 October 2012).

129. Morris, R.E. and Weigel, S.J. (1997) The synthesis of molecular sieves from non-aqueous solvents. *Chem. Soc. Rev.*, **26**, 309–317.

130. Gomes, M.F.C., Lopes, J.N.C., and Padua, A.A.H. (2010) in *Ionic Liquids*, Topics in Current Chemistry (ed. B. Kirchner), Springer, Berlin, pp. 161–183.

131. Cooper, R., Andrews, C.D., Wheatley, P.S., Webb, P.B., Wormald, P.W., and Morris, R.E. (2004) Ionic liquids and eutectic mixtures as solvent and template in synthesis of zeolite analogues. *Nature*, **430**, 1012–1016.

132. Morris, R.E. (2009) Ionothermal synthesis-ionic liquids as functional solvents in the preparation of crystalline materials. *Chem. Commun.*, 2990–2998.

133. Parnham, E.R., Wheatley, P.S., and Morris, R.E. (2006) The ionothermal synthesis of SIZ-6 – a layered aluminophosphate. *Chem. Commun.*, 380–382.

134. Xu, Y.P., Tian, Z.J., Xu, Z.S., Wang, B.C., Li, P., Wang, S.J., Hu, Y., Ma, Y.C., Li, K.L., Liu, Y.J., Yu, J.Y., and Lin, L.W. (2005) Ionothermal synthesis of silicoaluminophosphate molecular sieve in N-alkyl imidazolium bromide. *Chin. J. Catal.*, **26**, 446–448.

135. Han, L.J., Wang, Y.B., Li, C.X., Zhang, S.J., Lu, X.M., and Cao, M.J. (2008) Simple and safe synthesis of microporous aluminophosphate molecular sieves by ionothermal approach. *AIChE J.*, **54**, 280–288.

136. Han, L.J., Wang, Y.B., Zhang, S.J., and Lu, X.M. (2008) Ionothermal synthesis of microporous aluminum and gallium phosphates. *J. Cryst. Growth*, **311**, 167–171.

137. Fayad, E.J., Bats, N., Kirschhock, C.E.A., Rebours, B., Quoineaud, A.A., and Martens, J.A. (2010) A rational approach to the ionothermal synthesis of an AlPO4 molecular sieve with an LTA-type framework. *Angew. Chem., Int. Ed.*, **49**, 4585–4588.

138. Wragg, D.S., Slawin, A.M.Z., and Morris, R.E. (2009) The role of added water in the ionothermal synthesis of microporous aluminium phosphates. *Solid State Sci.*, **11**, 411–416.

139. Wragg, D.S., Fullerton, G.M., Byrne, P.J., Slawin, A.M.Z., Warren, J.E., Teat, S.J., and Morris, R.E. (2011) Solvothermal aluminophosphate zeotype synthesis with ionic liquid precursors. *Dalton Trans.*, **40**, 4926–4932.

140. Azim, M.M. (2011) Synthesis of porous solids using ionic liquids, Master Thesis, Institute of Chemical Technology, University of Leipzig.

141. Cai, R., Sun, M.W., Chen, Z.W., Munoz, R., O'Neill, C., Beving, D.E., and Yan, Y.S. (2008) Ionothermal synthesis of oriented zeolite AEL films and their application as corrosion-resistant coatings. *Angew. Chem., Int. Ed.*, **47**, 525–528.

142. Wang, S., Hou, L., Xu, Y., Tian, Z., Yu, J., and Lin, L. (2008) Effect of Al-containing precursors on ionothermal synthesis of aluminophosphate molecular sieve. *Chin. J. Process Eng.*, **8**, 93–96.

143. Leadbeater, N.E. and Torenius, H.M. (2002) A study of the ionic liquid mediated microwave heating of organic solvents. *J. Org. Chem.*, **67**, 3145–3148.

144. Hoffmann, J., Nuchter, M., Ondruschka, B., and Wasserscheid, P. (2003) Ionic liquids and their heating behaviour during microwave irradiation – a state of the art report and challenge to assessment. *Green Chem.*, **5**, 296–299.

145. Xu, Y.P., Tian, Z.J., Wang, S.J., Hu, Y., Wang, L., Wang, B.C., Ma, Y.C., Hou, L., Yu, J.Y., and Lin, L.W. (2006) Microwave-enhanced ionothermal synthesis of aluminophosphate molecular sieves. *Angew. Chem., Int. Ed.*, **45**, 3965–3970.

146. Pei, R.Y., Wei, Y., Li, K.D., Wen, G.D., Xu, R.S., Xu, Y.P., Wang, L., Ma, H.J., Wang, B.C., Tian, Z.J., Zhang, W.P., and Lin, L.W. (2010) Mixed template effect adjusted by amine concentration in ionothermal synthesis of molecular sieves. *Dalton Trans.*, **39**, 1441–1443.

147. Wang, L., Xu, Y.P., Wei, Y., Duan, J.C., Chen, A.B., Wang, B.C., Ma, H.J., Tian, Z.J., and Lin, L.W. (2006) Structure-directing role of amines in the ionothermal synthesis. *J. Am. Chem. Soc.*, **128**, 7432–7433.

148. Kirschhock, C., Feijen, E., Jacobs, P., and Martens, J. (2008) Hydrothermal zeolite synthesis, in *Handbook of Heterogeneous Catalysis* (eds G. Ertl, H. Knözinger, F. Schütz, and J. Weitkamp), Wiley-VCH Verlag GmbH, Weinheim.

149. Ma, H.J., Tian, Z.J., Xu, R.S., Wang, B.C., Wei, Y., Wang, L., Xu, Y.P., Zhang, W.P., and Lin, L.W. (2008) Effect of water on the ionothermal synthesis of molecular sieves. *J. Am. Chem. Soc.*, **130**, 8120–8121.

150. Parnham, E.R. and Morris, R.E. (2006) 1-Alkyl-3-methyl imidazolium bromide ionic liquids in the ionothermal synthesis of aluminium phosphate molecular sieves. *Chem. Mater.*, **18**, 4882–4887.

151. Parnham, E.R. and Morris, R.E. (2006) Ionothermal synthesis using a hydrophobic ionic liquid as solvent in the preparation of a novel aluminophosphate chain structure. *J. Mater. Chem.*, **16**, 3682–3684.

152. Xu, R.S., Zhang, W.P., Guan, J., Xu, Y.P., Wang, L., Ma, H.J., Tian, Z.J., Han, X.W., Lin, L.W., and Bao, X.H. (2009) New insights into the role of amines in the synthesis of molecular sieves in ionic liquids. *Chem. – Eur. J.*, **15**, 5348–5354.

153. Xu, R.S., Shi, X.C., Zhang, W.P., Xu, Y.P., Tian, Z.J., Lu, X.B., Han, X.W., and Bao, X.H. (2010) Cooperative structure-directing effect in the synthesis of aluminophosphate molecular sieves in ionic liquids. *Phys. Chem. Chem. Phys.*, **12**, 2443–2449.

154. Wragg, D.S., Le Ouay, B., Beale, A.M., O'Brien, M.G., Slawin, A.M.Z., Warren, J.E., Prior, T.J., and Morris, R.E. (2010) Ionothermal synthesis and crystal structures of metal phosphate chains. *J. Solid State Chem.*, **183**, 1625–1631.

155. Pei, R.Y., Tian, Z.J., Wei, Y., Li, K.D., Xu, Y.P., Wang, L., and Ma, H.J. (2010) Ionothermal synthesis of AlPO$_4$-34 molecular sieves using heterocyclic aromatic amine as the structure directing agent. *Mater. Lett.*, **64**, 2384–2387.

156. Pei, R.Y., Tian, Z.J., Wei, Y., Li, K.D., Xu, Y.P., Wang, L., and Ma, H.J. (2010) Ionothermal synthesis of AlPO$_4$ molecular sieves in the presence of quaternary ammonium cation. *Mater. Lett.*, **64**, 2118–2121.

157. Parnham, E.R. and Morris, R.E. (2006) The ionothermal synthesis of cobalt aluminophosphate zeolite frameworks. *J. Am. Chem. Soc.*, **128**, 2204–2205.

158. Wang, L., Xu, Y.P., Wang, B.C., Wang, S.J., Yu, J.Y., Tian, Z.J., and Lin, L.W. (2008) Ionothermal synthesis of magnesium-containing aluminophosphate molecular sieves and their catalytic performance. *Chem. – Eur. J.*, **14**, 10551–10555.

159. Ng, E.P., Sekhon, S.S., and Mintova, S. (2009) Discrete MnAlPO-5 nanocrystals synthesized by an ionothermal approach. *Chem. Commun.*, **1661–1663**.

160. Wheatley, P.S., Allan, P.K., Teat, S.J., Ashbrook, S.E., and Morris, R.E. (2010) Task specific ionic liquids for the ionothermal synthesis of siliceous zeolites. *Chem. Sci.*, **1**, 483–487.

161. Chauvin, Y., Mussmann, L., and Olivier, H. (1995) A novel class of versatile solvents for two-phase catalysis: hydrogenation, isomerization, and hydroformylation of alkenes catalyzed by rhodium complexes in liquid 1,3-dialkylimidazolium salts. *Angew. Chem., Int. Ed.*, **34**, 2698–2700.

162. Suarez, P.A.Z., Dullius, J.E.L., Einloft, S., de Souza, R.F., and Dupont, J. (1996) The use of new ionic liquids in two-phase catalytic hydrogenation reaction by rhodium complexes. *Polyhedron*, **15**, 1217–1219.

163. Geldbach, T.J. and Dyson, P.J. (2005) Searching for molecular arene hydrogenation catalysis in ionic liquids. *J. Organomet. Chem.*, **690**, 3552–3557.

164. Daguenet, C. and Dyson, P.J. (2006) Switching the mechanism of catalyst activation by ionic liquids. *Organometallics*, **25**, 5811–5816.

165. Favre, F., Olivier-Bourbigou, H., Commereuc, D., and Saussine, L. (2001) Hydroformylation of 1-hexene with rhodium in non-aqueous ionic liquids: how to design the solvent and the ligand to the reaction. *Chem. Commun.*, 1360–1361.

166. Stark, A., Ajam, M., Green, M., Raubenheimer, H.G., Ranwell, A., and Ondruschka, B. (2006) Metathesis of 1-octene in ionic liquids and other solvents: effects of substrate solubility, solvent polarity and impurities. *Adv. Synth. Catal.*, **348**, 1934–1941.

167. Stark, A. (2009) Ionic liquid structure-induced effects on organic reactions. *Ionic Liq.*, **290**, 41–81.

168. Suarez, T., Fontal, B., Reyes, M., Bellandi, F., Contreras, R.R., Ortega, J.M., Leon, G., Cancines, P., and Castillo, B. (2004) Catalytic hydrogenation of 1-hexene with RuCl$_2$(TPPMS)$_3$(DMSO). Part II: ionic liquid biphasic system. *React. Kinet. Catal. Lett.*, **82**, 325–331.

169. Stark, A., Behrend, P., Braun, O., Muller, A., Ranke, J., Ondruschka, B., and Jastorff, B. (2008) Purity specification methods for ionic liquids. *Green Chem.*, **10**, 1152–1161.

170. Fonseca, G.S., Domingos, J.B., Nome, F., and Dupont, J. (2006) On the kinetics of iridium nanoparticles formation in ionic liquids and olefin hydrogenation. *J. Mol. Catal. A: Chem.*, **248**, 10–16.

171. Widegren, J.A. and Finke, R.G. (2003) A review of soluble transition-metal nanoclusters as arene hydrogenation

catalysts. *J. Mol. Catal. A: Chem.*, **191**, 187–207.

172. Huang, J., Jiang, T., Han, B.X., Gao, H.X., Chang, Y.H., Zhao, G.Y., and Wu, W.Z. (2003) Hydrogenation of olefins using ligand-stabilized palladium nanoparticles in an ionic liquid. *Chem. Commun.*, 1654–1655.

173. Wasserscheid, P., Bosmann, A., and Bolm, C. (2002) Synthesis and properties of ionic liquids derived from the 'chiral pool'. *Chem. Commun.*, 200–201.

174. Pegot, B., Vo-Thanh, G., Gori, D., and Loupy, A. (2004) First application of chiral ionic liquids in asymmetric Baylis-Hillman reaction. *Tetrahedron Lett.*, **45**, 6425–6428.

175. Earle, M.J., McCormac, P.B., and Seddon, K.R. (1999) Diels-Alder reactions in ionic liquids – a safe recyclable alternative to lithium perchlorate-diethyl ether mixtures. *Green Chem.*, **1**, 23–25.

176. Machado, M.Y. and Dorta, R. (2005) Synthesis and characterization of chiral imidazolium salts. *Synthesis*, **15**, 2473–2475.

177. Gausepohl, R., Buskens, P., Kleinen, J., Bruckmann, A., Lehmann, C.W., Klankermayer, J., and Leitner, W. (2006) Highly enantioselective aza-Baylis-Hillman reaction in a chiral reaction medium. *Angew. Chem., Int. Ed.*, **45**, 3689–3692.

178. Ngo, H.L., Hu, A.G., and Lin, W.B. (2005) Catalytic asymmetric hydrogenation of aromatic ketones in room temperature ionic liquids. *Tetrahedron Lett.*, **46**, 595–597.

179. Geldbach, T.J. and Dyson, P.J. (2004) A versatile ruthenium precursor for biphasic catalysis and its application in ionic liquid biphasic transfer hydrogenation: conventional vs task-specific catalysts. *J. Am. Chem. Soc.*, **126**, 8114–8115.

180. Berthod, M., Joerger, J.M., Mignani, G., Vaultier, M., and Lemaire, M. (2004) Enantioselective catalytic asymmetric hydrogenation of ethyl acetoacetate in room temperature ionic liquids. *Tetrahedron: Asymmetry*, **15**, 2219–2221.

181. Feng, X.D., Pugin, B., Kuesters, E., Sedelmeier, G., and Blaser, H.U. (2007) Josiphos ligands with an imidazolium tag and their application for the enantioselective hydrogenation in ionic liquids. *Adv. Synth. Catal.*, **349**, 1803–1807.

182. Gavrilov, K.N., Lyubimov, S.E., Bondarev, O.G., Maksimova, M.G., Zheglov, S.V., Petrovskii, P.V., Davankov, V.A., and Reetz, M.T. (2007) Chiral ionic phosphites and diamidophosphites: a novel group of efficient ligands for asymmetric catalysis. *Adv. Synth. Catal.*, **349**, 609–616.

183. Zhou, Z.Q., Sun, Y., and Zhang, A.Q. (2011) Asymmetric transfer hydrogenation of prochiral ketones catalyzed by aminosulfonamide-ruthenium complexes in ionic liquid. *Cent. Eur. J. Chem.*, **9**, 175–179.

184. Zhao, Y.W., Huang, H.M., Shao, J.P., and Xia, C.G. (2011) Readily available and recoverable chiral ionic phosphite ligands for the highly enantioselective hydrogenation of functionalized olefins. *Tetrahedron: Asymmetry*, **22**, 769–774.

185. Sano, S., Beier, M.J., Mallat, T., and Baiker, A. (2012) Potential of ionic liquids as co-modifiers in asymmetric hydrogenation on platinum. *J. Mol. Catal. A: Chem.*, **357**, 117–124.

186. Dyson, P.J. and Geldbach, T. (2005) *Metal Catalysed Reaction in Ionic Liquid: Catalysis by Metal Complexes*, Springer, Berlin.

187. Daguenet, C. and Dyson, P.J. (2004) Inhibition of catalytic activity in ionic liquids: implications for catalyst design and the effect of cosolvents. *Organometallics*, **23**, 6080–6083.

188. Zeng, Y., Wang, Y.H., Xu, Y.C., Song, Y., Zhao, J.Q., Jiang, J.Y., and Jin, Z.L. (2012) Rh nanoparticles catalyzed hydroformylation of olefins in a thermoregulated ionic liquid/organic biphasic system. *Chin. J. Catal.*, **33**, 402–406.

189. Riisager, A., Fehrmann, R., Haumann, M., Gorle, B.S.K., and Wasserscheid, P. (2005) Stability and kinetic studies of supported ionic liquid phase catalysts for hydroformylation of propene. *Ind. Eng. Chem. Res.*, **44**, 9853–9859.

190. Lou, L.L., Dong, Y.L., Yu, K., Jiang, S., Song, Y., Cao, S., and Liu, S.X. (2010) Chiral Ru complex immobilized on mesoporous materials by ionic liquids

as heterogeneous catalysts for hydrogenation of aromatic ketones. *J. Mol. Catal. A: Chem.*, **333**, 20–27.

191. Virtanen, P., Salmi, T., and Mikkola, J.P. (2009) Kinetics of cinnamaldehyde hydrogenation by supported ionic liquid catalysts (SILCA). *Ind. Eng. Chem. Res.*, **48**, 10335–10342.

192. Berger, A., de Souza, R.F., Delgado, M.R., and Dupont, J. (2001) Ionic liquid-phase asymmetric catalytic hydrogenation: hydrogen concentration effects on enantioselectivity. *Tetrahedron: Asymmetry*, **12**, 1825–1828.

193. Virtanen, P., Salmi, T.O., and Mikkola, J.P. (2010) Supported ionic liquid catalysts (SILCA) for preparation of organic chemicals. *Top. Catal.*, **53**, 1096–1103.

194. Steffan, M. (2008) Heterogen katalysierte selektivhydrierung von citral in der flüssigphase, PhD thesis, Technischen Universität Darmstadt, Darmstadt.

195. Arras, J., Steffan, M., Shayeghi, Y., and Claus, P. (2008) The promoting effect of a dicyanamide based ionic liquid in the selective hydrogenation of citral. *Chem. Commun.*, 4058–4060.

196. Arras, J., Steffan, M., Shayeghi, Y., Ruppert, D., and Claus, P. (2009) Regioselective catalytic hydrogenation of citral with ionic liquids as reaction modifiers. *Green Chem.*, **11**, 716–723.

197. Arras, J., Ruppert, D., and Claus, P. (2009) Influence of ionic fluidity with functionalized cations in the palladium catalysed liquid phase of citral. *Chem. Ingen. Tech.*, **81**, 2007–2011.

198. Arras, J. (2010) Wirkungsweise ionischer flüssigkeiten auf Metall/Trägerkatalysatoren für die hydrierung von citral, PhD thesis, Technische Universität Darmstadt, Darmstadt.

199. Steffan, M., Lucas, M., Brandner, A., Wollny, M., Oldenburg, N., and Claus, P. (2006) Selective hydration of citral in organic solvents, in ionic fluid and in substance. *Chem.-Ing.-Tech.*, **78**, 923–929.

8
Gelling of Plant Based Proteins

Navam Hettiarachchy, Arvind Kannan, Christian Schäfer, and Gerhard Wagner

8.1
Introduction – Overview of Plant Proteins in Industry

The major storage proteins present in cereals, legumes, oilseeds, and vegetables having gelation properties are reviewed in this chapter. Table 8.1 lists major cereal plants and their storage proteins having gelation properties. Table 8.2 lists major leguminous/oilseed plants and their storage proteins having gelation properties. Table 8.3 lists vegetable plants and their storage proteins having gelation properties.

Among cereal plants the main storage proteins present are the glutelins of rice and wheat; prolamins of corn, sorghum, and barley; and the secalins and avenins of rye and oat respectively (Table 8.1). The glutelins and prolamins fall under different classes identified by their varying molecular sizes, and hence exhibit different structure–function properties, including gelation. The legumes and oilseed plants consist mostly of globulins as storage proteins (Table 8.2). These globulins are found in soybean, sesame, and sunflower, different types of peas, lentils, and cottonseed. Of these, soy bean proteins and their major storage fractions [conglycinin (7S) and glycinin (11S)] have been studied widely for functional gelation properties, particularly the glycoprotein β-conglycinin fraction. Lupin proteins (conglutin-storage proteins), on the other hand, represent a balance in essential amino acids and are being considered as a replacement for soy proteins in food formulations. The storage globulins present in the pea family and vegetable plants (Table 8.3) have generally been subjected to modifications involving either hydrolysis to obtain protein isolates or chemical modifications, and the fractions thus obtained have been shown to have good functional properties. Most of these storage proteins are able to participate in various reactions, including intramolecular disulfide bonding and other modification reactions such as deamidation due to the presence of specific amino acids such as cysteine and glutamine in their storage fractions, and hence can form gels of various structural integrities needed in various food formulations in industry. Since the structures of proteins play a major role in gelling reactions, the structure and formation of gels, factors influencing formation of gels, and determinations for studying gels are introduced in the following sections, followed by a review of the gelling properties and characteristics of formed gels for each of

Product Design and Engineering: Formulation of Gels and Pastes, First Edition.
Edited by Ulrich Bröckel, Willi Meier, and Gerhard Wagner.
© 2013 Wiley-VCH Verlag GmbH & Co. KGaA. Published 2013 by Wiley-VCH Verlag GmbH & Co. KGaA.

Table 8.1 Cereal proteins in gelation.

Plant	Botanical name	Main storage proteins	Application	Reference
Wheat	*Triticum* sp.	Gluten	Used to alter textural properties of mixed gels	[24, 25]
Rice	*Oryza sativa* L.	Glutelin	Can improve pasting and structural properties of rice. Applied as gels in several products	[35, 36][a]
Maize/corn	*Zea mays* L.	Zein	Gelling capacity and industrial application in coatings due to thermoplasticity	[38–40][a]
Sorghum	*Sorghum bicolor* L.	Kafirin	Gelling capacity with film forming abilities	[43–45]
Rye	*Secale cereale* L.	Secalins	Protein–carbohydrate interactions drive gelation process	[46]
Oat	*Avena sativa* L.	Avenins	Improved gelation properties with modification (deamidation)	[47]
Barley	*Hordeum vulgare* L.	Prolamins and glutelins	Good thermal stability for reinforcing gelling properties	[50]
Other potential cereal plant sources for gelation				
Hemp	*Cannabis sativa* L.	Edestin/ albumin	Potential modifications for functional properties	[51]

[a] Industrial application and research publications.

the major cereal, legume, oilseed, and vegetable plant storage proteins available in the current literature.

8.2
Structure and Formation of Protein Gels

Gelation refers to the transformation of proteins in the sol state into a gel-like structure through physical or chemical means. The formation of gels in food structures imparts important characteristics to the structure of food (food matrix) for carrying flavors, ingredients, and other properties, including water holding capacity. Protein gels can be divided into two types, gels formed by "random" aggregation and gels formed by association of molecules into strands in a more

Table 8.2 Leguminous/oilseed proteins in gelation.

Plant	Botanical name	Main storage proteins	Application	Reference
Soybean	*Glycine max* (L.) Merr.	Globulins	Good gelation properties induced with modifications or high-pressure processing, and application in several products[a]	[53–60][a]
Pea	*Pisum sativum* L.	Globulins	Gelling capacity in protein isolates and application in products	[92, 93][a]
Lupin	*Lupinus* sp.*Lupinus albus* L.*Lupinus angustifolius* L.	Conglutins	Potential functional properties with improved gelation using modifications	[63–68][a]
Sesame	*Sesamum indicum* L.	Globulins	Potential functionality particularly in foods showing hydrophobic nature	[69–71]
Sunflower	*Helianthus annuus* L.	11S globulins	Gelation properties applicable in modified (enzyme treated) protein isolates	[73–75, 113]
Canola	*Brassica* sp.	Cruciferin and napin	Gelation properties applicable in modified (enzyme treated) protein isolates	[79]
Bean	*Phaseolus* sp.	Phaseolin	Heat-, salt- induction favors gelation properties of bean protein isolates	[86–88]
Broad bean	*Vicia faba* L.	Legumin	Good for acidic foods where gel strength can be maintained	[89]
Chick pea	*Cicer arientum* L.	Globulins	Ultrafractionated isolates show improved gelling properties.	[92, 93]
Pigeonpea	*Cajanus cajan* (L.) Millspaugh	Globulins	Protein conctrates with less protein content can be used for functional properties owing to high solubility	[95]
Cowpea	*Vigna unguiculata* (L.) Walp.	Globulins	Heat-, salt- induction favors gelation properties in protein isolates	[96][a]
Lentil	*Lens culinaris* L.	Globulins	Ultrafractionated isolates show improved gelation	[93]

Table 8.2   (*Continued*)

Plant	Botanical name	Main storage proteins	Application	Reference
Peanut	*Arachis hypogaea* L.	Arachin	Gelation properties exhibited after extensive processing such as de-fatting, removing anti-nutritional factors, and enzyme hydrolysis.	[100]
Cottonseed	*Gossypium* sp.	Globulins	Processing and modifications needed for exhibiting gelling properties	[111]
Other potential legumes/oilseed sources for gelation				
Linseed	*Linum usitatissimum* L.			
Safflower	*Carthamus tinctorius* L.			

[a] Industrial application and research publications.

ordered way [1]. Owing to small physical changes affecting the thermodynamic characteristics, gels of both types can be formed from one protein and the transition from one type of gel structure to another can take place within 0.1 pH units [1].

Gelation of most proteins usually occurs as unfolding and aggregation of proteins. Heating causes proteins to unfold, exposing reactive groups from which intermolecular bonds can be formed with neighboring protein molecules. With sufficient intermolecular bonding, a three-dimensional network can be developed, forming a gel [2]. Factors affecting gelation properties of proteins include pH, ionic strength, heating and cooling rates, and the molecular forces that are influenced by these factors.

With various molecular forces driving the gelation processes, it is imperative to understand how proteins from different sources react to different forces and form gels. This chapter is thus an attempt to summarize findings on the gelation of plant proteins, possible rheological measurements, and physicochemical characteristics for a better understanding of the gelation phenomenon.

8.3
Factors Determining Physical Properties of Protein Gels

Protein gelation is the crosslinking of polypeptide chains to from a three-dimensional network. Crosslinking of proteins is caused by different molecular forces and may involve hydrogen, ionic, disulfide, and/or hydrophobic bonding [3].

Table 8.3 Vegetable/fruit proteins in gelation.

Plant	Botanical name	Main storage proteins	Application	Reference
Potato	*Solanum tuberosum* L.	Patatin	Moderate gelation with proteins isolates	[106]
Alfalfa/lucerne	*Medicago sativa* L.	Globulins	Moderate gelation observed with soft gels	[23]
Bitter melon	*Momordica charantia*	Albumin, globulin, glutelin fractions	Moderate functional properties–potential for modification to improve functional properties	[112, 113]
Other potential vegetable/fruit sources for gelation				
Crambe	*Crambe abyssinica* L.			
Cucumber	*Cucumis sativus*			
Luffa	*Luffa cylindrica*			
Pumpkin	*Cucurbita maxima*			
Avocado	*Persea americana*			

The molecular forces involved in the gel network are dependent upon the protein and the protein structure, which can be influenced by the method used for protein isolation [4]. The involvement of different interactive forces in the formation and structure of protein gels can be deduced from effects of pH, salts, reducing agents, and dissociating agents. Hydrogen bonds and hydrophobic interactions in protein can be de-stabilized by urea and guanidine hydrochloride (Gu·HCl). Urea denatures a protein molecule through preferential adsorption with charged protein solutes, dehydrating the molecules and causing repulsion between proteins, thus stabilizing the unfolded structures [5]. Consequently, urea can be involved with both hydrophobic interactions and hydrogen bonding by dehydrating the protein molecules and interacting through hydrogen bonds that might otherwise interact with the solvent surrounding the molecule. Sulfhydryl/disulfide exchange has been proposed to be involved in soy protein gelation based on the reaction of the gel (gel integrity) with several reagents: β-mercaptoethanol (β-ME or 2-ME) [6], dithiothreitol (DTT) [7], and N-ethylmaleimide (NEM) [8]. When electrostatic forces are involved in gel formation of whole plasma protein it was observed that the gel strength was affected by pH and salts [9, 10]. Therefore, involvement of electrostatic interactions can be determined by the effect of salts and pH.

The gel forming ability of proteins has been considered to be of great importance in rendering structure to foods. Heat induction to facilitate gel formation has long been practiced, and usually involves three steps: (i) unfolding of the protein by denaturation to expose residues buried in the core, (ii) aggregate formation by interaction of the exposed residues, and (iii) formation of a continuous network by arrangement of the aggregates. A balance of both protein–protein and protein–solvent interactions is essential for such heat-induced gel network formation [11]. In addition, pH and ionic strengths contribute as two main factors that influence these interactions, resulting in various types of network [12]. Among plant proteins, soy protein is the most widely studied. Renkema [13] investigated the effect of pH and ionic strength on the storage modulus, fracture strain, and permeability of soy protein isolate gels formed at 0, 0.2, and 0.5 M NaCl at pH 3.8, 5.2, and 7.6. The authors found that gels with a consistently higher storage modulus and lower fracture strain were formed at pH 3.8 compared with those formed at higher pH values, whereas ionic strength influenced the permeability of the gels (reflecting the pore size of the protein networks) more than the pH did. The authors concluded that, in addition to strand coarseness, information on the curvature of the strands was needed to relate the rheological properties to the network structures formed under different pH and ionic strength conditions. Since the strands that make up the gel network were composed of protein aggregates they concluded that understanding the influence of factors such as pH and ionic strength in affecting the gelation process (unfolding, aggregation, and gel network formation) was important.

8.4
Evaluating Gelation of Proteins

The gelation of proteins can be evaluated using various techniques and measurements mostly involving the strength of formed gels. Most of the measurements are based on the science of rheology-deformation and flow of matter, where a clear distinction can be made among small and large deformations due to load. Measurements such as rupture strength (RS), shear modulus, strain, and so on can be measured.

Small deformations are generally measured by the initial slope formed by shear modulus or rigidity (G) (stress/strain) and, usually, a simple compression test is carried out with a cylindrical probe. Large deformations – maximum load a gel can take – can give rise to the RS or gel strength. In this case, a probe will be inserted at constant velocity and the force at which the gel ruptures is recorded.

The viscoelasticity of gels is another property that can be measured based on applied stress. It is usually obtained by applying constant or sinusoidally oscillating stress/strain, determining creep compliance function (constant stress), denoted by $J(t) = \text{strain}(t)/\text{stress}$, and stress relaxation function (constant strain) denoted by $G(t) = \text{stress}(t)/\text{strain}$. Using these equations a *gel* can be defined as "a viscoelastic material for which G' is greater than G'' and for which both G' and G'' are almost independent of frequency."

In terms of non-rheological measurements, protein crosslinking, denaturation, extended protein chains, gel networks, opacity, and thermal reversibility are some of the properties that can be evaluated. Gel networks can form as random aggregation or string-of-beads aggregation, depending on the number of binding sites available per unfolded protein molecule, the spatial distribution of each protein, and the relative bonding strengths of each protein.

Other rheological techniques, typically non-destructive, can be used to obtain information on the aggregation process. When an oscillatory strain (deformation) of a fixed dimension is applied to the sample throughout the gelation process the stress developed over time can be measured. The stress developed is dependent upon the nature of the sample and its intrinsic material properties [14]. Because two structurally different gels could have the same stress response they cannot be identified by rheological techniques, and hence techniques such as microscopy can be beneficial for a complete understanding of gel network formation based on hypothesized structural features through rheology. Without heat-induced unfolding of globular proteins and exposure of buried residues, protein–protein interactions are unlikely to occur. The kinetics of unfolding and aggregation tend to result in the formation of an orderly assembled aggregate that depends on the rate of reaction. Usually, aggregation takes place at a slower rate than unfolding. When aggregation proceeds faster than unfolding, a non-orderly assembled aggregate forms [12]. A slow heating rate gives more time for aggregates to interact and assemble themselves in an orderly manner and an orderly arrangement of aggregates into a network is what creates a "fine-stranded network." Gels with transparent characteristics are observed for this type of network [15–17]. In contrast, when the processes occur too quickly, larger aggregate clusters are created that giving a more random arrangement of aggregates assembles into a network and turbid gels [18]. The structure of heat-induced gel networks is determined by the intermolecular interactions [19, 20]. Understanding these interactions is important in studying and modifying the gel texture of related foods.

8.5
Gelation of Proteins Derived from Plants

8.5.1
Gelation of Cereal Crop Plant Proteins

8.5.1.1 Wheat Proteins
Storage proteins form about 50% of the total protein in mature cereal grains and have nutritional impacts for humans and livestock with their functional properties in various food processing operations [21]. Gluten, a most abundant storage fraction of cereal protein obtained as a co-product of the wheat starch industry, is utilized as a functional protein due to its viscoelastic and structure-enhancing properties

[22, 23]. Wheat gluten is also used in meat systems as a source of proteins and amino acids and is a key ingredient owing to its functionality in terms of water and fat binding as well as texturizing properties [24]. Gluten, the storage protein, has amino acids containing amide groups (glutamine and asparagine) that are converted into glutamic and aspartic acids by deamidation. It was observed that the proteins formed weak gels below 90 °C but when autoclaved at 120 °C the proteins interacted to form a strong, elastic gel network. The proteins can be used to modify the textural properties of mixed gels. With the addition of only 1% gluten protein, a less brittle, smoother texture was noticed that, on further replacement with meat protein, became increasingly elastic and cohesive [25]. While some studies focus on the chemistry and strength of soy protein (owing to its functionality and popularity) gelation, others have used co-gelation processes with wheat gluten proteins as potential for a functional ingredient in gel-based (comminuted) muscle foods to enhance the texture [26]. Furthermore, studies on the effects of individual wheat glutenin subunits coded on specific chromosomes on the functional properties of wheat dough have demonstrated the role of certain structural features of proteins (size, number, and positions of cysteine residues) that are important in determining the gelation characteristics [27].

8.5.1.2 Rice Proteins

Rice proteins, owing to their hypoallergenic property, are being preferred in several food applications. However, due to poor solubility, applications are being limited. Therefore, several researchers have modified rice proteins in an attempt to improve their functionality. For example, Paraman *et al.* [28–30] used modifications to rice protein for enhanced functionality, and as potential gelling agents, too.

In general, these proteins play a central role in determining the pasting and textural properties of rice. Lim *et al.* [31] reported that reducing the protein content in rice flour increases its peak viscosity; and the three-dimensional network of the rice flour gel is weakened by the presence of prolamin in the matrix [32]. The glutelins, the major rice storage proteins, are soluble in dilute acid or alkaline solution. These globulin-type subunits tend to form large macromolecular complexes that can be stabilized by disulfide and hydrophobic bonding/interactions. Partial reduction is believed to open up the rice glutelin structural polymers via disulfide bond breakage. If a subsequent re-oxidation process is initiated it will restore the S–S bonding and enable rebuilding the polymer network. Besides hydrogen and disulfide bonding, extensive aggregation [33] and glycosylation [34] may also be partly responsible for the limited rice protein solubility. It is with this understanding that rice proteins are, generally, subject to different modification reactions to improve their functionalities. Crosslinking is one such modification that has been effective in achieving a desired structural network of glutelin proteins. This aspect is reviewed in Chapter 9. In addition, several applications are seeing the use of rice products as gels in puddings, and so on, and patents have also been issued on the use of rice products, possibly due to their increased functionality and other advantages, including hypoallergenicity [35].

8.5.1.3 **Maize/Corn**

Zein is the major storage protein of corn, comprising ≈45–50% of the protein in corn. The prolamins of maize (called *zeins*) consist of one major group of proteins (α-zeins) and several minor groups (β, γ, and δ-zeins) [36, 37].

Zein in solution is determined not only by its solubility but also by its gelling properties, and was utilized as early as the 1950s in industrial coatings owing to its thermoplasticity [38]. Zein gels form easily and this phenomenon, which is quite well documented, has been referred to as *"a troublesome characteristic of zein"* [38]. Factors that affect gelation are type of solvent, concentration of solvent (less water results in slower gel formation), temperature (higher temperatures promote gelation), pH, and mechanical factors (e.g., agitation). Gelation could also happen as a result of denaturation of zein and/or due to the presence of insoluble substances able to act auto-catalytically to precipitate or gel the zein [39]. Little work has been directed toward zein protein gelation but its high-temperature resistance can be used under various thermal processing conditions for food applications.

8.5.1.4 **Sorghum**

Kafirins, the major storage proteins in sorghum, are classified as prolamins, containing high levels of proline and glutamine [40]. They are classified as either α, β, γ, or δ based on molecular weight and solubility. The sorghum endosperm contains about 66–84% α-kafirin, 8–13% β-kafirin, and 9–21% γ-kafirin and low levels of a poorly characterized δ-kafirins [41]. The α-kafirins do not crosslink extensively and form mainly intramolecular disulfide bonds. The β-kafirins are rich in the sulfur-containing amino acids methionine and cysteine. Both β- and γ-kafirins form intermolecular and intramolecular disulfide bonds and are highly crosslinked. Overall, sorghum prolamins are rich in glutamic acid and non-polar amino acids (proline, leucine, and alanine), but almost devoid of the essential amino acid lysine [42]. A study conducted to compare gels formed by kafirins and zein observed that kafirin gels form more readily than those of zein and that the formed kafirin films show much higher tensile strength and lower extensibility than zein films. This could be attributed to the structures as disulfide crosslinks can be readily achieved in kafirins, owing to the presence of β- and γ-fractions rich in cysteine [43–45]. Gelling capacities with film-forming abilities can be used with the kafirin proteins under several applicable food-processing conditions.

8.5.1.5 **Rye, Oat, and Barley**

Very few studies have been carried out to evaluate the functional properties of rye, oat, barley, and hemp owing, possibly, to their carbohydrate rigid structures. In one study, the pentosan-protein material from rye was subjected to oxidative gelation. This consisted of fractions obtained by precipitating water-soluble from rye starch with ethanol and gelation was observed with addition of oxidizing agents, hydrogen peroxide, and horseradish peroxidase. It was evident that even though the protein component did not play a major role in the oxidative gelation process the pentosan material was required to drive the oxidation process. However, the mechanism of these observations is not clearly understood [46].

With oat, protein isolates were derived and even modified for functional properties. In a study by Ma and Khanzada [47], oat protein isolates were deamidated by mild acid hydrolysis. Amino acid analysis and gel filtration chromatography showed no significant cleavage of peptide linkages although the aggregated and oligomeric oat proteins were extensively dissociated. Deamidation led to marked improvement in solubility, emulsifying properties, and water and fat binding capacities. The pH for heat-induced gelation was lowered by deamidation, and a firm, elastic gel was produced by mixing egg white with deamidated oat isolates [47].

Barley endosperm contains mainly hordein (35–45%) and glutelin (35–45%) proteins, whereas cytoplasmic proteins (mainly albumin and globulin) are enriched in barley bran and germ [48]. These protein fractions might exhibit different functional properties for various food applications owing to different molecular structures [49] due mostly to their emulsifying properties. When adsorbed at the oil–water interface, the glutelin fraction aggregates to form a strong film due to the presence of high molecular weight polypeptides, which can prevent close contact between the oil droplets and decrease flocculation and coalescence. The good thermal stability of the glutelin-stabilized emulsion may be due to further gelation of the glutelin around the oil droplets during thermal treatment to form a reinforced film [50]. Table 8.1 lists major cereal plant proteins with gelation properties. In addition to the major cereal plant proteins, other sources like hemp can serve as a potential source of functional proteins. Owing to the ability to become functional when modified, as observed with proteolysis of its protein fractions [51], hemp proteins can be included in the area of functional proteins derived from cereal crops.

8.5.2
Gelation of Legume Plant Proteins

Of the several legume plant proteins, soybean protein and its fractions have been widely studied and used in the food industry for functional properties. Among their functional properties, heat-induced gelation is important in many food applications. Thermal or heat-induced gelation of globular proteins is a complex process, involving several steps such as denaturation, dissociation–association, and aggregation [1]. The final characteristics of the protein gels depend on many parameters, including the type of the proteins, their composition, concentration, and also other factors including pH, presence of any salts, and conditions for gelation, including time and temperature [52–54].

β-Conglycinin (7S) and glycinin (11S) are the major globulins in soybean, representing 80% of total protein in the soybean seed. β-Conglycinin is a trimeric glycoprotein composed of three subunits α, α', and β (68, 72, and 52 kDa, respectively). Glycinin is a hexameric protein composed of five possible types of subunit AB (58–69 kDa), which under reducing conditions produce the polypeptides A and B by cleavage of disulfide bonds. Soy proteins are generally added to food products as a high protein source such as soybean protein isolate (SPI), but glycinin and β-conglycinin have specific functional properties that might be of special interest to the food industry [55].

Several reports of soy protein gelation have been included in the literature, but lately the effects of salts and high-pressure processing have attracted attention. Calcium has a significant impact on soy protein functionality because of the protein interactions promoted by its presence [56, 57]. Calcium interacts with denatured soy proteins, (the basis of tofu making), and its presence promotes the formation of complexes with soybean protein hydrolysates [58]. Heat-induced gelation of soy protein isolate (SPI) and 7S or 11S globulins has been studied but the mechanism by which the gels form is poorly understood. Speroni *et al.* [57] observed that in the absence of calcium the glycinin-enriched fractions, 7S and 11S fractions, had the lowest and highest Q value, 0.050 and 0.124, respectively. Therefore, despite the fact that the T_{gel} of 7S differs from that of 11S (84.7 compared to 95.0 °C) and the denatured proteins in 7S have potentially more time to interact to establish a network, the greater proportion of the structure in 7S gels was formed upon cooling (Figure 8.1). Similar work has been published and the basis behind such phenomena of gelation is becoming understood.

Interest in high-pressure (HP) processing is mainly related to food preservation but this technology, which affects protein conformation and therefore modifies

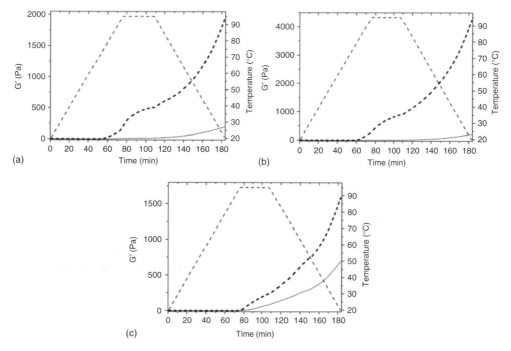

Figure 8.1 Typical gelation curves of 10% SPI (a), 7SEF (b), and 11SEF(c) in the absence (solid line) or presence of calcium (bolder dashed line – 25 mM for SPI and 7SEF; 2.5 mM for 11SEF). Temperature profile is indicated with the lighter dashed line. From Speroni *et al.* [57]; obtained with permission.

protein properties, is being used to confer unique attributes to food products. HP processing has been shown to promote soy protein aggregation and denaturation [59, 60]. Changes in the secondary, tertiary, and quaternary structure due to the pressure effects on hydrophobic and electrostatic interactions could influence the capacity of soy proteins to form three-dimensional gel networks [61]. The formation of soybean protein gels under pressure has been reported for a SPI dispersion with a protein concentration equal to or higher than 17% (w/w) [62, 63]. Figure 8.2 shows the textural properties of the HP-induced gels. The springiness and cohesiveness (Figure 8.2a,b) of the HP-induced gels of the three proteins, 7S, 11S, and SPI, showed slightly lower values than the heat-treated gel (control), differing significantly only in the case of SPI. Comparing the different pressures applied, most values were similar, with the only significant difference being in the springiness values of the 11S globulin, which increased with pressure up to 600 MPa, although at 700 MPa it decreased again. The adhesiveness (Figure 8.2c) showed higher values for the HP-induced gels of 7S and SPI compared to the control (HT gel). However, the 11S HP-induced gels, compared to the control, showed similar values of adhesiveness after treatment at 500, 600, and 700 MPa, but higher values after treatment at 300 and 400 MPa. The hardness (Figure 8.2d) of the HP-induced gels was lower than obtained for the HT-induced gels, showing that both the 7S and 11S yielded increased hardness with increasing pressure. In HT gels the hardness increased with increasing concentration of protein.

Although several studies have been reported on the gelation capabilities of soy proteins, the mechanisms of gel formation are being understood from a

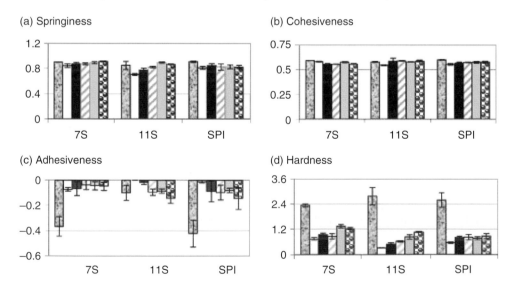

Figure 8.2 Textural properties of the HP induced gels of 20% (w/v) concentration of 7S, 11S, and SPI: (a) springiness, (b) cohesiveness, (c) adhesiveness, and (d) hardness (in Newton). Results are means of three different determinations, two different days: (▨) heated; pressure = (□) 300, (■) 400, (▧) 500, (▤) 600, and (▩) 700 MPa. From Molina *et al.* [63]; obtained with permission.

rheology point of view. Modifications are also being sought to understand gelation mechanisms and the phenomenon in general to be able to manipulate soy proteins or fractions for improved functionality.

8.5.2.1 Lupin

Lupin proteins are increasingly being used to replace soy proteins in low fat dairy and meat formulations. Lupin seeds consist of conglutins as storage proteins and are generally rich with a balance in essential amino acids, and are considered better than those of soybean proteins [64]. Lupin proteins have been found to not participate in crosslinking reactions as do the soy proteins, but appear to possess significantly high foaming and whipping capabilities thereby serving to replace egg white products. Their emulsion and gel forming capacities have also been documented [65–67]. In particular, protein isolates from lupin showed strong gelation properties in the presence of NaCl [66], and is believed that modification reactions to lupin proteins can render better functionality. Most of the lupin proteins can be recovered by ultrafiltration (UF) (and possibly a precipitation or coagulation step) and drying [68]. Potential functional properties with improved gelation using modifications are present for lupin proteins.

8.5.2.2 Sesame

Sesame seeds contain nearly 25% protein, consisting of mainly two fractions. One is the high molecular weight protein fraction, a globulin (nearly 60%), which has been isolated and characterized under various solution conditions [69]. Its quaternary structure is well established and is also well understood [70]. The other fraction, the low molecular weight protein fraction, a globulin, too, constitutes nearly 25% of the total protein. No information is available on this protein fraction. Very few studies have attempted to examine the functional properties of sesame proteins. In one case, factors affecting disulfide bonding contributing to gelation of sesame globulins were investigated. It was observed that hydrophobicity and hydrophobic interaction provided more stability to gels formed from sesame proteins [71, 72]. Potential functionality, particularly in foods showing a hydrophobic nature, exists for sesame proteins.

8.5.2.3 Sunflower

Sunflower protein concentrates/isolates have been identified with functional properties. It has been demonstrated that sunflower protein concentrates (with 11S globulin storage component) exhibit high water solubility, possibly enough to obtain foams and emulsions of different stability at different pH and ionic strength from these protein concentrates, as well as self-supporting gels produced by head induction [73].

The influence of several parameters on the gelation properties of trypsin hydrolysates of sunflower proteins has been studied by dynamic rheological methods. The degree of hydrolysis has very little effect on either storage modulus or gelation time. Sunflower protein gelation is strongly pH dependent. Gelation is only possible in the pH range 7–11. The storage modulus was shown to reach a maximum value at pH 8 with protein concentration. The gels formed at pH 7 or above pH 9 were very weak. Gelation time increased with pH and decreased with protein concentration.

It was also observed that gelation of sunflower proteins was not achieved by any of the usual treatments generally performed to induce gelation. However, after partial hydrolysis with trypsin, it was possible to obtain gels exhibiting the unusual property of deeply weakening when cooled below about 80 °C [74]. Gelation requires a "critical" macromolecular concentration. The macromolecular concentration also influences gelation time and the rheological behavior of the gels [75, 76]. Usually, protein gel properties are also strongly affected by pH [77] and other factors. Applicable gelation properties are currently found in modified (enzyme treated) protein isolates from sunflower proteins.

8.5.2.4 Canola

Canola protein, despite having an excellent amino acid profile, has found only marginal use in the food industry. Improvement of functional properties, such as gelation including modification could offer new markets for this protein [78].

Enzymatic hydrolysis has been shown to improve the functional properties of sunflower, peanut, soy, and rapeseed protein isolates [79–82]. The main storage proteins in canola are cruciferin and napin. Studies were carried out to improve canola protein gelation properties through enzymatic hydrolysis with trypsin, ficin, and bromelin. Proteolysis was found to reduce gel strength to below that of the control. This was accompanied by a decrease in molecular weight, observable through SDS-PAGE (sodium dodecyl sulfate polyacrylamide gel electrophoresis) analysis. However, limited proteolysis was shown to be a suitable pretreatment before applying modifications including crosslinking. Other research suggests that opening the protein structure prior to modifications can enhance its effectiveness. It was concluded that while limited proteolysis does not improve canola protein gelation on its own it can significantly improve gelation properties when combined with other modifications [78].

In a few other studies, the effects of pH, salt, guar gum, and protein concentrations on the gelling ability of canola protein isolate (CPI) were evaluated. Factorial and response surface optimization models were used to identify the optimum conditions (20%, w/v CPI, pH 10, 1.5% w/v guar gum, 0.05 m NaCl) that would simultaneously maximize G' ($\geq$28 000 Pa) and minimize $\tan \delta$ ($<$0.17) values of CPI-guar gum gels. Although a pH $>$8 is unconventional in food systems, strong and elastic CPI-guar gum gels were produced at pH 10, whereas gels prepared at pH 6 were less elastic. Under the optimum conditions, CPI alone formed a stronger gel than a CPI–guar gum mixture, suggesting that guar gum interfered with protein gelation [83].

8.5.2.5 Bean

Functional properties of the Great Northern bean (*Phaseolus vulgaris* L.), which has phaseolin as its main storage component, have been examined. One study observed that the proteins had the highest water and oil absorption capacity

(5.93 and 4.12 g g^{-1}, respectively) among several closely related samples [84]. The gelation characteristics of mucuna bean flour and protein isolates have been determined. The pH-dependent protein solubility profile of the flour and protein isolates indicated that the isoelectric point of the proteins was between 4.0 and 5.0. The lowest least gelation capacity (LGC) for the flours was obtained at a pH of either 4 or 5 depending on the *Mucuna* species. In the *Mucuna rajada* and *Mucuna cochinchinensis* flours, the lowest LGC of 14% w/v was obtained at a pH of 5, whereas the lowest LGC of 12% w/v was observed at pH 4 for *Mucuna veracruz* white. In addition, the LGC increased as the pH value moved away from the isoelectric region. A better gelation characteristic was observed for the protein isolates. *Mucuna veracruz* white protein isolate recorded the best LGC (8% w/v) at a pH of 4. The LGC in other isolates ranged from 10 to 12% w/v [85].

In another study, the influence of pH, ionic strength, and carbohydrates on the foaming and gelation properties of mucuna bean protein concentrate (MPC) were investigated. Gelation properties were reduced in alkaline and acidic conditions, except at pH 4, where the least gelation concentration endpoint (LGE) was 8. The gelation properties of MPC improved in the presence of carbohydrates in the mixture. The gel-forming properties also increased with increasing ionic strength from 0.1 to 0.4 M, while a further increase, from 0.6 to 1.0 M, reduced the gelation properties [86].

The heat-induced gelation properties of red bean globulin (RBG) and the viscoelastic properties of the gels were evaluated by small amplitude oscillatory tests and creep experiments. It was observed that gel structure developed progressively when the protein dispersion ($\approx$10% w/v) was heated from 25 to 95 °C, and both the storage modulus (G') and loss modulus (G'') increased rapidly during cooling. The addition of NaCl at lower concentrations led to increases in G' and G'' values, with decreased creep compliance, suggesting higher viscoelasticity. At higher salt concentrations, viscoelasticity was decreased; the optimum NaCl concentration to produce maximum gel rigidity was 0.2 M. The addition of sodium dodecyl sulfate and urea caused a more pronounced reduction of the gel moduli than did DTT and NEM. Such data suggest that, in the thermal-gelation process, hydrophobic interactions and hydrogen bonding play an important role, while electrostatic interactions and a balance of attractive–repulsive forces may also contribute [87].

8.5.2.6 Broad Bean

Protein fractions of broad beans (legumins – storage proteins) have been prepared and studied for gel formation characteristics. Gels were formed at acidic pH of the prepared protein isolate. The gels were fluid at high temperature and stronger when cooled. The gel strength increased with increasing pH, with increasing heat treatment, and with increasing protein concentration. Addition of calcium and dextrose also increased the gel strength whereas sucrose decreased it. The transparency of the gels decreased with increasing pH [88]. Although very few studies have been conducted using broad bean, it seems to be good in applications of acidic foods where gel strength can be maintained.

8.5.2.7 Pea, Chickpea, Lentil, and Pigeonpea (Pulses)

The major proteins found in pulses are globulins and albumins. Globulins represent roughly 70% of legume seed proteins and consist primarily of the 7S, 11S, and 15S proteins. The molecular sizes of these proteins range from 8000 to 600 000 Da [89]. These proteins generally have a minimum solubility at pH values between 4.0 and 5.0. By manipulating the solubility of the proteins and using filtration techniques that take advantage of their hydrodynamic properties, protein concentrates and isolates with varying purity and functionality can be obtained. The functional properties of pea, chickpea, and lentil protein concentrates were compared in a study that found, for each pulse, that the protein extract produced using UF had better gelling properties (i.e., lower LGC) than the one produced using isoelectric precipitation (IEP) [90]. This suggested that the processes used to produce the protein extracts had an impact on their gelling properties. Similar findings were reported by Papalamprou *et al.* [91]. In a study by Boye *et al.* [92], no gels were formed at concentrations of 2% and 4% (w/v) irrespective of the pulse type, variety, or processing technique. At 6% concentration the green lentil concentrate processed by UF formed a weak gel. A strong gel was formed at 8% (LGC); the GL-UF, thus, had the best gelling properties while the yellow peas, desi, and kabuli chickpeas processed by IEP had the lowest gelling properties (LGC of 14%, w/v). All other protein extracts demonstrated intermediate gelling properties [92].

The effects of pH and sodium and calcium chloride on the gelation properties of chickpea protein isolates were also investigated. Chickpea protein isolates dispersions with sodium and calcium salts showed different rheological behavior at different ionic strength and pH. Increasing the ionic strength of dispersions could strengthen the gelation properties of the protein isolate under acidic conditions; however, the elastic parameters of protein isolate were reduced at pH 7.0. In addition, it was deduced that the gel strength for samples with $CaCl_2$ was stronger than for those with NaCl at pH 3.0 since Ca^{2+} facilitated the electrostatic protein–protein interaction in favor of gel network formation [93].

Furthermore, the objective of another investigation was to study the functional properties of pigeonpea [*Cajanus cajan* (L.) Millsp.] flour and protein concentrate. The solubility of both samples was superior up to 70% at pH above 6.7 and below 3.5. The water and oil absorption were 1.2 and 1.07 ml g^{-1} of sample and 0.87 and 1.73 ml g^{-1} of flour and protein concentrate samples, respectively. The minimum concentration of flour and protein concentrate needed for gelation was 20% and 12%, respectively. Good emulsifying and foaming properties were also found. These properties indicate that the flour as well as the concentrate could have applications in various food systems [94].

8.5.2.8 Cowpea

The functional properties of protein isolates prepared from three cowpea varieties and two soybean varieties were determined. Gels were formed at 70 °C for 40 and 30 min for cowpea protein isolate (12%) and soy protein isolate (10%), respectively. A considerably higher protein concentration is usually required for the gelation of globular proteins compared with hydrocolloids. Cowpea solution formed a firm gel by heating a 12% solution at 70 °C for a minimum of 40 min and at 80 °C for a minimum of 20 min followed by cooling at 4 °C for 30 min, while a 10% soy protein solution formed a gel after heating at 70 °C for a minimum of 30 min and at 80 °C for a minimum of 10 min followed by cooling at 4 °C for 30 min. All of the isolates could form gels after heating for 10 min at 90 and 100 °C. Only 30 min cooling time was needed for firm gel formation [95]. In another study, the least gelation concentration of cowpea proteins was found to be 6% when the proteins were dissolved in 0.5 or 1.0 M NaCl [96]. These studies show that heat-, salt-induction favors gelation properties in protein isolates prepared from cowpea.

8.5.3
Oilseed Proteins

The functionality of oilseed proteins in relation to food applications has been reviewed by different authors, especially on industrial crops: (oilseeds) [97], sunflower [98], and peanut [99]. The importance of the protein structure in relation to the functional properties [100] and their interaction with polysaccharides has been reported by Braudo *et al.* [101]. Koumanov *et al.* [102] reviewed the electrostatic interactions of proteins in relation to their functional properties. Edible nut seeds contain 7–25% protein. The number and type of proteins present are dependent on the seed type, and typically these proteins are rich in the acidic amino acids glutamic acid and aspartic acid and the basic amino acid arginine. The protein content of defatted meals from dehulled oilseeds depends on the type of seed and ranges between 35 and 60%. Notably, the meals may contain anti-nutritional compounds, such as oligosaccharides, trypsin inhibitors, phytic acid, and tannins, and present low protein solubility, which could limit food applications. Processing with or without enzymes, or extracting the oil to make the nutrients available, can help to improve the functional properties of the proteins present in the oilseeds [103]. Table 8.2 lists oilseed proteins with gelation properties.

Probably owing to the need for additional processing steps in improving the functional properties of oilseeds, not many studies have been reported. A few studies have, however, compared the functional properties of several oilseeds, including gelation capacities. Gelation studies indicated the LGC (%, w/v) for almond, Brazil nut, cashew, hazelnut, macadamia, pine nut, pistachio, Spanish peanut, Virginia peanut, and soybean to be 6, 8, 8, 12, 20, 12, 10, 14, 14, and 16, respectively [104]. The data represent good solubility indexes of the isolated proteins. In cases where functional properties were not pronounced, the proteins were modified.

Enzyme modification was conducted through breakage of peptide bonds to produce peptides of the desired size, charge, and surface properties. Enzymatic hydrolysis was successful in increasing protein solubility. Trypsin treatment of protein products results in higher solubility and higher water hydration capacity in comparison with corresponding untreated products. It was observed that the foaming capacity and stability, and gelation capacity, of flour could be improved by protease hydrolysis [105, 106]. The co-precipitation of proteins from different oilseed protein mixtures showed higher nutritional values and better functional properties than those of their individual protein isolates. Application of genetic engineering to oilseed proteins can affect the functional properties to differing degrees depending on the protein [107–110]. Table 8.2 lists major legume and oilseed plant proteins with gelation properties. Linseed/flaxseed and safflower are other sources of proteins that could have potential functional properties, including gelation, based on the limited research conducted with their protein concentrates [105].

8.5.4
Vegetable/Fruit Proteins

The potential of vegetable/fruit proteins to form a stable network like a gel structure has been attributed to molecular properties, namely, their ability to unfold, which can be enhanced to a certain degree of thermal denaturation. Few studies have been conducted using vegetable/fruit proteins in terms of functionality. The gelation and interfacial properties of vegetable proteins were observed by van Vliet *et al.* [23]. In some cases, the functional properties were less pronounced, with LGEs as low as below 6% (e.g., potato protein isolate) (Table 8.3).

Moderate gelation was observed with protein isolates prepared from alfalfa, where acid-precipitated concentrates showed stress relaxation tests on the soluble leaf portion of 7/100 g, indicating softer gels with faster relaxation compared to whey protein isolate gels (13/100 g) [23].

Not many studies have been conducted using vegetable/fruit proteins or from prepared protein concentrates or isolates, in comparison to the wealth of data existing on the functional properties of proteins, protein fractions, or protein isolates derived from cereal, legumes, pulses, or oilseeds. It will be interesting to examine, in particular, extracts derived from vegetables/fruits or from vegetable leaves for proteins or peptides able to confer some functional property. Bitter melon, for instance, could be a potential source of proteins with functional properties. Horax *et al.* [111, 112] studied its seed protein fraction characteristics as well as some of the functional properties and found that bitter melon seed protein isolates exhibited functional properties including solubility, emulsifying, and foaming properties. It is possible that the protein isolates when modified can improve functional properties and could also serve as a potential gelling agent. Texturized vegetable proteins (reviewed in Chapter 9 by Schäfer) are being used as meat substitutes in several formulations owing to their flavor absorbing capacities, and may also confer functional properties. More research can be directed towards evaluating proteins derived from vegetables, including crambe, alfalfa/lettuce, bitter melon, pumpkin, luffa, avocado, and so on, for functional properties.

8.6
Protein Gels in Product Application

Being able to modify the appearance and textural properties of protein gels is important to food companies developing new products. Depending on the product, gels need a certain texture – from tough to brittle – high water-holding capacity, and an attractive appearance. Research into various protein systems is helping determine the ways that gels can be formed according to product need. The textural attributes of processed foods, such as cheese, custard, and meat products, are directly related to the type of protein used in their manufacturing and the way it is prepared. Gels made from proteins (dairy, egg, meat, and soy) or carbohydrates (pectin, carrageenan) can exhibit diverse mechanical and microstructural properties. In general, *gels* are defined as materials containing a continuous solid network that is assembled from particles or polymers embedded in an aqueous phase. Some of these networks can hold >99% water in their structures.

Gels can be divided into two major categories. The first is an opaque structure, typically formed by egg white proteins at their natural pH. This structure is composed of relatively large protein aggregates bound to one another to form the network. The second category is a fine-strand gel structure produced by the association of small diameter molecules to form an ordered network. These gels are usually clear and are more elastic than the opaque gels. Variation between these two groups exists, and mixed gels can also be formed. A typical egg white gel prepared with 5% protein at pH 8 (normal egg pH) will appear clearer when the egg white gel is prepared at pH 3.2 instead. The pH modification resulted in the formation of fine protein strands, which will allow light to pass through. By comparison, the opaque gel is composed of large aggregates that reflect most of the light and, therefore, the structure appears white/opaque.

Another major difference between the two gels is their water-holding capacity – an extremely important attribute to consumers. In products such as yogurt, cheese, and sliced meat, gels that possess a high water-holding capacity show no purge in the package. In terms of structure, opaque gels generally have poor water-holding capacity compared to clear gels. The reason for this is that the small capillary spaces within the structure of the clear gel can retain the water better than the large open spaces within the opaque gel structure. These spaces are generally very small, measured in nanometers. As they become larger, such as in opaque gels, water can no longer be tightly held within the structure. Obviously, consumers prefer to buy products that show no exudates in the package. Whey proteins, which are used to make certain cheeses or additives in various foods, can also be induced to form clear and opaque gels. In the case of whey proteins, the type of salt present can dictate the structure formed during heating.

A third type of semi-transparent gel can be produced at room temperature with a calcium salt. Using a new process called *"cold gelation,"* a protein solution is first preheated without any salt, and hence no gel is formed. The solution is then cooled and held in a refrigerator overnight prior to a slow calcium salt addition. This results in medium-size protein strand formation, which gives the gel a semi-transparent appearance. This process resembles preparing a gelatin-type dessert without the need for hot water. Such a process can be important in protecting heat-sensitive nutrients (e.g., vitamins, carotenoids) in fortified foods. Determining the optimal conditions for cold gelation is important to the food industry, which is interested in using the process to provide texture to products such as mayonnaise, and binding restructured meat products or onion rings prior to heating. It will be worth carrying out research into designs that confer such properties to gels based on industry's specific requirements.

8.7
Future Prospects and Challenges

Physicochemical, thermal, chemical, and enzymatic technologies applied to seeds, meals, concentrates, or isolates can be used to obtain products with desirable properties for food applications. The functional properties of oilseed proteins can be modulated by carefully selecting both original seed and the operational variables during extraction (pH, temperature, solvent, presence of salts, ionic strength, etc.). It will be interesting to look, in particular, into extracts derived from vegetables/fruits or from vegetable leaves, or industrial co-products including bran/meals that are most often used as feed, for proteins or peptide hydrolysates being able to confer functional property including gelation. Texturized vegetable proteins are being used as meat substitutes in several formulations owing to their flavor absorbing capacities, and they may also confer functional properties. Likewise, proteins/hydrolysates, if obtained from industrial co-products and having appropriate gelation characteristics or ones that can be modified to have better functional properties, can bring cheap sources of functional ingredients to the market and industry. More research can be directed towards evaluating proteins derived from vegetables, including crambe, alfalfa/lettuce, bitter melon, and so on, for functional properties.

References

1. Hermansson, A.M. (1979), in *Functionality and Protein Structure* (ed. A. Pour-EI), American Chemical Society, Washington, DC, pp. 81–103.
2. Lanier, T.C. (2000), in *Surimi and Surimi Seafood* (ed. J.W. Park), Marcel Dekker, New York, pp. 237–265.
3. Otte, J., Schumacher, E., Ipsen, R., Ju, Z., and Qvist, K.B. (1999) Protease induced gelation of unheated and heated whey proteins: effects of pH, temperature and concentration of proteins, enzyme and salts. *Int. Dairy J.*, **9**, 801–812.
4. Utsumi, S. and Kinsella, J.E. (1985) Forces involved in soy protein gelation: effects of various reagents on the formation, hardness and solubility of heat-induced gels made from 7S,

11S and soy isolate. *J. Food Sci.*, **50**, 1278–1282.
5. Wallqvist, A., Covell, D.G., and Thirumalai, D. (1998) Hydrophobic interactions in aqueous urea solutions with implications for the mechanism of protein denaturation. *J. Am. Chem. Soc.*, **120**, 427–428.
6. Wolf, W.J. (1993) Sulfhydryl content of glycinin: effect of reducing agents. *J. Agric. Food Chem.*, **41**, 168–176.
7. McKlem, L.K. (2002) Investigation of molecular forces involved in gelation of commercially prepared soy protein isolates, MS thesis, North Carolina State University.
8. Wang, C.H. and Damodaran, S. (1990) Thermal gelation of globular proteins:

weight-average molecular weight dependence of gel strength. *J. Agric. Food Chem.*, **38**, 1157–1164.

9. O'Riordan, D., Kinsella, J.E., Mulvihill, D.M., and Morrissey, P.A. (1988) Gelation of plasma proteins. *Food Chem.*, **33**, 203–214.

10. O'Riordan, D., Mulvihill, D.M., Kinsella, J.E., and Morrissey, P.A. (1988) The effect of salts on the rheological properties of plasma protein gels. *Food Chem.*, **34**, 1–11.

11. Arntfield, S.D. and Murray, E.D. (1990) Influence of protein charge on thermal properties as well as microstructure and rheology of heat-induced networks for ovalbumin and vicilin. *J. Texture Stud.*, **21**, 295–322.

12. Arntfield, S.D. and Murray, E.D. (1992) Heating rate affects thermal properties and network formation for vicilin and ovalbumin at various pH values. *J. Food Sci.*, **57**, 640–646.

13. Renkema, J.M.S. (2004) Relations between rheological properties and network structure of soy protein gels. *Food Hydrocolloids*, **18**, 39–47.

14. Ross-Murphy, S.B. (1988), in *Food Structure – Its Creation and Evaluation* (eds J.M.V. Blanchard and J.R. Mitchell), Butterworths, London, ch 21, pp. 387–400.

15. Tani, F., Murata, M., Higasa, T., Goto, M., Kitabatake, N., and Doi, E. (1995) Molten globule state of protein molecules in heat-induced transparent food gels. *J. Agric. Food Chem.*, **43**, 2325–2331.

16. Mine, Y. (1996) Laser light scattering study on the heat-induced ovalbumin aggregates related to its gelling property. *J. Agric. Food Chem.*, **44**, 2086–2090.

17. Matsudomi, N., Tomonobu, K., Moriyoshi, E., and Hasegawa, C. (1997) Characteristics of heat-induced transparent gels from egg white by the addition of dextran sulfate. *J. Agric. Food Chem.*, **45**, 546–550.

18. Tani, F., Murata, M., Higasa, T., Goto, M., Kitabatake, N., and Doi, E. (1993) Heat induced transparent gel from hen egg lysozyme by a two-step heating

method. *Biosci. Biotechnol. Biochem.*, **57**, 209–214.

19. Zheng, B.A., Matsumura, Y., and Mori, T. (1993) Relationship between the thermal denaturation and gelling properties of legumin from broad beans. *Biosci. Biotechnol. Biochem.*, **57**, 1087–1090.

20. Ikeda, S. and Nishinari, K. (2001) On solid-like rheological behaviors of globular protein solutions. *Food Hydrocolloids*, **15**, 401–406.

21. Shewry, P.R. and Halford, N.G. (2002) Cereal seed storage proteins: structures, properties and role in grain utilization. *J. Exp. Bot.*, **53**, 947–958.

22. Ahmedna, M., Prinyawiwatkul, W., and Rao, R.M. (1999) Solubilized wheat protein isolate: functional properties and potential food applications. *J. Agric. Food Chem.*, **47**, 1340–1345.

23. Van Vliet, T., Martin, A.H., and Bos, M.A. (2002) Gelation and interfacial behaviour of vegetable proteins. *Curr. Opin. Colloid Interface Sci.*, **7**, 462–468.

24. Maningat, C.C., Bassi, S., and Hesser, J.M. (1994) Wheat gluten in food and non-food systems. *Tech. Bull. Am. Inst. Baking Res.*, **16**, 1–8.

25. Comfort, S. and Howell, N.K. (2003) Gelation properties of salt soluble meat protein and soluble wheat protein mixtures. *Food Hydrocolloids*, **17**, 149–159.

26. Ramírez-Suárez, J.C., Addo, K., and Xiong, Y.L. (2005) Gelation of mixed myofibrillar/wheat gluten proteins treated with microbial transglutaminase. *Food Res. Int.*, **38**, 1143–1149.

27. Oszvald, M., Tömösközi, S., Tamás, L., and Békés, F. (2009) Effects of wheat storage proteins on the functional properties of rice dough. *J. Agric. Food Chem.*, **11**, 10442–10449.

28. Paraman, I., Hettiarachchy, N.S., Schaefer, C., and Beck, M.I. (2007) Hydrophobicity, solubility and emulsifying properties of enzyme modified rice endosperm protein. *Cereal Chem.*, **84** (4), 343–349.

29. Paraman, I., Hettiarachchy, N.S., and Schaefer, C. (2007) Glycosylation and deamidation of rice endosperm protein for improved emulsifying properties. *Cereal Chem.*, **84**, 593–599.

30. Paraman, I., Hettiarachchy, N.S., and Schaefer, C. (2008) Preparation of rice endosperm protein isolate by alkali extraction. *Cereal Chem.*, **85** (1), 75–81.

31. Lim, S.T., Lee, J.H., Shin, D.H., and Lim, H.S. (1999) Comparison of protein extraction solutions for rice starch isolation and effects of residual protein content on starch pasting properties. *Starch/Starke*, **51**, 120–125.

32. Baxter, G., Blanchard, C., and Zhao, J. (2004) Effects of prolamin on the textural and pasting properties of rice flour and starch. *J. Cereal Sci.*, **40**, 205–211.

33. Sugimoto, T., Tanaka, K., and Kasai, Z. (1986) Molecular species in the protein body II (PB-II) of developing rice endosperm. *Agric. Biol. Chem.*, **50**, 3031–3035.

34. Wen, T.N. and Luthe, D.S. (1985) Biochemical characterization of rice gluten. *Plant Physiol.*, **78**, 172–177.

35. Lanter, K., de Rodas, B., Miller, B.L., and Fitzner, G.E. (2012) Gel based livestock feed, method of manufacture and use. US Patent 8092853, Issued Jan. 10, 2012.

36. Coleman, C.E. and Larkins, B.A. (1999) in *Seed Proteins* (eds P.R. Shewry and R. Casey), Kluwer Academic Publishers, Dordrecht, pp. 109–139.

37. Leite, A., Neto, G.C., Vettore, A.L., Yunes, J.A., and Arruda, P. (1999) in *Seed Proteins* (eds P.R. Shewry and R. Casey), Kluwer Academic Publishers, Dordrecht, pp. 141–157.

38. Croston, B.C. and Evans, C.D. (1950) *Yearbook of Agriculture, 1950–1951*, U.S. Government Printing Office, Washington, DC, p. 607.

39. Shukla, R. and Cheryan, M. (2001) Zein: the industrial protein from corn. *Ind. Crops Prod.*, **13**, 3171–3192.

40. Swallen, L.C. (1941) Zein – a new industrial protein. *Ind. Eng. Chem.*, **33**, 394–398.

41. Dimitroglou, D.A. (1996) Factors affecting zein stability in aqueous ethanol or isopropanol solutions, MS thesis, University of Minnesota, St. Paul, MN.

42. Shewry, P.R. and Tatham, A.S. (1990) The prolamin storage proteins of cereal seeds: structure and evolution. *Biochem. J.*, **267**, 1–12.

43. da Silva, L. and Taylor, J.R.N. (2005) Physical, mechanical, and barrier properties of kafirin films from red and white sorghum milling fractions. *Cereal Chem.*, **82** (1), 9–14.

44. Belton, P.S., Delgadillo, I., Halford, N.G., and Shewry, P.R. (2006) Kafirin structure and functionality. *J. Cereal Sci.*, **44**, 272–286.

45. Mesa-Stonestreet, N.J., Alavi, S., and Bean, S.R. (2010) Sorghum proteins: the concentration, isolation, modification, and food applications of kafirins. *J. Food Sci.*, **75**, R90–R104.

46. Vinkx, C.J.A., Van Nieuwenhove, C.G., and Delcour, J.A. (1991) Physicochemical and functional properties of rye nonstarch polysaccharides. III. Oxidative gelation of a fraction containing water-soluble pentosans and proteins. *Cereal Chem.*, **68**, 617–622.

47. Ma, C.-Y. and Khanzada, G. (1987) Functional properties of deamidated oat protein isolates. *J. Food Sci.*, **52**, 1583–1587.

48. Lâsztity, R. (1984) *Chemistry of Cereal Proteins*, CRC Press, Boca Raton, FL, pp. 159–183.

49. Wang, C., Tian, Z., Chen, L., Temelli, F., Liu, H., and Wang, Y. (2010) Functionality of barley proteins extracted and fractionated by alkaline and alcohol methods. *Cereal Chem.*, **87**, 597–606.

50. Zhao, J., Tian, Z., and Chen, L. (2011) Effects of deamidation on aggregation and emulsifying properties of barley glutelin. *Food Chem.*, **128** (4), 1029–1036.

51. Yin, S.-W., Tang, C.-H., Cao, J.-S., Hu, E.-K., Wen, Q.-B., and Yang, X.-Q. (2008) Effects of limited enzymatic hydrolysis with trypsin on the functional properties of hemp (Cannabis sativa L.) protein isolate. *Food Chem.*, **106**, 1004–1013.

52. Hermansson, A.-M. (1986) Soy protein gelation. *J. Am. Oil Chem. Soc.*, **63**, 658–666.

53. Puppo, M.C. and Añón, M.C. (1999) Rheological properties of acidic soybean protein gels: salt addition effect. *Food Hydrocolloids*, **13**, 167–176.

54. Hua, Y., Cui, S.W., Wang, Q., Mine, Y., and Poysa, V. (2005) Heat induced gelling properties of soy protein isolates prepared from different defatted soybean flours. *Food Res. Int.*, **38**, 377–385.

55. Rickert, D.A., Johnson, L.A., and Murphy, P.A. (2004) Functional properties of improved glycinin and β-conglycinin fractions. *J. Food Sci.*, **69**, 303–311.

56. Scilingo, A.A. and Añón, M.C. (1996) Calorimetric study of soybean protein isolates: effect of calcium and thermal treatments. *J. Agric. Food Chem.*, **44**, 3751–3756.

57. Speroni, F., Jung, S., and De Lamballerie, M. (2009) Effects of calcium and pressure treatment of thermal gelation of soybean protein. *J. Food Sci.*, **75** (1), E30–E38.

58. Bao, X.L., Lv, Y., Yang, B.C., Ren, C.G., and Guo, S.T. (2008) A study of the soluble complexes formed during calcium binding by soybean protein hydrolysates. *J. Food Sci.*, **73**, 117–121.

59. Puppo, M.C., Chapleau, N., Speroni, F., De Lamballerie, M., Añón, M.C., and Anton, M. (2004) Physicochemical modifications of high-pressure treated soybean protein isolates. *J. Agric. Food Chem.*, **52**, 156–471.

60. Lakshmanan, R., De Lamballerie, M., and Jung, S. (2006) Effect of soybean-to-water ratio and pH on pressurized soymilk properties. *J. Food Sci.*, **71**, 384–391.

61. Balny, C. and Masson, P. (1993) Effects of high pressure on proteins. *Food Rev. Int.*, **9**, 611–628.

62. Dumoulin, M., Ozawa, S., and Hayashi, R. (1998) Textural properties of pressure-induced gels of food proteins obtained under different temperatures including subzero. *J. Food Sci.*, **63**, 92–95.

63. Molina, E., Defaye, A.B., and Ledward, D.A. (2002) Soy protein pressure-induced gels. *Food Hydrocolloids*, **16**, 625–632.

64. Cerletti, P. and Duranti, M. (1979) Development of lupine proteins. *J. Am. Oil Chem. Soc.*, **5**, 460–463.

65. Pozani, S., Doxastakis, G., and Kiosseoglou, V. (2002) Functionality of lupin seed protein isolate in relation to its interfacial behavior. *Food Hydrocolloids*, **16**, 241–247.

66. Kiosseoglou, A., Doxastakis, G., Alevisopoulos, S., and Kasapis, S. (1999) Physical characterization of thermally induced networks of lupin protein isolates prepared by isoelectric precipitation and dialysis. *Int. J. Food Sci. Technol.*, **34**, 253–263.

67. Mavrakis, C., Doxastakis, G., and Kiosseoglou, V. (2003) Large deformation properties of gels and model comminuted meat products containing lupin protein. *J. Food Sci.*, **68**, 1371–1376.

68. Sipsas S. (2008) Lupin products – concepts and reality, in *Lupins for Health and Wealth' Proceedings of the 12th International Lupin Conference, Fremantle, Western Australia, September 14–18, 2008* (eds J.A. Palta and J.B. Berger), International Lupin Association, Canterbury. ISBN: 0-86476-153-8.

69. Prakash, V. (1985) Hydrodynamic properties of a– globulin from Sesamum indicum L. *J. Biosci.*, **9**, 165–175.

70. Plietz, P., Damaschun, G., Zirwer, D., Gast, K., Schwenke, K.D., and Prakash, V. (1986) Shape and quaternary structure of alpha-globulin from sesame (Sesamum indicum L.) seed as revealed by small angle X-ray scattering and quasi-elastic light scattering. *J. Biol. Chem.*, **261**, 12686–12691.

71. Rajendran, S. and Prakash, V. (1988) Isolation and characterization of beta-globulin Low molecular weight protein fraction from sesame seed (Sesamum indicum L.). *J. Agric. Food Chem.*, **36**, 269–275.

72. Takenaka, Y., Arii, Y., and Masui, H. (2011) Network structure and forces involved in Perilla globulin gelation: comparison with sesame globulin. *Biosci. Biotechnol. Biochem.*, **75**, 1198–1200.

73. Salgado, P.R., Ortiz, S.E.M., Petruccelli, S., and Mauri, A.N. (2012) Functional food ingredients based on sunflower protein concentrates naturally enriched with antioxidant phenolic compounds. *J. Am. Oil Chem. Soc.*, **89**, 825–836.

74. Sánchez, A.C. and Burgos, J. (1995) in *Food Macromolecules and Colloids* (ed. E. Dickinson), Royal Society of Chemistry, London, pp. 426–430.

75. Oakenfull, P.G. and Morris, V.J. (1987) A kinetic investigation of the extent of polymer aggregation in carrageenan and furcellaran gels. *Chem. Ind.*, **16**, 201–202.

76. Clark, A.H. (1992) in *Physical Chemistry of Foods* (ed. H.G. Hartel), Marcel Dekker, New York, pp. 263–305.

77. Stading, M. and Hermansson, A.M. (1990) Viscoelastic behaviour of β-lactoglobulin gel structures. *Food Hydrocolloids*, **4**, 121–135.

78. Pinterits, A. and Arntfield, S.D. (2007) The effect of limited proteolysis on canola protein gelation. *Food Chem.*, **102**, 1337–1343.

79. Beuchat, L.R. (1977) Functional property modification of defatted peanut flour as a result of proteolysis. *Lebensm. Wiss. Technol.*, **10**, 78–83.

80. Hartnett, E.K. and Satterlee, L.D. (1990) The formation of heat and enzymes induced (plastein) gels from pepsin hydrolyzed soy protein isolate. *J. Food Biochem.*, **14**, 1–13.

81. Kim, S.Y., Park, P.S.W., and Rhee, K.C. (1990) Functional properties of proteolytic enzyme modified soy protein isolate. *J. Agric. Food Chem.*, **38**, 651–656.

82. Mahajan, A. and Dua, S. (1998) Role of enzymatic treatments in modifying the functional properties of rapeseed (Brassica campestris var. toria) meal. *Int. J. Food Sci. Nutr.*, **49**, 435–440.

83. Uruakpa, F.O. and Arntfield, S.D. (2005) The physico-chemical properties of commercial canola protein isolate-guar gum gels. *Int. J. Food Sci. Technol.*, **40**, 643–653.

84. Sathe, S.K. and Salunkhe, D.K. (1981) Functional properties of the Great Northern bean (Phaseolus vulgaris L.) proteins: emulsion, foaming, viscosity, and gelation properties. *J. Food Sci.*, **46**, 71–81.

85. Adebowale, Y.A. and Adebowale, K.O. (2008) Evaluation of the gelation characteristics of mucuna bean flour and protein isolate. *EJEAF Chem.*, **7**, 3206–3222.

86. Adebowale, K.O. and Lawal, O.S. (2003) Foaming, gelation and electrophoretic characteristics of mucuna bean (Mucuna pruriens) protein concentrates. *Food Chem.*, **83**, 237–246.

87. Meng, G.-T. and Ma, C.-Y. (2002) Thermal gelation of globulin from Phaseolus angularis (red bean). *Food Res. Int.*, **35**, 377–385.

88. Hartmann, B. and Schmandke, H. (1988) Gelation of a broad bean (Vicia faba L. minor) protein fraction (prepared at pH 2) in dependence on some influencing factors. *Food Nahrung*, **32**, 127–133.

89. Freitas, R., Ferreira, R., and Teixeira, A. (2000) Use of a single method in the extraction of the seed storage globulins from several legume species. Application to analyse structural comparisons within the major classes of globulins. *Int. J. Food Sci. Nutr.*, **51**, 341–352.

90. Makri, E., Papalamprou, E., and Doxastakis, G. (2005) Study of functional properties of seed storage proteins from indigenous European legume crops (lupin, pea, broad bean) in admixture with polysaccharides. *Food Hydrocolloids*, **19**, 583–594.

91. Papalamprou, E.M., Doxastakis, G.I., Biliaderis, C.G., and Kiosseoglou, V. (2009) Influence on preparation methods on physicochemical and gelation properties of chickpea protein isolates. *Food Hydrocolloids*, **23**, 337–343.

92. Boye, J.I., Aksay, S., Roufik, S., Ribéreau, S., Mondor, M., Farnworth, E., and Rajamohamed, S.H. (2012) Comparison of the functional properties of pea, chickpea and lentil protein concentrates processed using ultrafiltration and isoelectric precipitation techniques. *Food Res. Int.*, **43**, 537–546.

93. Zhang, T., Jiang, B., and Wang, Z. (2007) Gelation properties of chickpea protein isolates. *Food Hydrocolloids*, **21**, 280–286.

94. Mizubuti, I.Y., Biondo, J.O., Souza, L.W., da Silva, R.S., and Ida, E.I. (2000) Functional properties and protein concentrate of pigeon pea (Cajanus cajan

(I.) Millsp.) flour. *Arch. Latinoam. Nutr.*, **50**, 274–280.

95. Horax, R., Hettiarachchy, N.S., Chen, P., and Jalaluddin, M. (2004) Functional properties of protein isolate from cowpea (Vigna unguiculata L. Walp.). *J. Food Sci.*, **69**, 119–121.

96. Ragab, D.D.M., Babiker, E.E., and Eltinay, A.H. (2004) Fractionation, solubility and functional properties of cowpea (Vigna unguiculata) proteins as affected by pH and/or salt concentration. *Food Chem.*, **84**, 207–212.

97. Lampart-Szczapa, E. (2001) in *Chemical and Functional Properties of Food Proteins* (ed. Z.E. Sikorski), Technomic Publishing Company, pp. 407–436.

98. Sosulski, F.W. and Fleming, S.E. (1977) Chemical, functional, and nutritional properties of sunflower protein products. *J. Am. Oil Chem. Soc.*, **54**, 100–104.

99. El-Zalaki, L.M., Gomaa, E.G., and Abdel-Rahman, A.Y. (1955) Peanut protein: functional properties and nutritional studies. *Riv. Ital. Sostanze Gr.*, **72**, 505–508.

100. Schwenke, K.D. (2001) Reflections about the functional potential of legume proteins. *Nahrung*, **45**, 377–381.

101. Braudo, E.E., Plashchina, I.G., and Schwenke, K.D. (2001) Plant protein interactions with polysaccharides and their influence on legume functionality: a review. *Nahrung*, **45**, 382–384.

102. Koumanov, A., Ladenstein, R., and Karshikoff, A. (2001) Electrostatic interactions in proteins: contribution to structure–function relationships and stability. *Recent Res. Dev. Prot. Eng.*, **1**, 123–148.

103. Kumar, N.S.K., Nakajima, M., and Nabetani, H. (2000) Processing of oilseeds to recover oil and protein using combined aqueous, enzymatic and membrane separation techniques. *Food Sci. Technol. Res.*, **6**, 1–8.

104. Sharma, G.M., Su, M., Joshi, A.U., Roux, K.H., and Sathe, S.K. (2010) Functional properties of select edible oilseed proteins. *J. Agric. Food Chem.*, **58**, 5457–5464.

105. Wang, B., Wang, L.-J., Li, D., Bhandari, B., Wu, W.-F., Shi, J., Chen, X.D., and Mao, Z.-H. (2009) Effects of potato starch addition and cooling rate on rheological characteristics of flaxseed protein concentrate. *J. Food Eng.*, **91** (3), 392–401.

106. Hrckova, M., Rusnakova, M., and Zemanovic, J. (2002) Enzymatic hydrolysis of defatted soy flour by three different proteases and their effect on the functional properties of resulting protein hydrolysates. *Czech J. Food Sci.*, **20**, 7–14.

107. Taha, F.S. and Ibrahim, M.A. (2002) Effect of degree of hydrolysis on the functional properties of some oilseed proteins. *Grasas Aceites*, **53**, 273–281.

108. Heywood, A.A., Myers, D.J., Bailey, T.B., and Johnson, L.A. (2002) Functional properties of extruded-expelled soybean flours from value-enhanced soybeans. *J. Am. Oil Chem. Soc.*, **79**, 699–702.

109. Tandang, M.R., Adachi, M., Inui, N., Maruyama, N., and Utsumi, S. (2004) Effects of protein engineering of canola procruciferin on its physicochemical and functional properties. *J. Agric. Food Chem.*, **52**, 6810–6817.

110. Moure, A., Sineirob, J., Domíngueza, H., and Parajóa, J.C. (2006) Functionality of oilseed protein products: a review. *Food Res. Int.*, **39**, 945–996.

111. Horax, R., Hettiarachchy, N., Over, K., Chen, P., and Gbur, E. (2010) Extraction, fractionation and characterization of bitter melon seed proteins. *J. Agric. Food Chem.*, **58** (3), 1892–1897.

112. Horax, R., Hettiarachchy, N., Kannan, A., and Chen, P. (2011) Protein extraction optimization, characterisation, and functionalities of protein isolate from bitter melon (Momordica charantia) seed. *Food Chem.*, **124** (2), 545–550.

113. Sánchez, A.C. and Burgos, J. (1996) Thermal gelation of trypsin hydrolysates of sunflower proteins: effect of pH, protein concentration, and hydrolysis degree. *J. Agric. Food Chem.*, **44**, 3773–3777.

9
Enzymatically Texturized Plant Proteins for the Food Industry

Christian Schäfer

9.1
Introduction

Supplying today's growing world population of ~7 billion people with adequate and quality protein is becoming more and more critical. Opportunities for the production of sufficient proteins of animal origin as source for balanced amino acid provision to the population are already restricted in several areas due to limitations of available soil and water and for ethnic reasons. Proteins of plant origin in general, and from drought-resistant plants in particular, will become increasingly prominent in the future.

Owing to crises related to the production of animals as part of the food chain, such as bird flu, swine fever, and "mad cow" disease, plant proteins could benefit with regard to food safety and general perception. As a matter of course, the well-known allergenicity of certain plant proteins has to be taken into consideration, too.

The demand for high quality food proteins has increased over recent decades. Particularly regarding consumers' acceptance and preferences, methods for modifying the techno-functional properties and nutritional value of food ingredients, such as proteins, are of increasing interest in order to develop convenient and healthy foods [1].

Besides color, aroma, and flavor, texture is a key quality parameter in food technology. Proteins are food ingredients that contribute to the texture of food products. A wide range of proteins from animal and vegetable sources, including those of single-cell organisms, is available for the production of foods.

Texturized vegetable proteins (TVPs), mainly containing proteinous raw material from soybean such as concentrates or isolates, which differ in their protein content, are well known to the food industry. Many excellent scientific reports have been published over the last decades, including a recent review by Riaz [2]. Up to now, two processes, namely, extrusion and spinning, have predominantly been applied to manufacture texturized plant protein based food products [3]. Food products such as snacks, meat extenders, and several structured meat analogs can be found on the shelves of today's supermarkets.

Product Design and Engineering: Formulation of Gels and Pastes, First Edition.
Edited by Ulrich Bröckel, Willi Meier, and Gerhard Wagner.
© 2013 Wiley-VCH Verlag GmbH & Co. KGaA. Published 2013 by Wiley-VCH Verlag GmbH & Co. KGaA.

Cost-effectiveness, suitable functionality, and the demand for nutritional and health value are the challenges in today's food product development. In addition, beside gluten-free proteins, non-genetically modified and vegetarian products have become a key issue.

Therefore, vegetable proteins have gained increasing attention, being associated with high consumer acceptance and nutritional value. Cereals and grain legumes are the most important vegetable sources of protein. However, the allergenicity of cereal proteins and insufficient texture properties of proteins from plant origin have so far hindered the successful launch of such products. Thus, enhancing the texturizing properties of proteins from plant origin could open up new areas for the development of healthy, well accepted, and cost-effective food products.

Microbial transglutaminase (MTG) isolated from *Streptomyces mobaraensis* has been available on a commercial scale for several years and represents a powerful tool for the improvement of techno-functional properties of proteins in food systems, such as texture, gel firmness, viscosity, emulsifying properties, heat-resistance, water-binding, and oil-binding capacity. MTG catalyzes the formation of isopeptide bonds between γ-carboxylamide groups of glutamine residues and ε-amino groups of lysine residues in proteins that contribute to the formation of a stable protein network. The formation of these ε-(γ-glutamyl)lysine linkages catalyzed by MTG presents an effective method for achieving significant textural improvement of protein-containing food products.

Scientific publications concerning the suitability of transglutaminase (TG, protein-glutamine γ-glutamyl-transferase, EC 2.3.2.13) for enzymatic crosslinking of proteins to improve their techno-functional properties can be traced back at least three decades [4]. Application of MTG for the production of plant protein-based food products such as tofu, noodles, bread, and bakery products, however, is still largely focused on raw materials from soybean and wheat.

Recently, the application of a Ca^{2+} independent MTG has been demonstrated for leguminous protein isolates from soy, pea, and lupin [5].

Since the isolation of MTG from *Streptomyces mobaraensis* in 1989 [6], and its purification from *Streptomyces* sp. [7–9], commercial products containing MTG have been launched and found their way into the food industry. In particular, in manufacturing restructured meat and seafood products.

Various effects of MTG on the techno-functional properties of proteins such as improvement of texture, stability, increased water-binding, and utilization in the food industry have been reviewed by Dube *et al.* [10]. The authors reported specific applications for meat, fish, dairy, egg, bread, and bakery products, and in soybean processing. Furthermore, two independent patent searches from 1990 to 2007 [10] and from 1990 to 2008 [11] highlighted 165 and 62 relevant patent applications, respectively. In total, general applications of MTG, processing of dairy, fish, meat, other animal foods, bread and bakery, noodles and pasta, soybean, and other vegetable foods were listed. Whereas most inventions dealt with animal food products, MTG has rarely been used in the texturization of plant proteins, with the exception of tofu production from soybean, bread making, and noodle production from wheat. In addition, research papers in recent years reflect the potential of crosslinked proteins for developing novel food products and processing methodologies that allow the manufacture of products with high convenience and improved sensory and nutritional-physiological properties.

An increasing demand for vegetarian foods in many branches of the food industry has been observed over recent decades. Apart from the competitive price of plant proteins and their general security of supply as well as easy storage, ecological, and energetic advantages in comparison with their animal-derived counterparts, plant-derived food products are well accepted by consumers. The utilization of proteins from pea and lupin, which are major crops in the Mediterranean Basin and Central Europe, and proteins from so far underestimated plants, for example, sesame, rice, sorghum, and sunflower, as functional ingredients, seems promising.

For all these reasons, food products manufactured by usage of MTG-crosslinked vegetable proteins can be considered a reliable source for human diets based on sufficient and healthy protein with a balanced amino acid composition from plant sources.

9.2
Reactions Catalyzed by MTG

Ca^{2+}-independent MTG catalyzes an acyl-transfer between γ-carboxylamide groups of peptide- or protein-bound glutamine residues and primary amines (Scheme 9.1). In the absence of primary amines water is used as an acyl acceptor and as a result glutamine residues are deamidated (Scheme 9.1, reaction IV). The vicinity of the glutamine residue in the protein chain, in particular with regard to the nature and conformation of its amino acids, affects the reactivity of the acyl donor. In the early stage of the reaction, a sterically determined intermediate thioester between the proteinous glutamine residue and the cysteine residue of the reactive center of the enzyme is formed under release of ammonia (Scheme 9.1, reaction I). For this reason, the glutamine content and the bonding environment of this amino acid are considered the limiting parameters of the crosslinking reaction. Isopeptide bonds [ε-(γ-glutamyl)lysine crosslinks] are created when ε-amino groups of protein-bound lysine residues act as acyl acceptors (Scheme 9.1, reaction II). As a consequence, a stable protein network via inter- and intramolecular interactions is formed. Assuming that ε-amino side-chains from protein-bound lysine are present, the deamidation reaction is generally understood to be much slower than the linkage to primary amines and the formation of crosslinks, respectively.

The crosslinking reaction (Scheme 9.1, reaction II) facilitates modification of the techno-functional properties of proteins. Additionally, the reaction between γ-carboxylamide groups and primary amines (Scheme 9.1, reaction III) can be used to improve the nutritional value of proteins of plant origin, which are often deficient in specific amino acids such as lysine, cysteine, and so on. MTG crosslinking of proteins with peptides containing appropriate amino acids is a powerful tool for fortification of proteins in general.

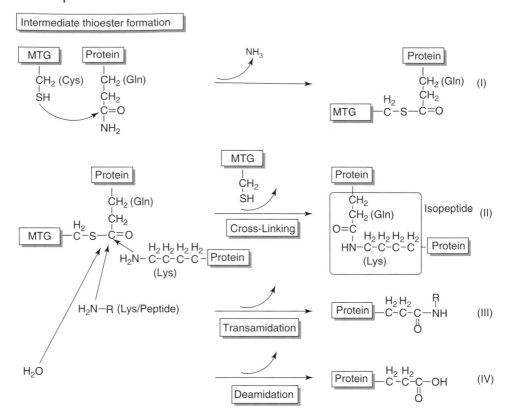

Scheme 9.1 Reactions catalyzed by Ca^{2+}-independent MTG (microbial transglutaminase), in particular crosslinking reaction (II) between protein-bound glutamine (Gln) and lysine (Lys) residues.

9.3
Current Sources of MTG

In general, appearance of TGs is widespread in nature. TG has been found in various tissues of animal and plant origin [10]. A prominent representative of TG is Factor XIII, also known as *Laki–Lorand factor*, which is involved in wound healing by crosslinking human fibrin during blood clotting. Isolation and extraction of TG and MTG has been attempted in multiple ways [10]. The most important milestone with regard to commercialization of MTG was reported in 1989. A Ca^{2+}-independent MTG from *Streptomyces mobaraensis* was extracted by Ando *et al.* [6]. Purification was rather easy, and the production of MTG containing enzyme preparations could be established commercially in subsequent years. Attempts to obtain MTG from other microorganisms by conventional fermentation or by means of genetic modification, with all the advantages and disadvantages in view of their usability on a commercial scale, are continuously being reported. Since food regulations with regard to mandatory labels apply – and acceptance by consumers, especially in Europe is limited – production of MTG by means of genetic modification is deemed to be less favorable.

At present, the only commercially available MTG is still isolated from *Streptomyces mobaraensis*.

9.4
Need for Novel Sources of MTG

All known MTGs have an optimum temperature in the range 40–60 °C [10]. For some food applications, crosslinking reactions at higher temperatures, for example, in soups and sauces, or at lower temperatures, for example, in deserts, would be advantageous. MTG variants with maintained activity at lower temperatures around 10 °C and mutants with increased stability at temperatures higher than 60 °C have been identified, isolated, and characterized from bacterial strains of *Escherichia coli* BL21Gold(DE3) [12]. As a consequence, inactivation of heat-sensitive variants could be carried out at lower temperatures, thus enabling strategies for energy saving in food processing. Furthermore, these MTGs could be applied to food products that need to be treated at low temperatures. A corresponding patent application with regard to increased heat resistance and enzyme activity in comparison to the MTG from *Streptomyces mobaraensis* has been filed quite recently [13]. In addition, successful isolation of MTG from *Streptomyces* sp. P20 [14] and from *Streptomyces hygroscopicus* [15] was reported. Cui *et al.* [15] described a variant of MTG from *Streptomyces hygroscopicus* that was stable up to 40 °C and the activity could be increased by the addition of polyols typically used as food ingredients, such as sorbitol, sucrose, and maltodextrin. Moreover, Yu *et al.* purified MTG from *Streptomyces netropsis* and achieved overproduction of a recombinant MTG in *Escherichia coli* [16]. To expand application areas in the food industry, efficient systems for the expression of TG from microbial sources, as exemplarily described in the above-mentioned publications, are a prerequisite and may result in further commercially available enzymes that are heat and cold sensitive.

9.5
Vegetable Proteins Suitable for Crosslinking with MTG

Besides proteins of animal origin, mostly proteins from soybean and wheat have been subjected to enzymatic crosslinking. In recent years, other crops have also been suggested as potential raw materials for the production of proteinous ingredients, for example, isolates from pea, lupin, rice, and sunflower [5, 10]. Depending on their botanical origin, the manufacture of proteins from cereals, pulses, and oilseeds includes specific processing such as dehulling, flaking, grinding, de-oiling, enzymatic extraction methods, and isolation via isoelectric precipitation. Anti-nutritive substances such as protease inhibitors, glucosinolates, and phytates are very often reduced to an acceptable level during aqueous extraction of the proteins. In addition, adsorption for the removal of unwanted substances such as polyphenols, enzymatic treatment for removal of sensory active molecules related to off-flavors such as degradation products from lipid oxidation, and drying steps may be used to obtain storable products. All these process steps are related to exposure of the proteins to elevated temperatures and thus have potential to influence their structure and as a consequence their techno-functional properties such as solubility, emulsification capacity, and water-binding.

The suitability of plant proteins for crosslinking with MTG depends on their amino acid composition, nutritive value, structure, availability, and utilization in human nutrition. In consideration of all these factors, successful product development of healthy and convenient food products is feasible. In addition, the quality of the proteins in general, and suitability for crosslinking with MTG in particular, is subject to factors such as harvesting period, climate, soil conditions, and so on, and the presence of secondary plant components such as polyphenols and alkaloids, respectively. Aqueous protein extraction processes include, partly, removal of the latter substances and are therefore considered advantageous.

The following vegetable proteins are listed with respect to their availability, utilization in the food industry, and suitability as substrate for MTG crosslinking technology.

9.5.1
Soy Protein

Soybeans [*Glycine max* (L.) Merr.] constitute the most important source of vegetable protein for the feed and food industry. Basically, a well-developed processing technology and their high protein content have allowed the manufacture of suitable products that could be commercialized and applied in food production. Additionally, beneficial effects on nutrition and health, such as lowering of plasma cholesterol and osteoporosis, have been reported [17]. The structures of soybean storage proteins, in particular the main fractions β-conglycinin and glycinin including their subunits and their contribution to gel formation, have been assayed *in extenso*. A recent review describes the structure–function relationship and its relevance for soy protein as a food ingredient in parallel to further physiologically active substances, for example, isoflavones [18]. Already in 1994, Nonaka *et al.* [19] studied the basic properties of soy protein treated with MTG. The rheological characteristics of MTG-treated protein gels from soy protein isolate (SPI) and soy protein concentrates (SPCs) were compared. Breaking strength and deformation of MTG-treated gels from SPI and SPC increased. Gels obtained from SPI were more rigid and more elastic than gels from SPC. The presence of higher carbohydrate contents in SPC was deemed to be the cause of this effect. The amounts of isopeptide increased with rising enzyme dosage and MTG-induced gels were more resistant to large deformation and had a higher rigidity in comparison with thermally induced glycinin gels. The inter- and intramolecular association of protein molecules through ε-(γ-glutamyl)lysine crosslinks was considered to be more regular than in thermally induced gels. Furthermore, a stronger contribution to protein–protein interactions via the irreversible formation of covalent isopeptide bonds was reconsidered [20]. Kang *et al.* [21] reported that the nature of surface lysine and glutamine residues of glycinin molecules affected the gel properties of soybean glycinin in addition to the amount of enzymatically formed crosslinks. The content of ε-(γ-glutamyl)lysine crosslinks could be increased as a result of heating the glycinin solution prior to the MTG reaction.

9.5.2
Wheat Protein

Since gluten from wheat (*Triticum* sp.) lacks in the essential amino acids lysine and threonine, the protein is considered of inferior quality especially with regard to its nutritive value [22]. Because of its higher glutamine content, gluten is assumed to serve as a better acceptor in the MTG reaction than other food proteins, although its low lysine content limits its participation as an amine donor substrate [23]. Superior reactivity of wheat glutenins and gliadins in comparison with prolamines from oats, maize, and rice was reported by Porta *et al.* [24]. During the MTG reaction the high-molecular weight glutenin subunits were the most crosslinked ones [25, 26]. MTG-treated gluten appeared to be less sensitive to heat treatment than its native counterpart with regard to its viscoelastic properties [27] and formed gels with considerably enhanced elasticity [28]. Numerous studies have reported, both positive and negative effects, of MTG utilization in the manufacture of bakery products. In essence, distinct optimum enzyme levels are required to obtain maximum improvement of the techno-functional properties [29].

9.5.3
Rice Protein

The annual production of rice (*Oryza sativa* L.) is nearly as high as that of wheat. In contrast, reports about enzymatic crosslinking of proteins from rice, in particular rice bran protein and rice endosperm protein, are scarce in research papers and patent applications. In principle, rice proteins can be considered interesting substrates for MTG crosslinking. Today, the commercial availability of rice bran concentrates and isolates is still limited. The complex nature of rice bran proteins and especially their high content in phytate (1.7%) and fiber (12%) may serve as explanation. The latter components were shown to bind with proteins. Additionally, the protein content of rice is lower than the level found in most other cereals. Interestingly, the lysine content of rice proteins exceeds that of wheat by more than 50%. Rice proteins cannot form the network necessary to hold gas produced during the fermentation process [30]. It was assumed that applications of rice flour could be broadened if rice proteins were crosslinked to produce a stable network [31]. Doughs with improved viscoelastic behavior resulted from crosslinking with MTG. Rice constitutes a major food source for a large part of the world's population. For this reason, extensive efforts have been made as to the production of rice protein products and improvement of their functional and nutritive properties by the use of MTG. Rice endosperm protein is accessible through various methods, such as alkali, salt, and enzymatic extraction [32] and, especially, alkaline extraction followed by isoelectric precipitation and fractionation by ultrafiltration. Since the physicochemical properties were shown to be positively affected by these extraction methods [33], proteins from plants that were earlier considered poor candidates, such as rice endosperm and in addition rice bran protein [34], may serve as techno-functional ingredients using MTG crosslinking technology.

9.5.4
Pea Protein

More than 90% of the pea (*Pisum sativum* L.) production in Europe is used for animal feed. Peas and other legumes contain various anti-nutritive compounds, such as phytic acid, that interact with their proteins. Processes for dephytinized proteins have been developed for soy and pea protein. Larré *et al.* [35] studied the action of MTG on polypeptides of pea legumin. Whereas the α-polypeptides were polymerized in the native globular conformation, the β-polypeptides were inaccessible to MTG. Legumin was a rather poor substrate for MTG although its content of glutamine residues was relatively high. The compact globular structure of these storage proteins was considered to be the main factor restricting access of MTG to the reactive residues. Acylation was applied to modify the protein conformation and was shown to be a useful technique for converting 11S seed globulins into suitable substrates for MTG [36]. The substrate conformation of pea legumin is one key factor for successfully applying MTG crosslinking technology. Potential phytate–protein interactions in MTG application to pea proteins should be kept at a minimum level by the use of dephytinized proteinous substrates.

9.5.5
Lupin Protein

Lupin seeds (*Lupinus* sp.) even grow on poor soils that are unsuitable for soybeans and combine high protein content with high nutritive value. Cultivars from *Lupinus albus* L. and *Lupinus angustifolius* L., so-called sweet lupins, low in alkaloids have found their way into the food industry. Owing to their high contents of toxic quinolizidinic alkaloids, the use of bitter lupins is restricted in human foods. Removal of the water-soluble alkaloids during the aqueous extraction and production of lupin protein isolates may overcome this problem. The addition of lupin flour of up to 5% provided optimal bread quality, whereas at higher levels adverse effects appeared. More recently, lupin proteins have found their way into successful food applications, for example, ice cream, and contribute to meeting consumer demand for more vegetarian food products. Owing to their high nutritive value, high lysine content, and techno-functional properties the potential of lupin proteins for food applications on the basis of MTG-catalyzed crosslinking is evident. The suitability of protein isolates from lupin for enzymatic crosslinking was demonstrated in model systems [37]. MTG-induced formation of ε-(γ-glutamyl)lysine crosslinks was determined via HPLC-MS (high-performance liquid chromatography–mass spectrometry) after proteolytic digestion of the crosslinked protein. The amounts of generated isopeptide bonds correlated with increasing gel strengths due to the formation of protein networks and gel structures.

9.5.6
Sesame Protein

Without observing adverse effects on bread sensory properties, protein isolate obtained from sesame seeds (*Sesamum indicum* L.) has been used in bread making. The addition to wheat flour, up to a protein level of 18%, and the admixture of concentrates or flours up to a protein level of 16% were possible. Dough development time was lowered and the water absorption was increased. Furthermore, commercially available sesame protein isolate was used as protein source in a liquid nutritional supplement [38]. The emulsion stability of a prototype formulation was similar to that of commercial beverages. A sensory panel preferred the product with sesame protein in comparison with counterparts prepared with a soybean isolate and a commercial brand containing dairy proteins. Broader utilization of sesame protein seems to be unlikely due to the low content of the limiting amino acid lysine, although it appears to be suitable for crosslinking with MTG.

9.5.7
Sunflower Protein

Soybean, wheat, lupin, and several other storage proteins display a considerable allergenic potential. Sensitization by sunflower (*Helianthus annuus* L.) proteins has rarely been described. Moreover, in contrast to leguminous proteins, for example, from soybean or pea, sunflower is devoid of green and beany flavors, especially those induced by lipoxygenase activity. However, sunflower flour was found to contain phenolic compounds at significant levels: approximately 4% were detected, mainly chlorogenic acid. Pierpoint studied the reactions of amino acids and peptides with *o*-quinones produced by enzymatic and non-enzymatic oxidation of chlorogenic and caffeic acid [39]. Except for lysine and cysteine, amino acids react through their α-amino groups to give red or brown products. Secondary reactions may follow and absorb oxygen, resulting in black or green colored products. This effect is still the major hindrance in the use of sunflower protein in food products. Several processes have been suggested to provide sunflower proteins low in chlorogenic acid.

A combination of sunflower protein concentrate and texturized soy protein resulted in stronger and less elastic doughs with an increased water-binding capacity. The potential of textured sunflower protein for the use as a meat extender was described by Rossi [40]. Ground-beef patties showed higher juice retention and enhanced chewiness when 15–45% of their meat content was replaced with sunflower protein. However, the use of sunflower protein isolates is still limited due to their gelation behavior. Sunflower protein based gels cannot be manufactured by controlled acidification or heating. Helianthinin, the main storage protein of sunflower, aggregated similarly to other 11S globulins at different pH and ionic strengths, but it remained in a trimeric form at pH 11. The thermal stability was higher than that of other 11S globulins. Partial protein hydrolysis with trypsin was successfully applied and gels were obtained that became notably weaker when cooled below 80 °C. MTG crosslinking may be a promising tool for gelation of sunflower protein, avoiding protein hydrolysis and its disadvantage in forming bitter peptides. In the presence of phenolic compounds such as chlorogenic acid, MTG activity may be negatively affected due to enzyme inhibition.

Recently, an optimized method for the extraction of high-quality sunflower protein in terms of light-colored protein isolates using mild-acidic extraction has been reported [41]. Colorless protein isolates from sunflower low in phenolic compounds can be considered a valuable ingredient for texturized foods.

9.5.8
Canola Protein

Canola (*Brassica* sp.) is primarily grown because of its ability to store high amounts of oil in its seeds and represents, at 14%, the second important source of oilseed meal following soybean, which is the most important with 59% for the world oilseed market [42]. After oil removal via pressing and extraction, significant amounts of protein remain in the meal. To date, canola proteins primarily make their way into animal feed production, whereas they are not widely used in the food industry. However, at least three different Canadian companies have commercialized products based on canola proteins for the food industry. Solubility, water- and fat-binding, emulsification, foaming, gelation, and film forming are the functional properties that determine their potential for application, for example, in beverages, baked goods, dairy, dressings and sauces, and desserts. Especially in areas where texture is a prerequisite for successful product development, canola proteins may benefit from MTG-crosslinking technology. Since canola proteins have not traditionally been consumed to a significant extent by humans, regulatory approval as a novel food ingredient or the designation Generally Recognized as Safe (GRAS) may be required according to relevant regulations. Adequate petitions have either been granted by or submitted to authorities in the Unites States of America, Canada, and Europe. However, its well balanced composition with regard to high contents of indispensible amino acids makes canola protein attractive for use in food products and in particular for use in MTG-crosslinking as lysine and glutamine contents are in general high [42].

9.5.9
Potato Protein

Following maize, rice, and wheat, potato (*Solanum tuberosum* L.) is ranked fourth amongst world staple crops. They are used directly as part of the daily menu and in addition for the extraction of starch [43]. As a co-product from industrial starch processing a valuable protein fraction is available. Since no significant protein modification such as the formation of lysino-alanine could be demonstrated and low levels of glycoalkaloids and residual sulfite from processing could be achieved, respective commercial products have obtained the GRAS status in the Unites States of America and have been approved as Novel Food in Europe. Potato proteins show good emulsifying and foaming properties. Therefore, the usage of commercially available potato proteins in the food industry is recommended for dressings, emulsion-based beverages, ice cream and sorbet-type ice creams, and desserts. Nutritional parameters such as PER (protein efficiency ratio), BV (biological value), and PDCAAS (protein digestibility corrected amino acid score) of protein isolates from potato are high in comparison with, for example, soy and wheat and therefore indicate high potential for their use in enzymatically texturized food products in particular and in human foods in general [43].

9.5.10
Sorghum Protein

Grain sorghum (*Sorghum bicolor* L.) is an important food crop in many countries, used for food as grain and in sorghum syrup or "sorghum molasses," fodder, the production of alcoholic beverages, as well as for the manufacture of biofuels. Most varieties are drought tolerant and heat tolerant, and are especially important in arid regions where the grain is staple, or one of the staples, for poor and rural people. They form an important component of pastures in many tropical regions. Sorghum is an important food crop in Africa, Central America, and South Asia and is the fifth most important cereal crop grown in the world. The protein content of sorghum is usually between 9% and 15% and ranks high in nutritional quality.

Sorghum endosperm proteins are recognized as hypoallergenic and can, thus, be a suitable alternative source of protein. However, high insolubility, hydrophobicity, and poor functionality of sorghum endosperm protein at neutral pH limit its industrial application as a functional ingredient in food.

The extracted proteins are highly insoluble in nature and the conditions used in protein isolation further decrease their solubility. High-protein sorghum products can be obtained from sorghum flour by alkali extraction followed by precipitation at the isoelectric pH of the protein. Starch-hydrolyzing enzymes such as α-amylase, glucoamylase, and pullulanase are often used to separate proteins in sorghum flour by solubilizing and removing starch. In addition to starch hydrolyzing enzymes, cellulase and hemicellulase enzymes have been used to further increase the protein content in sorghum protein concentrate. However, information on suitable extraction methods and functionalities of such isolates is limited. Efficient extraction methods using approved food-grade enzymes and chemicals are essential for commercial production and application of sorghum protein. Several promising protein modification studies such as fermentation, deamidation, enzymatic hydrolysis, and protein–polysaccharide conjugation have been undertaken in attempting to overcome these problems. Therefore, sorghum has recently been classified as an attractive raw material and a good source of protein due to the neutral flavor and color of specific varieties, low allergenicity, and its ability to grow in drought-like conditions. The potential for usage in food applications is deemed to be huge [44]. MTG-crosslinking technology may represent a promising tool to modify techno-functional properties of sorghum proteins in the desired direction. An improvement in solubility and emulsifying properties of sorghum proteins as a result of protein modification using MTG purified from *Streptoverticillium cinnamoneum* subsp. *cinnamoneum* IFO12852 has already been described by Babiker and Kato [45].

9.5.11
Various Other Vegetable Protein Sources

As shown by the cited examples, the suitability of vegetable proteins for crosslinking with MTG does not only depend on protein quality, nutritive value, sensory acceptance, and chemical properties, such as amino acid composition and structural conformation, but also on widespread commercial availability, healthiness, and ethnic familiarity based on traditional usage. For these reasons, extension of MTG-crosslinking technology to proteins from commonly grown crops as listed in Chapter 8, and many more, is deemed advisable.

9.6
Strategies to Modify and Improve Protein Sources for MTG Crosslinking

9.6.1
Hydrolysis

Enzymatic and acidic hydrolyses, followed by MTG treatment, have been reported to improve foaming and emulsifying capacities, solubility, and bitterness of soy [46–48] and wheat protein [49]. In addition, MTG-induced changes in foaming and emulsifying properties of hydrolyzed wheat gluten prepared by partial hydrolysis were attributed to pH-dependent solubility changes, amphiphilic nature of gluten proteins, and increased electrostatic repulsion resulting from deamidation [50]. Ionic strength (0 and 0.6 M NaCl), pH (4.0 and 6.5), and temperature (5 and 20 °C) were the major parameters influencing the use of MTG in food processing. The applicability of controlled enzymatic hydrolysis to rice endosperm protein to enhance solubility and emulsifying properties has also been described [51] and represents a tool to make proteins from rice and other plants an interesting source for application in enzymatically texturized foods. Crosslinking proteins in rice flour with MTG have already been shown to improve the hardness, gumminess, and rheological properties of rice gels [52]. Restricted enzymatic hydrolysis of protein isolates from pea, soy, and lupin to a degree of hydrolysis (DH) in the range of 1–3% clearly enhanced solubility and emulsifying properties. Subsequent MTG crosslinking of all three partially hydrolyzed proteins resulted in increased gel strengths as compared to the unmodified isolates [5]. Recently, a positive effect was reported for combined enzymatic hydrolysis with chymotrypsin and MTG crosslinking on the techno-functional properties, such as solubility at pH levels from 2 to 12, emulsifying stability, and emulsifying activity of proteins from two different sorghum cultivars [53].

As described for proteins from soy, wheat, rice, sorghum, and lupin, moderate enzymatic hydrolysis may represent a useful tool to make further vegetable proteins available for texturized products.

9.6.2
Maillard Reactions

Techno-functional properties can be improved via Maillard-type protein–polysaccharide conjugation, which is, thus, a promising method for broadening the use of plant proteins as food ingredients. Allergenicity and bitterness, a widespread phenomenon resulting from enzymatic hydrolysis, can also be reduced by covalent attachment of polysaccharides [52]. Glycosyl units incorporated via MTG crosslinking increased the solubility of pea legumin and wheat gliadins within the range of their isoelectric points [54]. Furthermore, denser and finer gel networks, in comparison with control gels, were obtained when MTG crosslinking and subsequent Maillard-type reactions by heating SPI in the presence of ribose were applied [55, 56].

9.6.3
Deamidation and Glycosylation

Paraman *et al.* [57] investigated the impact of modification methods on the functional properties of rice endosperm protein. Controlled alkaline deamidation increased the solubility, emulsion activity, and emulsion stability. Glycosylation improved the emulsion activity and emulsion stability. Furthermore, enzymatic hydrolysis slightly improved the solubility, emulsion activity, and emulsion stability. Transfer of these findings to other vegetable protein sources should be the subject of further research.

9.6.4
Solubilization and Hydrothermal Treatment

The availability of reaction sites of vegetable proteins for MTG crosslinking may be enhanced by preheating. Improved formation of the MTG induced network could be observed [5] for protein isolates from lupin, pea, and soy, when proteins were solubilized for 30–60 min at 50–60 °C. Liu *et al.* [58] reported that hydrothermal treatment in combination with MTG crosslinking enhanced the water absorption of wheat gluten.

9.6.5
Removal of Undesired Substances from Vegetable Proteins

The preparation of model foods basedd on MTG crosslinked protein isolates from Soy Protein Isolate (SPI) and Pea Protein Isolate (PPI) resulted in an undesirable beany off-flavor [5]. Flavor active lipids from protein isolates can be removed by their enzymatic conversion into water-soluble degradation products and subsequent aqueous extraction as described in related patent literature [59]. An adsorption process for the elimination of undesirable compounds, particularly flavor and pigments from vegetable proteins, was described in a recent patent application [60]. On the whole, MTG-catalyzed crosslinking of several vegetable proteins can be enhanced by enzymatic and physical treatments, such as partial enzymatic hydrolysis, glycosylation, preheating, deamidation, and adjustment of pH, temperature, and ionic strength to improve the functional properties of proteins. Since all these methods are based on aqueous systems they may also serve for the removal of undesired water-soluble compounds such as polyphenols, phytic acid, and alkaloids.

9.6.6
Improving the Nutritive Value of Plant Proteins

Variation of the amino acid composition of plant proteins and consequently influence on their nutritive value is widely observed. Furthermore, factors such as plant diversity, harvesting conditions, cultivation area, allocation of proteins in the organism, and process technology including extraction (e.g., isoelectric precipitation), purification (e.g., membrane filtration), and especially drying (temperature profile) applied during manufacture significantly contribute to techno-functional properties and BV. WHO/FAO has provided a standard for a well-balanced amino acid composition in human nutrition. Since plant proteins frequently lack one or more amino acids, methods for improving the nutritive value of food proteins are desirable. Blending of different protein sources and supplementation with amino acids may disqualify usage of plant proteins in various applications especially in infant nutrition. The incorporation of amino acids into food proteins through TG-catalyzed acyl-transfer between peptide-bound glutamine residues and primary amino groups was investigated by Ikura *et al.* [61]. Whereas free amino acids, with the exception of D- and L-lysine, did not serve as amine substrates, most of the methyl or ethyl esters and amides of L-amino acids were suitable amine substrates. In comparison with the starting material an increase in the methionine content to 250% was achieved via incorporation of methionine ethyl ester into soybean proteins by usage of TG. Moreover, lysine was successfully incorporated into wheat gluten. In another study, successful incorporation of lysine into citraconylated α_{s1}-casein using MTG was demonstrated by Nonaka *et al.* [62].

Incorporation of protein fractions from the same source, proteins from different sources, or suitable dipeptides containing the limiting amino acids using MTG-crosslinking technology may represent a promising approach for improving the nutritive value of food proteins derived from plants. In parallel to achieving a well-balanced amino acid composition for human nutrition, the techno-functional properties may be improved.

The strategies described in this section, including combinations thereof, for example, moderate enzymatic hydrolysis followed by glycosylation may all have an effect on the suitability of various plant proteins for MTG crosslinking. The selection of optimum parameters for protein modifications to increase the availability of reaction sites for MTG is deemed a prerequisite for successful development of products based on vegetable proteins.

Several studies have reported MTG crosslinking of salt-extracted proteins from pea [63], protein digests from sorghum [53], and heat treated SPIs in the presence of ribose and sucrose [64]. In general, interest in MTG crosslinking technology has increased and, in parallel, attention is being paid to novel and modified protein sources.

9.7
Applications of MTG in Processing Food Products Containing Vegetable Protein

Texturization of food products of animal origin has been shown to represent the major application areas for MTG crosslinking technology. The following examples have already been highlighted by Dube *et al.* [10] and illustrate the usage of MTG application in the processing of food products containing plant proteins, especially from soybean and wheat.

9.7.1
Soybean Products

The improvement of texture, sensory properties, and shelf life of tofu are the main drivers for MTG application. Traditional tofu (moisture content ~90%) kept at temperatures below 10 °C has a shelf life of two to four days. Large amounts of coagulants, such as magnesium chloride, are required when packaged stable tofu is produced from sterilized soybean milk, resulting in non-uniform coagulation. A process for producing packaged tofu, in which MTG was added to sterilized soybean milk either simultaneously or after addition of glucono-δ-lactone, was disclosed by Matsuura *et al.* [65]. The tofu was evaluated as excellent in both hardness and flavor.

Nonaka *et al.* [66] described in a patent application that MTG-treatment generated a softer and smoother tofu, which was stable for six months or longer at 25 °C, whereas sterilized soybean curds (e.g., 100–125 °C; 10–30 min) were generally too hard and dry. MTG crosslinking was responsible for a higher resistance to heat exposure and the tofu lost less of its weight through cooking for 30 min.

A process for industrially produced tofu products with excellent taste and freeze resistance was described by Oomura *et al.* [67]. Control of the coagulation reaction was easier and a smooth texture could be produced by application of MTG.

A method described in a patent application by Sawano and Sawano [68] enabled the utilization of soybean curd lees, which are usually discharged during the production of soybean milk and curd, as a food product. The plant fiber of pulverized curd lees was mixed with water and macerated by different tissue disintegrating enzymes, such as pectinases and cellulases. MTG was added to the soybean curd lees slurry to produce functional curd lees "milk."

9.7.2
Bread and Bakery Products

Several studies revealed beneficial effects on baking properties and product quality when MTG was applied. Positive effects of MTG on the rheological properties of doughs from wheat flours and bread quality were shown by Basman *et al.* [29]. Interestingly, decreasing flour–water absorption was observed with increasing enzyme levels. Parameters such as crust and crumb characteristics were improved at low levels of MTG, whereas higher MTG levels had detrimental effects. Another study of Basman *et al.* [69] dealt with the influence of MTG on the admixture of barley or soy flour to wheat flour. Higher loaf volumes as well as better crumb and crust characteristics were reported for bread from MTG-treated soft wheat flour when barley flour was added up to 30%. MTG-treated products were softer than those from non-treated samples. In contrast, slight decreases in the loaf volume at all levels of barley flour supplementation were recorded for MTG treatment of hard flour. This effect was attributed to the presence of high proportions of fibers like bran or rye, which interfered with the balanced ratio of starch, gluten, and pentosans in the dough, thus reducing baking capacity. In another study, higher homogeneity and improved stability of mixed doughs was achieved when MTG was added to mixed flour systems with rye flour and/or fiber components [70].

The quality of bakery products with regard to their freshness could be enhanced by the use of starch-degrading and non-starch-degrading enzymes. Breads with softer and less chewy fresh crumbs were obtained when MTG was added to α-amylase-supplemented doughs [71].

Gerrard *et al.* reported that MTG was able to improve the lift of puff pastry and the volume of croissants [72]. MTG-treated products had an excellent texture, mouth-feeling, and layer mainly due to an increase in size in comparison with non-treated samples. These effects persisted even after storing the pastry and croissant doughs frozen for periods of up to 90 days. An additional study by Gerrard *et al.* [73] revealed that specific high-molecular weight glutenin subunits appeared to be crosslinked by MTG.

Food products such as pastry or tart can become soggy as a result of filling them with creams and fruits containing high amounts of water. Respective products were found to be crispier and well-protected against dough-softening when treated with MTG [74].

Cereal products, especially bread, are the basic components of peoples' diet in many countries. The demand for gluten-free breads to serve the needs of persons with celiac disease has increased in recent years. Gluten is the major structure-forming protein in bread manufacturing. For this reason, Moore *et al.* [75] tried to mimic the viscoelastic properties by utilization of MTG in breads that consisted mainly of white rice flour, potato starch, corn flour, and other minor ingredients. It was possible to produce a stable network within a gluten-free bread system and to improve loaf volume and crumb characteristics when soy flour, skim milk powder, and egg powder were used as protein sources.

9.7.3
Noodles

A drastic increase in both storage and loss modulus of fresh noodle sheets was reported by Wu and Corke [76] after application of MTG. The effect decreased with higher levels of enzyme due to the limited content of lysine in gluten. The elasticity and breaking strength of cooked noodles was positively affected by MTG crosslinking.

The loss of the glutinous taste of noodles caused by freezing, drying, or heat treatment was identified as the main problem. Approaches to improving the properties by adding starch, egg white, gliadin, glutenin, or MTG were reported. Several patent applications [10, 11] describe the addition of MTG to wheat flour to obtain noodles with enhanced elasticity, texture, and sensory properties.

9.8
Applications of MTG Crosslinked Leguminous Proteins in Food Models and Realistic Food Products

9.8.1
Crosslinking Protein Isolates from Pea, Lupin, and Soybean in Food Models

Publications on the suitability of MTG crosslinking technology in real food products containing vegetable proteins other than from soy and wheat are still rare. However, MTG crosslinked leguminous proteins from soy, pea, and lupin were recently demonstrated to be suitable for the preparation of real foods [77]. The effects of food ingredients such as salts and oil on the enzymatic texturization of protein isolates from soy, pea, and sweet lupin were evaluated in simplified aqueous model systems. Protein crosslinking with MTG was significantly improved when NaCl was added. In comparison, the addition of $CaCl_2$ resulted in reduced gel strengths. For this reason, the utilization of decalcified water was suggested for the preparation of food products based on MTG crosslinked proteins. Emulsification of corn oil into aqueous suspensions of protein isolates from pea and soy with oil/protein ratios of 1 and $2\,g\,g^{-1}$ prior to enzymatic crosslinking enhanced gel formation depending on the emulsification technique. As generally shown by simple O/W emulsion model systems, the effect was more pronounced when ultrasonication was used in comparison with a rotor–stator system. Furthermore, variation could be observed among the protein isolates with regard to the influences of NaCl and oil on the obtainable gel strengths and optimum doses of these ingredients. MTG crosslinking technology was applied to leguminous proteins in complex food models such as a thickened soup (liquid model), a mousse (foamed model), and a sausage-like substitute (solid model). General applicability was deduced from their performance in comparison with their counterparts based on milk, gelatin, and meat. Whereas unfavorable grittiness was reported for the liquid soup model, attractive texture was achieved in the foamed and solid food models. Members of a trained sensory panel were asked about their general perception of food products based on enzymatically texturized leguminous proteins in comparison with their reference products. Deviations were significant especially for the soup model and acceptance of the liquid food model was poor because of unfavorable grittiness. Within the same panel, acceptance rates with good prospects of 60% and 80% were recorded for the mousse-type food model and the sausage-like substitute, respectively. Taste masking by addition of suitable flavors during product development or usage of proteinous raw material low in off-flavors suggest themselves. Enzymatically crosslinked leguminous proteins were found to be suitable for application in food products in terms of visual appearance, texture, and color [77].

9.8.2
Methods for Monitoring the Enzymatic Texturization

Methods for controlling the enzymatic reaction are a prerequisite for successful development of products based on texturized plant proteins. A reliable HPLC-MS method for the quantification of ε-(γ-glutamyl)lysine isopeptide in lyophilized proteolytic digests of crosslinked plant protein samples has been developed in addition to several alternative analytical approaches [37]. Levels of isopeptide crosslinks, detected via HPLC-MS analysis, correlated well with gel strength of texturized proteins from soy, pea, and lupin. The determination of gel strength was shown to be a fast and practicable method for controlling MTG-induced gelation [78].

9.9
Safety of MTG and Isopeptide Bonds in Crosslinked Plant Proteins

Transglutaminase is a naturally occurring ubiquitous enzyme and isopeptide bonds occur in nature. Nevertheless, both the enzyme and isopeptide bond resulting from MTG crosslinking technology have to be contemplated when incorporated in human foods.

9.9.1
Safety of the Isopeptide

Isopeptide bonds are formed as a result of posttranslational modification of proteins. Additionally, ε-(γ-glutamyl)lysine is formed during heating of proteins or by enzymatic reaction catalyzed by MTG. The biodistribution and catabolism of ^{18}F-labeled N^{ε}-(γ-glutamyl)-L-lysine isopeptide have been investigated by Hultsch et al. [79]. According to this study, isopeptides resulting from MTG application in food biotechnology are readily cleaved by peptidases after absorption and do not pose a risk to the health of consumers.

9.9.2
Safety of MTG

Pedersen et al. [80] have evaluated the allergenic potential of MTG using a decision scheme of the FAO/WHO. No IgE-mediated allergy to bacterial proteins has been reported and the enzyme was fully degraded after pepsin treatment. Homology with known allergens down to a match of six contiguous amino acids could not be detected, and cross reactivity between MTG and the major codfish allergen Gad c1 could be excluded. In this study, no safety concerns regarding the allergenic potential of MTG from *Streptomyces mobaraensis* were identified and any allergenicity of MTG was thought to be unlikely.

Furthermore, the safety of MTG has been evaluated in conventional safety studies, and no obvious effects with regard to acute oral toxicity and its mutagenic potential were found in rats, in Slc:dd male mice, *Salmonella typhimurium*, *Escherichia coli*, and in a Chinese hamster lung cell line [81].

9.9.3
Allergenicity of MTG Crosslinked Plant Proteins

Celiac disease is a chronic enteropathy caused by intolerance to gluten proteins [82]. The effect of MTG crosslinked gluten proteins on allergenicity is controversial.

The involvement of tissue transglutaminase (TTG) in the pathogenesis of celiac disease has been reported when MTG was applied to wheat gluten [83].

Cabrera-Chávez *et al.* [84] reported the formation of new epitopes in prolamins of wheat and maize similar to those endogenously formed by TTG in peptides of prolamins inside the epithelial cells of celiac patients, when MTG was added during bread manufacture.

In an earlier study by Gerrard and Sutton [85], the addition of MTG to cereal products was shown to generate the epitope responsible for celiac disease and the authors recommended stopping the use of MTG, although there has been no experimental evidence that cereal grain food products baked in the presence of MTG may trigger symptoms of celiac disease.

A study from Berti *et al.* [86] revealed increased immunoreactivity of IgA anti-gliadin antibodies to gliadins from gluten after MTG-catalyzed deamidation.

In contrast, earlier reports claimed beneficial effects of MTG in cereal products on functional properties [26, 72, 73, 87–89], and the assumption of reduced allergenicity following MTG treatment has been reported [90].

In 2008 Rossi *et al.* [91] claimed, in a patent application, that MTG treatment reduces or completely eliminates the pathological immune response of gluten or similar products derived from cereals such as maize and rice in patients affected by celiac disease.

In summary, the evidence as to whether MTG crosslinked proteins may influence gluten-related intolerance is not consistent.

9.9.4
Allergenicity of Plant Proteins

Since they are likely to cause adverse reactions in susceptible individuals, labeling of several food ingredients, for example, protein isolates from cereals, soybean, sesame seeds, and lupin, is mandatory, according to European Union commission directive 2007/68/EC [92].

Enzymatic hydrolysis of soy proteins [93] and enzymatic hydrolysis followed by ultrafiltration of soy flour protein [94] are considered to reduce allergenic potential.

Synergistic effects may occur by combination of enzymatic hydrolysis and MTG crosslinking. Plant proteins in general might benefit from reduced or eliminated anti-nutritive properties. To the best of the author's knowledge, increased allergenicity or toxicity as a result of crosslinking leguminous protein isolates with MTG has not been reported so far.

9.10
Conclusions

Enzymatically crosslinked vegetable proteins have a high potential as ingredients for texturized plant-based food products, including pastes and gel-like structures. Adequate extraction, including pre- and post processing, of the proteins should be applied especially to remove undesired compounds and, consequently, increase the diversity of healthy and suitable protein sources for MTG crosslinking technology.

The suitability of proteins from a huge variety of plant sources for crosslinking with MTG mainly depends on the protein source, its nutritive value, the conformational structure of the protein, and the accessible amount of lysine and glutamine residues. With regard to the provision of alternative protein sources of plant origin in comparison to SPI, numerous research activities during the past two decades have revealed interesting candidates such as proteins from lupin, rice, sunflower, pea, and sorghum. Further research activities will hopefully open up new horizons for many more proteinous substrates, facilitating the development of innovative plant-based food products with paste or gel-like structures.

Figure 9.1 visualizes results from a literature search using SciFinder of the American Chemical Society for the time period 1991–2011. The search revealed an increasing number of publications, including patent applications, in the field of "plant proteins" and "vegetable proteins," each in combination with the search term *"food."* Within the decade starting in 1991, the number of publications tripled from 50 to 150 in 2001 and doubled again within the following decade to 345 publications in 2011. Altogether, 3736 (Figure 9.1: dark gray area) publications were related to the search terms *"vegetable protein"* and *"food"* (1415) and *"plant protein"* and *"food"* (2321) over the last 20 years. Within the same time period, the number of publications related to the search terms *"transglutaminase"* and *"food"* (including food products from animal and plant origin) also increased by a factor of 10, from 8 publications in 1991, 94 in 2009, and 73 in 2011 (Figure 9.1: light gray area). This observation can be attributed mainly to publications dealing with food products containing animal protein. Surprisingly, the search terms *"plant protein"* and *"vegetable protein"* in combination with *"transglutaminase"* resulted in only 63 publications over the last 20 years (Figure 9.1: black area). Twenty-seven of these publications were issued during 2008–2011, indicating that crosslinking proteins of plant origin have recently gained more attention.

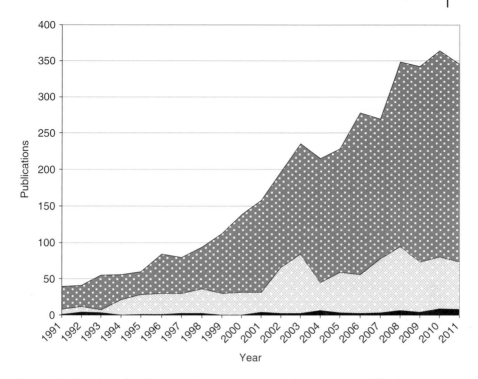

Figure 9.1 Overview of publications: Data source: SciFinder, American Chemical Society, February 2012 with search terms "plant proteins" and "food" and "vegetable proteins" and "food" (dark gray area); "transglutaminase" and "food" (light gray area); "plant proteins" and "transglutaminase" and "vegetable proteins" and "transglutaminase" (black area).

Considering the increasing demand of consumers for convenient and healthy food products of vegetable origin, MTG has high potential as a processing aid for the manufacture of innovative plant-based food products. Several opportunities to modify and improve the functional properties of plant proteins and their nutritive value by crosslinking with MTG should be subject to future research activities and might broaden application areas further.

References

1. Zhu, Y., Bol, J., Rinzema, A., Tramper, J., and Wijngaards, G. (1999) Transglutaminase as a potential tool in developing novel food proteins. *Agro-Food-Ind. Hi-Tech.*, **10**, 8–10.
2. Riaz, M.N. (2011) in *Handbook of Food Proteins* (eds G.O. Phillips and P.A. Williams), Woodhead Publishing Limited, Cambridge, pp. 393–418.
3. Belitz, H.-D., Grosch, W., and Schieberle, P. (2004) *Food Chemistry*, Springer-Verlag, Berlin, pp. 87–88.
4. Motoki, M., Nio, N., and Takinami, K. (1984) Functional properties of food proteins polymerized by transglutaminase. *Agric. Biol. Chem.*, **48**, 1257–1261.
5. Schäfer C (2011) *Enzymatic Texturisation of Leguminous Proteins and their*

Application in Food Models, Shaker-Verlag, Aachen.

6. Ando, H., Adachi, M., Umeda, K., Matsuura, A., Nonaka, M., Uchio, R., Tanaka, H., and Motoki, M. (1989) Purification and characteristics of a novel transglutaminase derived from microorganisms. *Biol. Chem.*, **53**, 2613–2617.

7. Gerber, U., Jucknischke, U., Putzien, S., and Fuchsbauer, H.L. (1994) A rapid and simple method for the purification of transglutaminase from Streptoverticillium mobaraense. *Biochem. J.*, **299**, 825–829.

8. Motoki, M., Okiyama, A., Nonaka, M., Tanaka, H., Uchio, R., Matsuura, A., Ando, H., and Umeda, K. (1989) Novel transglutaminase. JP Patent 1027471.

9. Ando, H., Matsuura, A., and Hirose, S. (1992) Process for producing a transglutaminase derived from Streptomyces. JP Patent 4108381.

10. Dube, M., Schäfer, C., Neidhart, S., and Carle, R. (2007) Texturisation and modification of vegetable proteins for food applications using microbial transglutaminase. *Eur. Food Res. Technol.*, **225**, 287–299.

11. Santos, M. and Torné, J.M. (2009) Recent patents on transglutaminase production and applications: a brief review. *Recent Pat. Biotechnol.*, **3**, 166–174.

12. Marx, C.K., Hertel, T.C., and Pietzsch, M. (2008) Random mutagenesis of a recombinant microbial transglutaminase for the generation of thermostable and heat-sensitive variants. *J. Biotechnol.*, **136**, 156–162.

13. Hertel, T.C., Pietzsch, M., and Marx, C.K. (2009) Thermostable transglutaminases. Patent Appl. PCT WO 2009/030211.

14. Macedo, J.A., Sette, L.D., and Sato, H.H. (2008) Optimization studies for the production of microbial transglutaminases from a newly isolated strain of Streptomyces sp. *Food Sci. Biotechnol.*, **17** (5), 904–911.

15. Cui, L., Du, G., Zhang, D., and Chen, J. (2008) Thermal stability and conformational changes of transglutaminase from a newly isolated Streptomyces hygroscopicus. *Bioresour. Technol.*, **99**, 3794–3800.

16. Yu, Y.-J., Wu, S.-C., Chan, H.-H., Chen, Y.-C., Chen, Z.-T., and Yang, M.-T. (2008) Overproduction of soluble recombinant transglutaminase from Streptomyces netropsis in Escherichia coli. *Appl. Microbiol. Biotechnol.*, **81**, 523–532.

17. Friedmann, M. and Brandon, D.L. (2001) Nutritional and health benefits of soy proteins. *J. Agric. Food Chem.*, **49**, 1069–1086.

18. Fukushima, D. (2011) in *Handbook of Food Proteins* (eds G.O. Phillips and P.A. Williams), Cambridge, Woodhead Publishing Limited, pp. 393–418.

19. Nonaka, M., Toiguchi, S., Sakamoto, H., Kawajiri, H., Soeda, T., and Motoki, M. (1994) Changes caused by microbial transglutaminase on physical properties of thermally induced soy protein gels. *Food Hydrocolloids*, **8**, 1–8.

20. Chanyongvorakul, Y., Matsumura, Y., Nonaka, M., Motoki, M., and Mori, T. (1995) Physical properties of soy bean and broad bean 11S globulin gels formed by transglutaminase reaction. *J. Food Sci.*, **60**, 483–493.

21. Kang, I.J., Matsumura, Y., Ikura, K., Motoki, M., Sakamoto, H., and Mori, T. (1994) Gelation and gel properties of soybean glycinin in a transglutaminase-catalyzed system. *J. Agric. Food Chem.*, **42**, 159–165.

22. Friedmann, M. and Finot, P.A. (1990) Nutritional improvement of bread with lysine and γ-glutamyllysine. *J. Agric. Food Chem.*, **38**, 2011–2020.

23. Iwami, K. and Yasomoto, K. (1986) Amine-binding capacities of food proteins in transglutaminase reaction and digestibility of wheat gliadin with ε-attached lysine. *J. Sci. Food Agric.*, **37**, 495–503.

24. Porta, R., Gentile, V., Esposito, C., Mariniello, L., and Aurrichio, S. (1990) Cereal dietary proteins with sites for crosslinking by transglutaminase. *Phytochemistry*, **29**, 2801–2804.

25. Mujoo, R.a. and Ng, P.K.W. (2003) Identification of wheat protein components involved in polymer formation on incubation with transglutaminase. *Cereal Chem.*, **80**, 703–706.

26. Bauer, N., Köhler, P., Wieder, H., and Schieberle, P. (2003) Studies on effects of microbial transglutaminase on gluten proteins of wheat. I. Biochemical analysis. *Cereal Chem.*, **80**, 781–786.

27. Larré, C., Denery-Papini, S., Popineau, Y., Deshayes, G., Desserme, C., and Lefebvre, J. (2000) Biochemical analysis and rheological properties of gluten modified by transglutaminase. *Cereal Chem.*, **77**, 121–127.

28. Larré, C., Deshayes, G., Lefebvre, J., and Popineau, Y. (1998) Hydrated gluten modified by transglutaminase. *Nahrung*, **42**, 155–157.

29. Basman, A., Köksel, H., and Ng, P.K.W. (2002) Effects of increasing levels of transglutaminase on the rheological properties and bread quality characteristics of two wheat flours. *Eur. Food Res. Technol.*, **215**, 419–424.

30. Gujral, H.S., Guardiola, I., Carbonell, J.V., and Rosell, C.M. (2003) Effect of cyclodextrinase on dough rheology and bread quality from rice flour. *J. Agric. Food Chem.*, **51**, 3814–3818.

31. Gujral, H.S. and Rosell, C.M. (2004) Functionality of rice flour modified with a microbial transglutaminase. *J. Cereal Sci.*, **39**, 225–230.

32. Paraman, I., Hettiarachchy, N.S., Schäfer, C., and Beck, M.I. (2006) Physicochemical properties of rice endosperm proteins extracted by chemical and enzymatic methods. *Cereal Chem.*, **83** (6), 663–667.

33. Paraman, I., Hettiarachchy, N.S., and Schäfer, C. (2008) Preparation of rice endosperm protein isolate by alkali extraction. *Cereal Chem.*, **85** (1), 76–81.

34. Wang, M., Hettiarachchy, N.S., Qi, M., Burks, W., and Siebenmorgen, T. (1999) Preparation and functional properties of rice bran protein isolate. *J. Agric. Food Chem.*, **47**, 411–416.

35. Larré, C., Chiarello, M., Dudek, S., Chenu, M., and Gueguen, J. (1993) Action of transglutaminase on the constitutive polypeptides of pea legumin. *J. Agric. Food Chem.*, **41**, 1816–1820.

36. Larré, C., Kedzior, Z.M., Chenu, M., Viroben, G., and Gueguen, J. (1992) Action of transglutaminase on an 11S seed protein (pea legumin): influence of the substrate conformation. *J. Agric. Food Chem.*, **40**, 1121–1126.

37. Schäfer, C., Schott, M., Brandl, F., Neidhart, S., and Carle, R. (2005) Identification and quantification of ε-(γ-glutamyl)lysine in digests of enzymatically crosslinked leguminous proteins by high-performance liquid chromatography–electrospray ionization mass spectrometry (HPLC-ESI-MS). *J. Agric. Food Chem.*, **53**, 2830–2837.

38. López, G., Flores, I., Gálvez, A., Quirasco, M., and Farréz, A. (2003) Development of a liquid nutritional supplement using a Sesamum indicum L. protein isolate. *Lebensm. Wiss. Technol.*, **36**, 67–74.

39. Pierpoint, W.S. (1969) o-Quinones formed in plant extracts: their reactions with amino acids and peptides. *Biochem. J.*, **112**, 609–616.

40. Rossi, M. (1988) Textured sunflower protein for use as a meal extender. *Lebensm. Wiss. Technol.*, **21**, 267–270.

41. Pickardt, C., Neidhart, S., Griesbach, C., Dube, M., Knauf, U., Kammerer, D.R., and Carle, R. (2009) Optimisation of mild-acidic protein extraction from defatted sunflower (Helianthus annuus L.) meal. *Food Hydrocolloids*, **23**, 1966–1973.

42. Arntfield, S.D. (2011) in *Handbook of Food Proteins* (eds G.O. Phillips and P.A. Williams), Woodhead Publishing Limited, Cambridge, pp. 289–315.

43. Alting, A.Z. and Pouvreau, L. (2011) in *Handbook of Food Proteins* (eds G.O. Phillips and P.A. Williams), Woodhead Publishing Limited, Cambridge, pp. 316–334.

44. de Mesa-Stonestreet, N.J., Alavi, S., and Bean, S.R. (2010) Sorghum proteins: the concentration, isolation, modification, and food applications of kafirins. *Food Sci.*, **75** (5), R90–R104.

45. Babiker, E.E. and Kato, A. (1998) Improvement of the functional properties of sorghum protein by protein-polysaccharide and protein–protein complexes. *Nahrung*, **42** (5), 286–289.

46. Babiker, E.E., Khan, M.A.S., Matsudomi, N., and Kato, A. (1997) Polymerization of soy protein digests by microbial transglutaminase for improvement of the

functional properties. *Food Res. Int.*, **29** (7), 627–634.

47. Babiker, E.E. (2000) Effect of transglutaminase treatment on the functional properties of native and chymotrypsin-digested soy protein. *Food Chem.*, **70** (2), 139–145.

48. Walsh, D.J., Cleary, D., McCarthy, E., and FitzGerald, R.J. (2003) Modification of the nitrogen solubility properties of soy protein isolate following proteolysis and transglutaminase crosslinking. *Food Res. Int.*, **36**, 677–683.

49. Babiker, E.E., Fujisawa, N., Matsudomi, N., and Kato, A. (1996) Improvement in the functional properties of gluten by protease digestion or acid hydrolysis followed by microbial transglutaminase treatment. *J. Agric. Food Chem.*, **44**, 3746–3750.

50. Agyare, K.K., Addo, K., and Xiong, Y.L. (2009) Emulsifying and foaming properties of transglutaminase-treated wheat gluten hydrolysate as influenced by pH, temperature and salt. *Food Hydrocolloids*, **23**, 72–81.

51. Paraman, I., Hettiarachchy, N.S., Schäfer, C., and Beck, M.I. (2007) Hydrophobicity, solubility, and emulsifying properties of enzyme-modified rice endosperm protein. *Cereal Chem.*, **84** (4), 343–349.

52. Kato, A., Babiker, E.E., Fujisawa, N., and Matsudomi, N. (1998) in *Plant Proteins from European Crops* (eds J. Gueguen and Y. Popineau), Springer, Berlin, pp. 146–151.

53. Mohamed, M.A., Ahmed, I.A.M., Babiker, E.E., ElGasim, A., and Yagoub, A. (2010) Polymerization of sorghum protein digests by transglutaminase: changes in structure and functional properties. *Agric. Biol. J. N. Am.*, **1**, 47–55.

54. Colas, B., Caer, D., and Fournier, E. (1993) Transglutaminase-catalyzed glycosylation of vegetable proteins: effect on solubility of pea legumin and wheat gliadins. *J. Agric. Food Chem.*, **41** (11), 1811–1815.

55. Gan, C.-Y., Cheng, L.-H., and Easa, A.M. (2008) Physicochemical properties and microstructures of soy protein isolate gels produced using combined

crosslinking treatments of microbial transglutaminase and Maillard crosslinking. *Food Res. Int.*, **41**, 600–605.

56. Gan, C.-Y., Cheng, L.-H., and Easa, A.M. (2009) Assessment of crosslinking in combined crosslinked soy protein isolate gels by microbial transglutaminase and Maillard reaction. *J. Food Sci.*, **74** (2), 141–146.

57. Paraman, I., Hettiarachchy, N.S., and Schaefer, C. (2007) Glycosylation and deamidation of rice endosperm protein for improved solubility and emulsifying properties. *Cereal Chem.*, **84** (6), 593–599.

58. Liu, Z., Zhang, D., Liu, L., Wang, M., Du, G., and Chen, J. (2010) Enhanced water absorption of wheat gluten by hydrothermal treatment followed by microbial transglutaminase reaction. *J. Sci. Food Agric.*, **90**, 658–663.

59. Schäfer, C., Bahary-Lashgary, S., Wäsche, A., Eisner, P., and Knauf, U. (2003) Vegetable protein preparations and use thereof. Patent Application PCT WO 03/106486.

60. Ruf, F., Sohling, U., Hasenkopf, K., Eisner, P., Müller, K., Pickardt, C., and Bez, J. (2009) Elimination of unwanted accompanying substances from vegetable protein extracts. Patent Application PCT WO 2009/043586.

61. Ikura, K., Yoshikawa, M., Sasaki, R., and Chiba, H. (1981) Incorporation of amino acids into food proteins by transglutaminase. *Agric. Biol. Chem.*, **45**, 2587–2592.

62. Nonaka, M., Matsuura, Y., and Motoki, M. (1996) Incorporation of lysine dipeptides into αs1-casein by Ca^{2+}-independent microbial transglutaminase. *Biosci. Biotechnol. Biochem.*, **60**, 131–133.

63. Sun, X.D. and Arntfield, S.D. (2011) Gelation properties of salt-extracted pea protein isolate catalyzed by microbial transglutaminase crosslinking. *Food Hydrocolloids*, **25**, 25–31.

64. Gan, C.-Y., Latiff, A.A., Cheng, L.-H., and Easa, A.M. (2009) Gelling of microbial transglutaminase crosslinked soy protein in the presence of ribose and sucrose. *Food Res. Int.*, **42**, 1373–1380.

65. Matsuura, M., Sasaki, M., Sasaki, A., and Takeuchi, T. (2000) Production for producing packed tofu. US Patent 6042851.

66. Nonaka, M., Soeda, T., Yamagiwa, K., Kowata, H., Motogi, M., and Toiguchi, S. (1991) Process of preparing shelf-stable "tofu" at normal temperature for long term. US Patent 5055310.

67. Oomura, H., Adachi, T., Nakatani, S., and Akasaka, T. (2002) Tofu products excellent in freeze resistance and process for producing the same. US Patent 6342256.

68. Sawano, E. and Sawano, H. (2003) Process for producing functional okara milks and functional tofus. US Patent 6582739.

69. Basman, A., Köksel, H., and Ng, P.K.W. (2003) Utilization of transglutaminase to increase the level of barley and soy flour incorporation in wheat flour breads. *J. Food Sci.*, **68**, 2453–2460.

70. Poza, O.D. (2002) Transglutaminase in baking applications. *Cereal Foods World*, **47**, 93–95.

71. Collar, C. and Bollaín, C. (2005) Impact of microbial transglutaminase on the staling behaviour of enzyme-supplemented pan breads. *Eur. Food Res. Technol.*, **221**, 298–304.

72. Gerrard, J.A., Newberry, M.P., Ross, M., Wilson, A.J., Fayle, S.E., and Kavale, S. (2000) Pastry lift and croissant volume as affected by microbial transglutaminase. *J. Food Sci.*, **65**, 312–314.

73. Gerrard, J.A., Fayle, S.E., Brown, P.A., Sutton, K.H., Simmons, L., and Rasiah, I. (2001) Effects of microbial transglutaminase on the wheat proteins of bread and croissant dough. *J. Food Sci.*, **66**, 782–786.

74. Ishii, C. and Soeda, T. (1997) Method of manufacturing baked products. EP Patent 0760209.

75. Moore, M.M., Heinbockel, M., Dockery, P., Ulmer, H.M., and Arendt, E.K. (2006) Network formation in gluten-free bread with application of transglutaminase. *Cereal Chem.*, **83**, 28–36.

76. Wu, J. and Corke, H. (2005) Quality of dried white salted noodles affected by microbial transglutaminase. *J. Sci. Food Agric.*, **85**, 2587–2594.

77. Schäfer, C., Neidhart, S., and Carle, R. (2011) Application and sensory evaluation of enzymatically texturised vegetable proteins in food models. *Eur. Food Res. Technol.*, **232**, 1043–1056.

78. Schäfer, C., Zacherl, C., Engel, K.-H., Neidhart, S., and Carle, R. (2007) Comparative study of gelation and crosslink formation during enzymatic texturisation of leguminous proteins. *Innov. Food Sci. Emerg. Technol.*, **8**, 269–278.

79. Hultsch, C., Bergmann, R., Pawelke, B., Pietzsch, J., Wuest, F., Johannsen, B., and Henle, T. (2005) Biodistribution and catabolism of 18F-labelled isopeptide Nε-(γ-glutamyl)-L-lysine. *Amino Acids*, **29**, 405–413.

80. Pedersen, M.H., Hansen, T.K., Sten, E., Seguro, K., Ohtsuka, T., Morita, A., Bindslev-Jensen, C., and Poulsen, L.K. (2004) Evaluation of the potential allergenicity of the enzyme microbial transglutaminase using the 2001 FAO/WHO decision tree. *Mol. Nutr. Food Res.*, **48**, 434–440.

81. Bernard, B.K., Tsubuku, S., and Shioya, S. (1998) Acute toxicity and genotoxicity studies of a microbial transglutaminase. *Int. J. Toxikol.*, **17** (6), 703–721.

82. Dewar, D., Pereira, S.P., and Ciclitira, P.J. (2004) The pathogenesis of coeliac disease. *Int. J. Biochem. Cell Biol.*, **36**, 17–24.

83. Stenberg, P., Roth, E.B., and Sjöberg, K. (2008) Transglutaminase and the pathogenesis of coeliac disease. *Eur. J. Int. Med.*, **19**, 83–91.

84. Cabrera-Chávez, F., Rouzaud-Sández, O., Sotelo-Cruz, N., and Calderón de la Barca, A.M. (2008) Transglutaminase treatment of wheat and maize prolamins of bread increases the serum IgA reactivity of celiac disease patients. *J. Agric. Food Chem.*, **56**, 1387–1391.

85. Gerrard, J.A. and Sutton, K.H. (2005) Addition of transglutaminase to cereal products may generate the epitope responsible for coeliac disease. *Trends Food Sci. Technol.*, **16**, 510–512.

86. Berti, C., Roncoroni, L., Falini, M.L., Caramanico, R., Dolfini, E., Bardella, M.T., Elli, L., Terrani, C., and Forlani, F. (2007) Celiac-related properties of chemically and enzymatically modified gluten proteins. *J. Agric. Food Chem.*, **55**, 2482–2488.

87. Gerrard, J.A., Fayle, S.E., Wilson, A.J., Newberry, M.P., Ross, M., and Kavale, S. (1998) Dough properties and crumb strength as affected by microbial transglutaminase. *J. Food Sci.*, **63**, 472–475.

88. Bauer, N., Koehler, P., Wieser, H., and Schieberle, P. (2003) Studies on effects of microbial transglutaminase on gluten proteins of wheat II. Rheological properties. *Cereal Chem.*, **80**, 787–790.

89. Collar, C. and Bollaín, C. (2004) Impact of microbial transglutaminase on the viscoelastic profile of formulated bread doughs. *Eur. Food Res. Technol.*, **218**, 139–146.

90. Watanabe, M., Susiki, T., Ikezawa, Z., and Arou, S. (1994) Controlled enzymatic treatment of wheat proteins for production of hypoallergenic flour. *Biosci. Biotechnol. Biochem.*, **58**, 388–390.

91. Rossi, M., Gianfrani, C., and Siciliano, R.A. (2008) Treatment of cereal flour and semolina for consumption by celiac patients. Patent Application PCT WO 2008/053310.

92. EU (2007) Commission Directive 2007/68/EC of 27 November amending Annex IIIa to directive 2000/13/EC of the European Parliament and of the Council as regards certain food ingredients. *Off. J. Eur. Union*, **L 310**, 11–14.

93. Eun-Hee, K., Sang-Il, L., and Sangsuk, O. (2006) Effect of enzymatic hydrolysis of 7S globulin, a soybean protein, on its allergenicity and identification of its allergenic hydrolyzed fragments using SDS-PAGE. *Food Sci. Biotechnol.*, **15** (1), 128–132.

94. Calderón de la Barca, A.M., Wall, A., and López-Díaz, J.A. (2005) Allergenicity, trypsin inhibitor activity and nutritive quality of enzymatically modified soy proteins. *Int. J. Food Sci. Nutr.*, **56** (3), 203–211.

10
Design of Skin Care Products

Wilfried Rähse

10.1
Product Design

The manufacture of designed products, according to the customer's wishes, requires trained teams of chemists, chemical engineers, and marketing people. The product includes the packaging. It is possible to change product properties within wide limits or to set superior features. Novel forms, applicators, and increased performances, developed in cooperation with the customer, characterize product design (Figure 10.1). This new way of thinking starts from a desired product, but not from an existing process. In addition, especially for consumer awareness, the brand (not discussed here) provides a key for sales [1].

Many measurable parameters characterize fine solid particles, such as the particle sizes and distributions, form factors, colors, surfaces and release properties, fragrances, and bulk densities. Suitable, coordinated, and directed processes may define their product design [2]. In addition, liquid products, in particular liquid mixtures, are subject to the laws of product design. To illustrate the diversity of product design, liquid skin care products are particularly suitable, because there are numerous ways to fix the product performance, handling convenience, and aesthetics. The following is a detailed discussion in the form of examples to prove the statements made, based on two papers in German [3, 4].

For the design settings, the recipe is responsible, which is composed of auxiliaries, additives, and active ingredients. Figure 10.2 lists medical and physical-chemical parameters that play a role in product development, and are adjustable by appropriate substances.

Further degrees of freedom exist by producing different physical-chemical emulsions. The possible emulsions consist of oil or water droplets, in the form of a (macro-) emulsion with droplet sizes in the micrometer range or as a mini- (or nano-) emulsion with droplet sizes of around 100–600 nm [6]. Additionally, the selection of an appropriate packaging optimizes, in another way, the product design.

Product Design and Engineering: Formulation of Gels and Pastes, First Edition.
Edited by Ulrich Bröckel, Willi Meier, and Gerhard Wagner.
© 2013 Wiley-VCH Verlag GmbH & Co. KGaA. Published 2013 by Wiley-VCH Verlag GmbH & Co. KGaA.

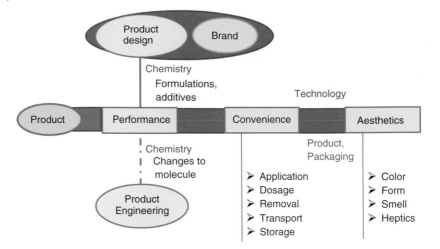

Figure 10.1 Characteristics and composition of product design.

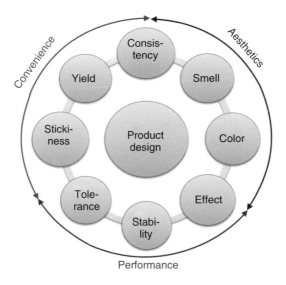

Figure 10.2 Elements of product design for cosmetic products.

10.2
Skin Care

Dermatological products are developed for the care, health and healing of skin, hair, and nails, and are classified, according to the task, as cosmetics, cosmeceuticals, and pharmaceuticals. Cosmeceuticals play a major role in medical skin care. The skin has an area of about 2 m², weighs 3.5–10 kg, and is 0.3–5 mm thick. As the largest organ in people's body it characterizes appearance and fulfills numerous functions, such as protection, regulating the heat balance, and the perception of contact stimuli.

The drying out and roughness of the skin prevents the hydro-lipid coat, which is also called the *acid mantle*, that must be available on the skin in the right composition and thickness. The skin is daily exposed to many stress factors (too much sun, cleaned too intensely, the ambient air is too dry, environmental influences). The strain under prolonged exposure damages the film on the skin. Therefore, skin care is of great importance for the well-being of people, today and in the future.

The cosmetic industry provides suitable products for all aspects of skin caring (Table 10.1). After application, the cream remains for the most part on the surface, smoothing the skin. Two cosmetic product lines are available: perfumed creams for beauty (cosmetics in the strict sense) differ from those for maintaining health (cosmeceuticals). Furthermore, there are products to relieve symptoms (cosmeceuticals,

Table 10.1 Features of dermatological creams.

Parameter	Cosmetics (skin care)	Cosmeceuticals (medical skin care)	Dermatics (medicament)
Manufacturer	Cosmetic industry	Cosmetic industry	Pharmaceutical industry
Tasks	Beauty: • care • moisture • scent	Health: • care/prevention/ concomitant therapy; • skin rejuvenation/ moisture; • itching; • acne; • sun damage; • medical treatment of diseased skin	Healing of skin diseases: • eczema, • dermatitis/psoriasis, • seborrheic keratosis, • microbial infestation, inflammation, • rashes, • wounds
Active ingredients:			
• number	1 (advertised)	>2	mostly 1
• concen- trations	low	medium to high	effective-and age dependent (children, adults)
Additives:			
• perfume	yes	no	no
• dye	in some cases yes in some cases	no	no
Emulsion type (preferred)	O/W	O/W	W/O
Water (%)	70–>80	50–70	60–80
Container	(Pump) bottles crucibles	Dispensers	Tubes

dermatics) as well as for healing damaged skin (dermatics, pharmaceuticals). A precise definition is not possible as there are overlapping areas.

Pharmaceutical products are used to treat and heal sick skin. The healing ingredients remain not only on the skin surface but also penetrate into the epidermis. Dermatics treat primarily pathological disorders of the skin such as psoriasis, eczema, inflammation, cancer, rashes, and deficiencies, and many others [7]. They are dermatological therapeutics. The galenic composition supports the active material, transport into the skin, and release (duration) of the therapeutic substance.

10.2.1
Cosmetic Products for Beautifications

The care of healthy skin around the eyes and lips, at the face, hands, and the total body is possible with commercially available creams and lotions. The corresponding products represent the great diversity found in the cosmetic industry. The nature of skin depends on, among other things, genetic factors, gender, type, age, and weight. The formulations used also differ, depending on the one hand on the thickness of the skin (feet, hands, face) and the skin condition (dry, normal, bold) and, on the other hand, on the intended use (skin texture, sun protection, anti-aging). Cleaning, revitalizing, perfuming, or deodorizing are the main functions of the cosmetics. In addition, there are color cosmetics, which underline the beauty with rouge, eye shadow, mascara, lipstick, and nail polish, and further with hair styling products.

One performance parameter stands out in the development of usual cosmetics. Mostly, people wish to improve skin moisture or the surface structure and barrier function; people of a mature age want effective ingredients against skin aging (Figure 10.3). Improvements in structure play a role in the case of dry skin. A cleaning agent reduces problems associated with oily skin. A weak level of sun

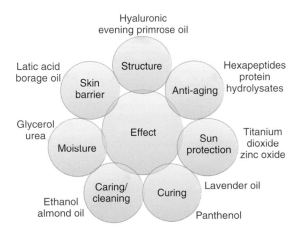

Figure 10.3 Benefits of skin care products (cosmetics, cosmeceuticals) with examples of suitable ingredients.

protection could be beneficial in products used daily, because UV light causes increased skin aging. Bright sunlight requires special products with high sun protection factors (SPFs). The SPF of the lotion depends on skin type and the intensity of solar radiation.

Typical commercial products for the skin have a pleasant consistency and provide short-term care. From a physical point of view, the creams preferably consist of oil in water (O/W) emulsions. Normal (usual) cosmetics contain only a few active ingredients, formulated in low concentrations, but in relation with many auxiliary materials. Traditional products contain about 25–40 ingredients, including many substances for preserving, emulsifying, and thickening, besides the active and base materials.

Almost all cosmetic care products are scented, and spread a pleasant fragrance. The effect only lasts a short time. A product sale takes place because of, on the one hand, the fragrance and, on the other hand, an interesting substance (silk proteins, urea) or because of a particular function (smoothing, moisturizing). According to the EU Cosmetics Directive [8], first the products must do what is written on the packaging (moisturizer, reduces wrinkles, etc.) and secondly, the products require a declaration of nanoparticles (solids), even if they are safe, on the packaging.

10.2.2
Active Cosmetics for a Healthy Skin

Depending on the recipe, cosmetics cure slight dermatological diseases [9]. For such products, between cosmetic care and healing with medicines, dermatologists in the United States created the term *"cosmeceuticals."* The word, formed from cosmetics and pharmaceuticals, means "active cosmetics." These designed skin care products ensure not only beauty but also a healthy condition of the skin with essential ingredients [10].

Cosmeceutical creams or lotions are, preferably, O/W emulsions. They combine effective cosmetic ingredients with "healing" natural oils and vitamins/provitamins in amounts that provide a rich supply to the skin and show visible success. By using selected ingredients, physiological processes occur more than usual and support the skin's own regeneration. To avoid undesired effects it is necessary to reduce the auxiliaries to a minimum. This affects not only the amounts of emulsifiers and preservatives but also waxes, synthetic oils, and other synthetic adjuvants.

To minimize the risk of allergic reactions, cosmeceuticals should not contain perfumes or dyes. To increase acceptance, some creams contain intensively tested fragrances that do not cause allergic reactions. The possible addition of (auxiliary) materials and their maximum levels, as well as the identification of substances on the packaging (INCI = International Nomenclature of Cosmetic Ingredients), are subject to the EU Cosmetics regulations [8, 11].

Cosmeceuticals contain biologically active substances in effective concentrations. The first group of these substances comes from plants/plant parts, obtained through different processes such as grinding/pressing and extracting with solvents or with the aid of superheated steam. Depending on the origin of applied substances, the mode of action on the skin surface can be described as healing, caring, moisturizing, lipid-supplying, smoothing, refreshing, rejuvenating, revitalizing, promoting circulation, antiseptic, anti-inflammatory, antioxidant, astringent, and UV-protective. The second group includes the vitamins and provitamins. These substances act on and in the skin in many ways. Some can work with synergistic effects, especially the antioxidants (vitamins C, E), the stimulators of circulation (vitamin B_3), regulators of sebum production (vitamin B_6), anti-aging (vitamin A), and healing ingredients (vitamin B_5). The third group of substances, proteins and peptides derived from microorganisms or synthesized in analogy with nature, provides improvements in skin hydration, firmness, elasticity, and skin protection. Examples are products that can reduce small facial wrinkles, improve peripheral circulation, reduce hair growth in unwanted places, or fight a fungal infection [12].

Cosmeceuticals [13] not only serve for the maintenance of health but also often for the support and supplement of a therapy. These problems require compositions other than those normally used in the cosmetic industry. The active substances work simultaneously in effective concentrations, partially with specific synergistic actions. Among the most important materials are some unsaturated natural oils that boost, depending on the amount, the sheen on the skin. This is undesirable for usual, daily cosmetic products but is acceptable for active cosmetics. The oils, in combination with moisturizers and vitamins, enable medical actions besides the usual care. The primary target is to maintain or restore a healthy skin. The inclusion of novel peptides [11, 14] provides firming effects. A medical application of cosmetics has demonstrated the calming impact on a dry, itchy skin, bringing the skin back to a healthy state. Studies of volunteers and/or the descriptions of many applications document the improvements achieved by cosmeceuticals.

For cosmeceuticals, there are no legally binding rules. They are subject to the same laws that apply to the usual cosmetics. Table 10.2 gives some aspects of the delimitation between normal and active cosmetics used for therapeutic agents.

10.2.3
Differences between Cosmeceuticals and Drugs

Cosmeceuticals are cosmetic products and not medicines. The good manufacturing practice (GMP) cosmetic guidelines represent the basis of manufacturing. Marketing requires a qualified security assessment of the recipe, executed in accordance with the EU cosmetic regulations. Offices of the federal states publish the formal procedures for marketing. For pharmaceuticals, however, there are legal provisions. The Medicines Act [15] contains the definition, requirements, production, licensing, registration, clinical testing, delivery assurance, and quality control aspects. For marketing, authorization (§ 21 AMG) is needed. The permit

Table 10.2 Legal basis of dermatological products in Europe [5].

	Cosmetics	Cosmeceuticals	Dermatics (pharmaceuticals)
Legal basis	LMBG Food and commodities act	No legal basis; functions proposed by the US dermatologists	AMG Medicines act
Tasks	• Externally applied, • care, cleaning, • appearance, • odor	• Externally applied; • improve, prevent, and eliminate forms of minimal dermatological diseases; • physiological effects are limited to the skin; • no side effects; • no systemic effect	• Inside and outside; • heal, alleviate symptoms of diseases and pathological symptoms • render harmless pathogens, parasites, and foreign substances; • systemic[a] effects possible

[a] Possible effects throughout the body via the blood circulation.

is granted by the competent federal authority or the Commission of the European Communities or the Council of the European Union.

After AMG §1 (abridged) pharmaceuticals are defined as "substances or preparations of substances that are intended for use in the human body and show properties for treating or alleviating or preventing diseases or pathological symptoms ...." The results of clinical studies represent the basis for the approval process of drugs, especially the proof of claimed effects (indications proof). Furthermore, the product must be safe in application. The emulsions or ointments have a healing effect in the case of skin lichen, eczema, inflammations, skin rashes, and nutritional deficiencies. Pharmaceuticals contain not only the actual active ingredient but also different consistency regulators, emulsifiers, and preservatives. They are usually "water in oil (W/O) emulsions." A fatty layer on the skin and a very slow penetration of water-soluble active ingredients characterizes this type of emulsion after application. Single-phase systems are predominantly known as an "*ointment*," which consists, for example, of a base substance (Vaseline®) and an active ingredient. Dermatologists prescribe recipes to treat and heal diseased skin with pharmaceuticals, while cosmeceuticals are on sale without prescription.

To check the effectiveness of therapeutic agents, they usually only contain one medicinal substance, at optimized concentration. Normally, a special galenic with suitable excipients guarantees the pharmaceutical availability [16] of the active ingredient. The pharmacological potency and duration of action depend on the concentration, additives, and the formulation (pharmaceutical chemistry and technology). The active substance exists as a pure or chemically modified natural material, or is synthesized completely, or produced biotechnologically. Except for α-hydroxy- and salicylic acid and urea, the drugs contain completely different agents as formulated in the cosmeceuticals. Examples are potent cortisone, antibiotics, antifungals, and retinoids.

10.3
Emulsions

The optimal supply of lipophilic and hydrophilic active substances in larger quantities requires the formation of an emulsion. An emulsion offers the best physical attributes for skin care products to act. The preparations are, preferably, O/W emulsions because in this way the 1–10 μm large oil drops can quickly penetrate into the skin, without leaving an oily film on the skin. In contrast, on applying W/O emulsions, oil remains for several minutes, often as sticky film, which is undesirable in cosmetics. Polymers can control the release of substances in both phases. The water as the continuous phase partially evaporates on the skin, leaving a thin film with the non-diffusible substances in this environment and reinforcing the hydro-lipid film (acid mantle). With a buffer in the water phase, the pH on the skin is set to 5. The substances smooth the skin, and provide protection against pathogens with a suitable, partially water/oil soluble preservative.

10.3.1
Basics (Definition, Structure, Classification)

Emulsions are thermodynamically unstable systems of two immiscible liquids. A surfactant distributes the first one evenly in the form of droplets in the other, continuous phase. The liquid is the external or continuous phase, while the internal or dispersed phases are the droplets. In case of oil droplets an O/W emulsion arises, otherwise a W/O emulsion with water droplets is formed [17]. According to the Bancroft rule [18], the HLB (hydrophilic–lipophilic balance) number of the emulsifier (Figure 10.4) decides the type of emulsion. The HLB value indicates the hydrophilic–lipophilic balance of the system and can be determined, analogously to Griffin [19], from the molecular weight of the hydrophilic portion (MG_{hydro}) in relation to the total molecular weight (MW).

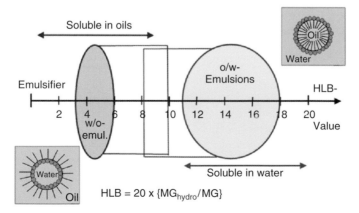

Figure 10.4 Control of the emulsion type by the HLB value of the emulsifier (insets: lipophilic parts are shown by lines, the hydrophilic groups by circles).

Emulsifiers with HLB values between 3 and 6 (up to 8), which are dissoluble in oils, form W/O emulsions. Moreover, surfactants with HLB values of 11–18 produce O/W emulsions. HLB values of 8–11 characterize the transition states between the lipophilic and hydrophilic behavior [20]. In this region arises an O/W- or W/O-emulsion, depending on the type of production and temperature, but preferably the O/W type. The percentage distribution of the phase components (water and oil) does not affect the type. W/O droplets are often smaller than oil droplets in water (Figure 10.5).

The stability of the system depends crucially on the emulsifier. Some nonionic surfactants stabilize emulsions very effectively and are therefore preferred. The choice of emulsifier depends on the chemistry of the oil, because oil and emulsifier should be compatible. High quality natural oils such as borage, hemp, and evening primrose consist mainly of C_{18} triglycerides. Clearly, therefore, it is best to use for the manufacture of an O/W emulsion a nature-based C_{18}-emulsifier.

Ethoxylated fatty alcohols are one of the most effective nonionic surfactants, creating optimal results with about 30 molecules of ethylene oxide (MW about 1600, HLB = 16.6). Other emulsifiers are in most cases less effective (higher quantity is required, larger droplets). A macro-emulsion, containing high quality natural oils, is already stable for several years with 0.5–0.7% of a fatty alcohol with 30-ethylene oxide units. Such a stabilized cream, about four weeks old at the time of measurement, is displayed in Figure 10.6. With this small amount of surfactant in combination with natural oil, the argument that skin drying out is caused by EO-containing nonionic surfactants is unfounded. A real assessment is possible, by testing the complete recipe in comparison with other less effective surfactants, which require an addition of 2–5%.

While the drop size distributions of commercial products are usually in the micron range, nano-sized droplets of about 80–600 nm characterize a mini-emulsion. To generate such a nano-emulsion two surfactants are required, namely, a hydrophilic and a lipophilic emulsifier. The preferred surfactant combination

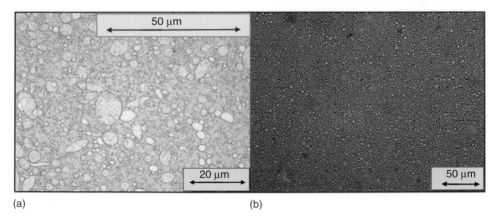

(a)   (b)

Figure 10.5 Microscopy images of commercial skin care products: (a) O/W-emulsion with small and large droplets; (b) W/O-emulsion type with some large drops.

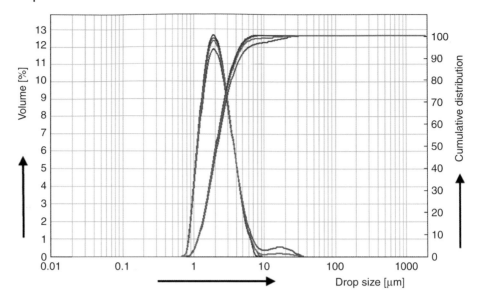

Figure 10.6 Droplet size distribution of a nonionic surfactant-stabilized O/W skin cream with a d_{50} of 1.8 μm; triple measure after dilution by laser diffraction/equipment: Malvern.

with HLB values of 16.6 and 5 results in small, stable droplets. The temperature stability of the emulsion depends on the molecular weight of the emulsifiers, the higher the better. The best results are found with a polymeric silicone-based emulsifier, which is even suitable for multiple emulsions, namely, cetyl PEG/PPG-10/1-dimethicone (Abil® EM 90 from Evonik) in quantities of 0.5% – especially for a stable O/W mini-emulsion applied in combination with about 2% Eumulgin® B 3 (COGNIS/BASF). The droplet size distribution also depends on the homogenizer and the operating conditions. A dual-piston high-pressure homogenizer (Niro Soavi) gives, after three runs, a close and reproducible drop distribution, with $d_{10} = 80$ nm, $d_{50} = 122$ nm, and $d_{90} = 200$ nm (Figure 10.7).

Micro-emulsions are clear and transparent. Because of the high amounts of emulsifiers (about 15–35%) needed for the production, a use for skin care seems less appropriate; similar restrictions apply to multiple emulsions. A multiple emulsion can be produced by the emulsification of an O/W emulsion in oil, or a W/O emulsion in water. The additional costs for multiple emulsions (higher amounts of surfactant, manufacturing, and fewer degrees of freedom in the formulation) result in no benefit for skin care today.

10.3.2
Stability of Emulsions

An essential quality feature of skin care products is the time to use after manufacture and after opening the package. The stability of the emulsion [17], the constancy

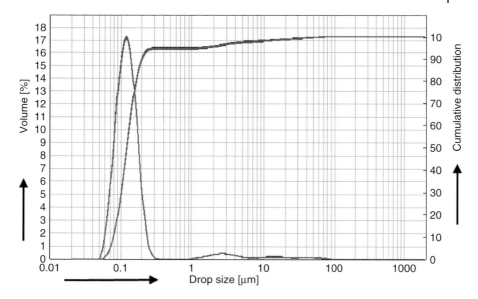

Figure 10.7 Stable cream containing nano-droplets (50–300 nm) after three runs through a high-pressure homogenizer; triple measure after dilution by laser diffraction/equipment: Malvern.

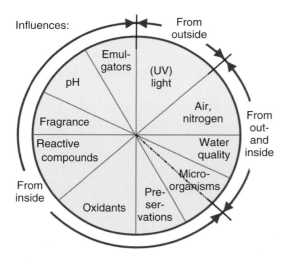

Figure 10.8 Influences on the stability of skin care products.

of color, flavor, and consistency (such as immediately after preparation) are the guidelines for product design. Several steps, regarding the recipe, manufacturing and filling conditions, and the packaging material (Figure 10.8), can optimize the lifetime of the product.

For oxidation-sensitive ingredients such as vitamins and natural oils, protection is possible by introducing chemical groups and by the addition of antioxidants.

Vitamin A palmitate and vitamin C phosphate are examples of the chemical modification of sensitive components. Tocopherols (especially in the presence of citric acid) protect the precious natural oils, with a high content of γ-linolenic acid (18-3), against rancidity. Both smell and color are inconspicuous. The use of substances with oxidizing power leads to significant changes in color, smell, and effect.

The quality of water is of crucial importance, because almost all infections in the product come from the water. Ion-exchange columns remove heavy metals and lime-forming ions prior to use. Membrane plants, supported by UV lamps, sterilize the water. Preservatives in the product not only inhibit the growth of microorganisms but also destroy them. An agent used in low concentration is sufficient, when working in the optimum pH range. Sensitive ingredients also need a stable, constant pH. Various reactions such as hydrolysis, oxidation, esterification, and ester cleavage can alter on the one hand the pH, thus reducing the effectiveness of preservation, and on the other hand create a pleasanter environment for microorganisms. Therefore, the use of a buffer system (pH 5.0–5.5) has certain advantages.

During manufacture, deviations in the temperature profiles, in the order of recipe ingredients, or in the dissolution of substances as well as an insufficient energy input lead to a destabilization of the emulsion. Non-optimal type and amount of emulsifiers, reactions supported by heavy metals, UV radiation and or oxygen, temperature fluctuations, changes in pH, and microbial contaminations over time contribute to the disintegration. On the one hand, the stability of an emulsion depends on creaming/sedimentation as well as the Ostwald ripening [24]. This describes the disappearance of small droplets, because bigger ones are more stable. On the other hand, after some time aggregations and coalescences stimulate the destabilization.

Careful selection of formulation components, emulsifiers, and preservatives, as well as setting and maintaining a constant pH by a buffer system, guarantees the stability of the cream/lotion. Production and filling run under optimized conditions in clean, disinfected equipment, taking into account the recipe-dependent temperature limits. The flushing of all containers with nitrogen is recommended during production and bottling. Furthermore, a packaging (bottle) is needed that is impermeable to UV light, water vapor, and oxygen, and equipped with a removal system without contact to the air and to the product.

Depending on the chemistry and technology, the distribution expresses the quality of the emulsion. The lower the d_{50} and the narrower the monomodal distribution the better should be the stability. The stability is checked with a centrifuge test (40 °C, 30 000g, and 5 min). All narrow distributed emulsions that pass the test remain stable over a long period (some years).

In a few cases, the formulation contains fine solids (powder). The particles settle mainly on the surface of the droplets and assist in the stabilization (Pickering emulsions). These and other types of emulsions, like the phase inversion temperature (PIT) and micro- and multiple emulsions, are fully described in the literature [25, 26] and are not discussed here. They do not play a major role in skin care.

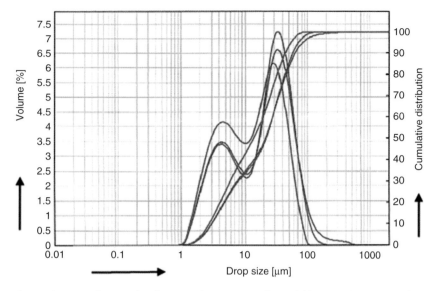

Figure 10.9 Droplet size distributions of a commercially available O/W-skin cream, a few weeks after production; triple measure after dilution by laser diffraction/equipment: Malvern.

Well-known competing products from the market (creams) have been measured with laser diffraction, to gain an impression of the emulsion stability a few weeks/months after manufacture. Remarkably, the four investigated samples of different producers show bimodal distributions. The corresponding O/W-droplet size distribution is depicted in Figure 10.9, and the associated micrograph (Figure 10.5a) shows the visible drops. In addition, a measured W/O-market product also displays a bimodal distribution (Figure 10.5b). The larger droplets in the emulsions are in the range of 20 to about 60 μm (maximum 100 μm); they point to an already incipient destabilization. In other images, some big bubbles were considerably more visible. Amazingly, several market products begin to disintegrate after only a few months. Probably, the emulsifiers are inadequately stabilized. The consumer notices the segregation only after complete disintegration, and thereafter pumps only water out of the bottle.

10.3.3
Preparation of Emulsions in the Laboratory

A manufacturing process is developed in a laboratory/pilot plant before transfer to production in accordance with the scale-up conditions. The following laboratory instructions are used to manufacture the emulsion, often described in this or a similar form, but with higher temperatures. First, in a stirred vessel successively dissolve the hydrophilic components at about 45–50 °C in sterile water. Depending on the ingredients and the consistency of the added viscosity regulators, the dissolution process may vary and definitely can take an hour or more to complete.

Next, acids or bases are added to adjust the pH to 5.0–5.5. AHAs (α-hydroxy acids) such as lactic or citric acid are suitable for this purpose.

To limit the required temperature it is recommended to formulate the recipes without high-melting organic substances, and also without the use of hard paraffin and waxes. In a separate vessel under a nitrogen atmosphere, the lipophilic components dissolve at about 60 °C to give a clear phase. The perfume oil runs in at the end. Immediately after dissolving all components, the oil phase flows into the nitrogen-layered aqueous solution at 45 °C. Here, the homogenizer runs for several minutes with varying speeds. A temperature control system ensures that the temperature in the liquid remains constant at 45 °C. After homogenization and cooling down, the addition of shear- and temperature-sensitive materials as aqueous solutions (e.g., peptides) under slow stirring completes the procedure. Finally, the product is filled, at about 25 °C, under nitrogen into the dispensers.

The manufacture of mini-emulsions from macro-emulsions requires, after addition of another surfactant, more than one run through a high-pressure homogenizer at 45–50 °C.

10.4
Structure of a Skin Care Cream

Creams/lotions are thickened emulsions; they consist of excipients, additives, and active ingredients. The nature, number, concentrations, and interactions of these substances determine the product design (performance, handling, esthetics), and can be influenced greatly.

10.4.1
Excipients

Excipients are the substances needed for the manufacturing, application to the skin, and stabilization. The number of excipients listed in Table 10.3 shows that a stable cream needs at least nine, and in some cases contains up to 20, agents. These include the thickener for the two phases, emulsifiers, solvents, and spreading agents. In addition, in a cream, pH regulators are required to set the pH of the skin to 5. A buffer system (acid/base) preferably executes the pH constancy. Furthermore, the buffer, the absence of oxygen, moderate temperatures, and UV-light impermeable dispensers prevent undesirable chemical reactions. The strength of the preservatives depends on the concentration and on the pH (Section 10.4.2).

The emulsifier system is crucial for the stability of an emulsion, and for the effect on the skin. This persists in O/W emulsions of one or two emulsifiers (surfactants) with HLB values of about 12–17 as well as viscosity regulators for both phases. Non-ionic surfactants such as ethoxylated fatty alcohols are effective already at low concentrations (0.5–1.5%). Other emulsifiers need 2.0–5.0% for stability and a low limit of allowable amounts of salt.

The stabilization of the emulsion happens not only with emulsifiers but also with the thickeners. It is recommended that there are at least two substances as viscosity

Table 10.3 Excipients of a cream according to their functions.

Excipients	Number of common substances	Number of required/recommended substances	Used concentration ranges (%)[a]	Typical substances
(a) Consistency (viscosity control)				
Organic phase	>60	1	2–5	Fatty alcohols (FAs)
Aqueous phase	>40	1	0.2–3	Acrylates, cellulose ethers, xanthan
(b) Emulsifiers				
O/W macro-emulsions	>80	1	0.7–2	FA ethoxylates
W/O macro-emulsions	>50	1	1–3	Sorbitan oleate
Mini (nano) emulsions	? (<20[b])	2	0.5–3	
(c) Preservatives	56	1	0.1–0.4	Sorbic, salicylic acid, parabens
(d) Complexing	12	0 or 1	0.2–0.6	Sodium citrate
(e) Antioxidants	15	1	0.2–2.5	Vitamin E
(f) pH control	31	2	0.1–1	Citric acid, sodium citrate
(g) Solvents, solubilizer	>14	1	0.5–4	Ethanol, propylene glycol
(h) Spreading	>17	1	0.5–3	Silicones
(i) Liposomes	>5	0 or 1	<2	Lecithin

[a] Material and application dependent.
[b] Estimated.

regulators, a lipophilic and hydrophilic one. The viscosity of the water phase can be adjusted by synthetic polymers, like modified polyacrylates, as well as numerous natural or semi-synthetic polymers such as starch, xanthan, and cellulose ethers. The hyaluronic acid thickens the water phase extremely, so even tiny amounts are enough to produce, with this well-known skin ingredient, an increase in viscosity. Lipophilic compounds of high molecular weight, such as long-chain fatty alcohols, set the viscosity of the oil phase. The final viscosity reached in the emulsification process depends on the optimum amount of thickening agents as a function of type and amount of the ingredients, which is estimated in a series of experiments. The amount and nature of excipients determine the structure of the cream. Furthermore, extensive experience is of help in choosing the most suitable active substances, of the best combinations for the viscosity, for the spreadability and haptic feeling.

The stabilization of polyunsaturated natural lipids requires an addition of antioxidants. Even a slight oxidation of the natural oils results in a clearly perceptible, rancid odor. This undesired reaction, catalyzed by heavy metal ions, should be additionally suppressed by the exclusion of oxygen and the use of complexing agents. After applying the cream, the larger water-soluble molecules remain on the skin. For their transport into the epidermis, proteins encapsulated in liposomes are suitable. Dissolved lecithin in water forms hollow spheres (vesicles), in which the proteins can be stored. The loaded vesicles are able to diffuse along the hair shaft and through the pores into the skin and transport active ingredients to the epidermis.

It is assumed that the classes of non-ionic fatty alcohol ethoxylates make the skin more permeable and support the introduction of (harmful) substances, further mobilizing the skin fats. A quote of Paracelsus (1493–1541) may be the right comment on these findings: "All things are poison and nothing is without poison, only the dose makes that a thing is no poison." In particular, the low concentrations, but also interactions and the influence of the environment (pH, salts), thickening agent, and lipids impinge on the effect of surfactants on the skin. Alcohol ethoxylates have the advantage of emulsifying at low concentrations very well, even in the presence of high salt loads. During the application of such cosmeceuticals no adverse effects on the skin are observed or measured.

10.4.2
Preservations

A cream is manufactured at a low level of bacterial counts. It is not sterile, and so germs are present in minimal amounts. These few germs are usually not a problem. (**Caution**: with organic solids, microorganisms can be introduced into the water phase.) After opening the package until complete consumption some months later, bacteria contaminate the content, which renders it useless. Therefore, the utilization of a preservative is required. First, the preservative prevents further multiplication of introduced microorganisms. Formulated in higher concentrations, it is able to reduce the number of germs. The required concentration of a preservative lies mainly in the range 0.1–0.35%, and depends particularly on the efficacy of the recipe and the packaging. The permissible limits are set from the European Cosmetics regulations [8].

In cosmeceuticals, some preservatives are preferred such as the *para*-hydroxybenzoic esters (parabens), sorbic acid, and phenoxyethanol. Phenoxyethanol represents an ether of phenol with ethylene glycol, which has only a moderate activity spectrum, especially against Gram-negative bacteria, and is therefore either formulated in relatively high concentrations (about 0.8%; allowed: <1%) or in combination with other agents. The esterified hydroxybenzoic acid, with a methyl, ethyl, propyl, butyl, or benzyl group, works effectively in mixtures (0.1–0.8%), preferably in a combination of methyl with propyl 4-hydroxybenzoate. The optimal effect lies in the pH range 6–8. The propyl ester is more effective against molds, and is also hydrolytically stable. From the perspective of chemistry, on the one hand, hydrolysis occurs under basic conditions with a pH shift and, on the other hand, the strong oil solubility (= inactivation) reduces possible applications.

Parabens rarely cause allergies but show, compared to sorbic acid, a higher allergy potential. Analyses of breast tumor tissues of women revealed traces of benzyl esters. They are suspected to cause cancer and not allowed to use longer [27]. Furthermore, butyl ester possibly triggers hormonal changes in children. Methylparaben may promote skin aging in sunlight (UV rays). According to present knowledge, documented in numerous studies, the parabens preserve safely. Nevertheless, they provide a degree of risk under intense usage.

In the pH range of skin (<5.0–5.5) the physiologically acceptable sorbic acid (hexadienoic) applies as the best preservative. This acid disassembles in the body similar to the fatty acids via β-oxidation [28]. Above pH 5.5, equilibrium exists between the undissociated acid and the salt, with too little free acid, so that the effectiveness decreases. The hexadienoic acid, which is poorly soluble in oil and in water, dissolves satisfyingly in organic solvents. These are present in cosmetic formulations as solubilizers and, therefore, sorbic acid is particularly suitable for cosmetic creams. Sufficient sorbic acid-preserved cosmeceuticals pass the microbiological stress test described in the European Pharmacopeia [29]. The bacterial count in the inoculated samples is reduced by the preservative within seven days by at least 3 orders of magnitude.

Sorbic acid is sensitive to oxidation and builds up gradually in the presence of air over time. Three measures help to suppress the reactions: the exclusion of oxygen, the addition of antioxidants, and stabilizers. Therefore, the cream runs in a container under nitrogen. By using "airless" dispensers, there is no gas atmosphere above the cream. No growth of microorganisms happens in buffered products in a long-time application.

The water-soluble potassium sorbate in combination with the buffer system citric acid/citrate (Figure 10.10) preserve well at a pH of about 5. The salt releases sorbic acid, which is stabilized by the buffer system. The buffer provides not only for a

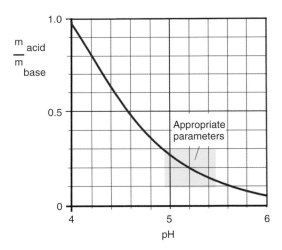

Figure 10.10 Mass ratio of citric acid monohydrate to sodium citrate dihydrate, the best suitable buffer system for skin creams.

constant pH of the cream but also for the correct pH on the skin for a long time, and is important for the effectiveness and durability of the ingredients.

10.4.3
Additives

Perfume is a pleasant-smelling mixture of many odorants, often containing essential oils. Scents enhance the well-being and have a positive effect on the psyche. Depending on the composition, they exert a calming or stimulating effect on the user's and on others nearby. Manifold scents are available. Therefore, many people use a cosmetic cream with a perfume of their choice, for personal characterization. To create a perfume, a perfumer selects 30–200 scents [30] out of more than 2500 natural, nature-identical, and synthetic fragrances, in which a scent may consist of several components. The perfumer's art consists in the compilation of individual oils in kind and amount to a pleasing scent composition with respect to the impression, intensity, and duration in a skin cream or in another consumer product ("product design"). Because of their allergenic effects, 26 fragrances should be used only in very small quantities, or better not at all. According to the annex of the Cosmetics Directive [8], they must be declared on the product package, when exceeding a certain amount (100 ppm for rinse-off and 10 ppm for leave-on materials). Many customers tolerate fragrance oils and choose cosmetic products for the scent impression.

Sensitive people search for well-tolerated, fragrance-free products. Patch test studies showed that 15–20% of the German population is sensitized to at least one of the most important contact allergens (studies of IDU: Information Network of Departments of Dermatology [31]). For more than 7% this is a major problem [32], contributing to fragrance allergy. Additives (Table 10.4) increase the risk of an allergic reaction on sensitive skin. Therefore, substances for scent and for staining are typical ingredients in "normal" cosmetic products, but are usually undesirable in cosmeceuticals.

Several essential fragrant oils show positive effects on the skin. Examples are chamomile, carrot, rose, rosemary, sage, and cedar. Only special products contain these oils. Their typical, intense smell is often not perceived as pleasant, despite

Table 10.4 Additives for color and fragrance.

	Additives	Number of common substances	Number of required/recommended substances
(a)	Fragrances, including essential oils	>2500	0
(b)	Perfumes	>1000	0
(c)	Dyes	145	0

their good efficacy. In contrast to cosmetics, cosmeceuticals normally do not contain any additives. After several positive experiences, without waxes, all creams should be free of substances of solid or semi-solid consistency, such as natural and synthetic waxes, solid vegetable fats, paraffin, or Vaseline.

10.4.4
Groups of Active Substances

The active ingredients divided into groups (Table 10.5). In groups 1 and 2 are the natural oils and vitamins/provitamins listed, in 3 and 4 the moisturizing agents and the AHAs. Other added substances (group 5) enhance or generate a desired effect. Examples include itch-relieving, anti-inflammatory, keratolytic, or especially rejuvenating substances.

Table 10.5 Substances with specific effects on and in the skin.

Group	Active ingredients	Number of common substances	Number of required/ recommended substances	Used concentration ranges (%)	Typical substances
1	Lipids				
	Natural oils	>40	1–3	5–20	Hemp, evening primrose, borage
	Synthetic oils	12	1	<3	Silicone
	Natural or hardened waxes	>20	0	<2	Beeswax, lanolin
	Synthetic waxes	42	0	<2	Artificial jojoba
2	Vitamins/provitamins	13	2 or 3	0.2–2.5	Vitamins A, E, B_3, B_5
	Panthenol			1.5–5	
3a	Natural moisturizing factors (NMFs)	16	2–5	0.2–6	Urea, lactate, amino acids
3b	Further NMF, salts, proteins	>14	2 or 3	0.2–4	Dead sea salts, sorbitol, protein hydrolysates, hyaluronic acid
4	AHA (α-hydroxy acids, fruit acids, salts)	9	3 or 4	<1	Lactic, citric acid
5a	Plant extracts (drugs)	>100	0 or 1	<2	Green tea
5b	Caffeine substances	8	0	<0.3	Caffeine
5c	Essential oils	>125	0 or 1	0.3–0.8	Geranium
5d	Biotechnological proteins	>3	0–2	<0.1	Antarcticine®
5e	Synthetic peptides	>20	0–2	<0.1	Hexa-, tetra-peptides

These active materials determine the effect on the skin, depending on the concentrations and together with the remaining components. Examples are the AHAs, urea, vitamin A and natural product extracts, and the peptides. In relatively low concentrations, the substances are very useful, in high may lead to undesirable effects. The patch test [33] checked the recipe for compatibility (contact allergy) in susceptible individuals. Low concentrations and/or avoidance of critical ingredients lead to a positive safety assessment [34].

10.4.5
Typical Effects of Cosmetics

Products for the skin protect and care for the body and face, eyes and lips, and hands and feet. The cosmetic industry offers most of these specialties for children, teenagers, women, and men. Some examples of the advertised benefits of cosmetic skin care products, especially the facial care, are listed in Table 10.6.

Cosmeceuticals have higher concentrations of active ingredients, and they operate fruitfully in combination with a drug therapy. In general, they are suitable for all people. Table 10.7 shows some examples of product applications that clarify the differences from cosmetic care.

10.5
Essential Active Substances from a Medical Point of View

Each skin cream should contain at least three essential active ingredients in effective concentrations to maintain a healthy skin: each one selected out of the group of natural oils, moisturizers, and vitamins. The following subsections describe the effect of selected substances on the skin.

Table 10.6 Cosmetic skin care, O/W- and W/O- creams, and lotions ("milk").

Skin type	Body part	Advertised benefits
All	Body, face, hands, feet	General skin care; sun protection
All	Eyes	Care in the areas around the eyes
Normal skin (also sensitive and mixed skin)	Face (body)	Moisture, vitalization; cleaning with body milk; (matting) care with UV protection; care with self-tanners
Dry skin	Face, body, hands	Against rough, dry skin feeling; moisture supply
Older skin	Face (body)	For firmer body and/or facial skin; against wrinkles, reduces wrinkle depth
Impure skin	Face	For clear, beautiful skin

Table 10.7 Effects of cosmeceuticals [21].

Skin type	Body part	Offered effect
All (also for sensitive and dry skin)	Body, face, hands, feet	Care by simultaneous supply of moisture, lipids, and (pro-)vitamins
Acne	Face	Reduces the impact
All	Nose	Cares for and reduces snoring
Neurodermitic areas	Body, face, hands, feet	Care by simultaneous supply of moisture, lipids, and specific (pro)vitamins; contains substances to destroy microorganisms
All	Feet	Destroys fungi
All	Face	Reduces lady beard
Older skin	Face	Care and firming by the simultaneous supply of moisture, lipids, (pro)vitamins, and several anti-aging ingredients
Older skin	Body	For the prevention of pressure ulcers

10.5.1
Linoleic and Linolenic

First, among the essential ingredients are natural oils that contain bound linoleic and γ-linolenic fatty acids. The human body cannot produce either of these fatty acids. Therefore, they must be supplied from outside. A deficiency leads to a disruption of the barrier function of the skin with a significant increase in transepidermal water loss. Linoleic acid is a di-unsaturated (C18:2, ω-6) fatty acid that is an essential part of the epidermis. The top layer of skin, the stratum corneum, contains ceramides, free fatty acids, and phospholipids. Ceramides exist as lipid bilayers and regulate the water balance in the skin. The very important linoleic acid represents the largest share of the essential ceramide 1. Secondly, linoleic acid supports the elimination of skin irritation after topical application (contact dermatitis), and reduces light damage to the skin and age spots.

At least as important is the triple-unsaturated (C18:3, ω-6) γ-linolenic acid, which is missing in neurodermitic skin and represents a raw material for the group of tissue hormones (prostaglandins [5]). A lack of these hormones, which are involved in cell metabolism, leads to rough, dry, cracked, and itchy skin. Natural oils such as borage, evening primrose, hemp, and black currant, and from the seed of pomegranate, supply the skin with linoleic and γ-linolenic. One of these natural oils (triglycerides) in amounts of 5–20% should be included in good skin cosmetics. The composition of fatty acids bound to hemp oil corresponds most closely to the skin's fatty acids.

10.5.2
Urea

In dermatology, one of the most important natural moisturizing factors is urea. It binds water in the upper layers of the skin, reduces transepidermal water loss, and contributes to the elasticity of the horny layer. Urea acts as anti-inflammatory, anti-bacterial, and in higher concentrations also as anti-pruritic and keratolytic. On application to open wounds, this substance – as acids – initiated a short burn. Urea is non-toxic and well tolerated.

Essentially, compared to healthy skin, urea is lacking in dry skin by up to 50%, in psoriatic skin up to 60%, and in neurodermatitic up to 70%. These missing amounts must be supplied from outside to the skin. Not only in supporting a therapy, but also as a precaution, active creams contain 2–4% urea, and thus are also suitable for children.

10.5.3
Panthenol

The water-soluble provitamin panthenol [35] penetrates into the skin and reacts in the epidermis to give pantothenic (vitamin B_5). Pantothenic is the main component of coenzyme A, which controls in the skin some metabolic reactions of fats, carbohydrates, and proteins. Therefore, topically applied panthenol enhances the formation of new skin cells and thus measurably promotes regeneration of the skin. Furthermore, this "panacea" improves moisture retention and elasticity of the skin, soothes itchiness and shows anti-inflammatory and wound healing properties. For use in cosmetics, apply the recommended amount of 1.5–3%, and in specialty products for wound healing apply up to 5%. Panthenol is stable at pH 5 up to 45 °C.

10.6
Penetration into the Skin

To optimize skin care products, it is essential to study the penetration of lipophilic and hydrophilic substances, to examine the evidence, and to derive rules for the formulation. In this way the product design is controlled by measurements.

10.6.1
Skin Structure

An explanatory picture of the structure of skin is given in the cross sections shown in Figure 10.11. The thickness of each layer depends on location, age, weight, and gender. Values for the skin thickness, such as those found on the forearm, characterize typical skin areas. Depending on the activity of the fat and sweat

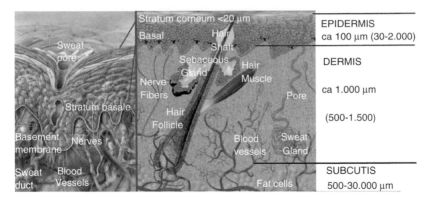

Figure 10.11 The structure of skin. Unlabeled photos: courtesy of COGNIS, Skin Care Forum 29, 34.

glands, and on the current cell condition, the surface of the skin is covered by a film made of water, electrolytes, polar substances, lipids, urea, and amino acids. This film shows acidic properties and, therefore, is called the *"acid mantle,"* or because of its moisture-preserving function it is also known as the "natural moisturizing factor." It provides an additional protective barrier against microorganisms.

The cornea consists of a network of lipid double membranes and dead cells (keratinized corneocytes) and renews permanently from the basal cells of the epidermis. The basal membrane delimits to the directly underlying dermis. Figure 10.12 shows a conceptual model of the construction of the stratum corneum. This concept shows that lipophilic substances easily penetrate through the lipid phase ("mortar") into the lower layers of the epidermis. The water-soluble substances must diffuse through the corneocytes ("stones") and meet, thereby, again on lipid phases, making it difficult to penetrate deeper. A simpler option is diffusion through sweat-filled channels and pores.

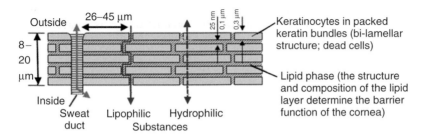

Figure 10.12 Proposed models for the structure of the stratum corneum. Based on Reference [36].

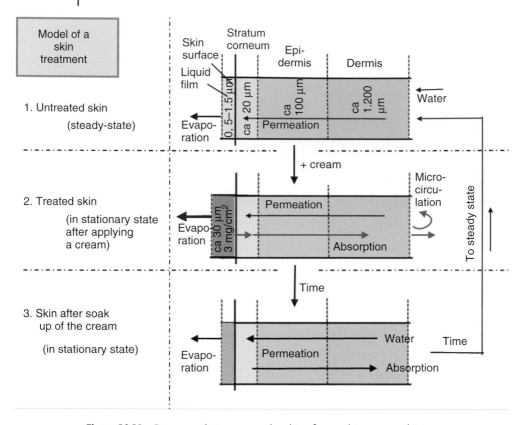

Figure 10.13 Processes that occur on the skin after applying an emulsion.

10.6.2
Applying the Emulsion

The emulsion is applied to the cornea. The surfactants of the cream emulsify the acid mantle, so that the agents are directly in contact with the skin and diffuse more easily. The processes on the skin that expire after application of the emulsion, especially three actions, are shown schematically in Figure 10.13.

Process 1 stands for the stationary state of the skin, showing the acid mantle and a slow drying. In the second process, the applied cream forms a layer on the skin, heated to body temperature. Notably, the surface area is increased greatly by applying the cream. Parts of the film dry by circulated air, whereby the emulsion breaks down. Dehydration (process 3), negatively affects diffusion into the skin. Without changes on the skin, the steady state of the film lasts about 12–24 h, depending on the recipe. Movement, sweating, the rubbing of laundry on the skin, and cleaning the body all disturb or remove the acid mantle temporarily.

10.6.3
Proof of Performance

Comparison of different skin parameters before and after application and over a longer period proves the performance (power) of a skin cream/lotion. Any distortion of results is avoided through a statistical design with about 15–25 (in special cases 50) persons. First, a performed test for skin tolerance excludes primary irritant properties. This happens either in patch tests or in the open application tests. The test aims to exclude the possibility of allergic reactions over a period of one to about five weeks, under realistic conditions or intensified by a provocation test. This test is particularly sensitive and is important for allergic skin types [33].

Various standardized measurement methods are available (Table 10.8) to assess the barrier function, moisture, surface structure, and lipid content in the skin. On one hand, the values obtained allow a precise statement about the skin condition. On the other hand, the results disclose improvements after use of the care product with statistical certainty.

Figure 10.14 displays an example for the determination of skin moisture after a single application by a capacitive measurement over a period of 48 h with over 20 volunteers. Suitable water-soluble substances improve the skin moisture significantly. Charge-free, low molecular weight substances can penetrate into the skin, especially in the case of a hydrated cornea, although the cornea constitutes a barrier to water. The diffusion paths run transepidermal, transfollicular along the hair follicles, or transglandular through pores.

The fine chemicals manufacturer and commodity traders usually give some evidence of the effects of formulated substances, for concentrations and incompatibilities with other substances, pH values, and temperatures. The desired effect is targeted and the recipe is optimized towards the measured skin parameters

Table 10.8 Determination of skin characterizing parameters with different apparatus [23].

Measuring device	Measurement
Corneometer	Skin moisture
Sebumeter	Skin surface lipids
Cutometer	Skin elasticity
FOITS (fast optical *in vivo* topometry of human skin)	Skin surface, structure, wrinkles
TEWL (transepidermal water loss)	Barrier function
pH	pH on the skin surface
Ultrasonic	Skin thickness
3D image analysis of silicone replicas	Microrelief (fold width and depth)
Confocal laser scanning microscopy (CLSM) and optical coherence tomography (OCT) [22]	Structural changes with age (fibrous structure)

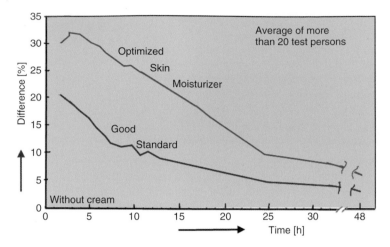

Figure 10.14 Measurement of skin moisture difference (%) with a Corneometer. The cream was applied at time $t = 0$; capacitive measurements of the treated skin minus measured values of the untreated skin.

("product design"), including the exclusive use of natural products. For each desired effect, there are several more or less effective alternative substances from nature.

10.6.4
Penetration of Lipophilic Substances

A specialized NMR (nuclear magnetic resonance) spectroscopy [37], performed *in vivo* on volunteers, tracks the diffusion of thin liquid, natural oils into the skin. Particularly suited is the mobile NMR-Mouse® [38] in conjunction with a newly developed and assembled device to hold up the arm. The equipment changes the position of the measuring arrangement in small steps according to the desired measuring depth. The measuring areas used are on the left and right forearm. Each measurement field produces slightly different values for the untreated skin. The measuring system produces data approximately every 15–30 μm over the skin depth. Each measurement takes about 1 min (65 s) to complete. More recently, both the step size and the measurement times have been reduced.

Evaluation of the NMR measurements succeeds, on the one hand, by calculating the proton densities, which are based on the maximum value of the untreated skin in the epidermis. On the other hand, the response time t_2 of the signal identifies, after calibration, the substance (water: 12.2 ms; natural oil: 88.1 ms). The relevant test results come from curves of the treated minus untreated skin. One example for supplying the skin with pure natural oils (mixture of almond oil with borage) is displayed in Figure 10.15. After 10 min, the supernatant oil on the skin was removed. The measured values [3] indicate that these oils diffuse very quickly in considerable amounts into the epidermis. About 40 min after application there is still a significant effect in the epidermis. In the meantime, most parts of the oil have penetrated into the deeper layers of the dermis.

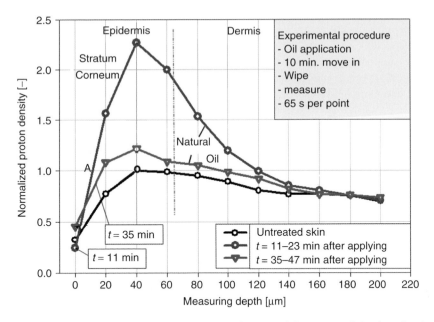

Figure 10.15 Normalized proton densities as a function of the measured depth in the skin after treatment with a mixture of natural oils (triglycerides).

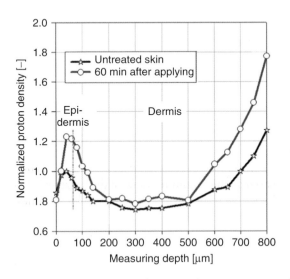

Figure 10.16 Treatment of the skin with an O/W mini-emulsion; differences between the two curves caused by diffused natural oils on the left forearm are clearly visible in the stratum basal of the epidermis and in the deeper layers of the dermis. (The measuring depth zero for the first value can be approximately 25 μm inside the skin.)

Figure 10.16 shows, graphically, typical responses, first for the untreated skin and secondly for the skin after applying a mini-emulsion. In sections, calculated according to Fick's law, the curves emerged under the assumption of a thermodynamic equilibrium at the basal membrane. Already after a few minutes, high values of the oil appear in the epidermis near the basal layer and in the bottom layers of the dermis (reticular dermis, elastic fiber network layer). In the lower layers of the dermis, the values increase for some hours before the oil phase values slowly drop to normal levels. The measurements show reproducibly the same patterns, in the case of mini-emulsions ($d_{50} = 120$ nm) with open, as in practical application, as well as with covered application without water evaporation for theoretical considerations. Comparable values for the untreated skin originated from magnetic resonance imaging (MRI) measurements [39].

The examined inter- and intracellular water contents in the epidermis depend on the measuring depth in the skin [40]. The maximum values are in the stratum spinosum and basal, with inter-and intracellular water contents of comparable sizes.

On covering the skin area after applying the cream, for several formulations both macro-emulsions (d_{50} about 2 μm) and mini-emulsions (d_{50} about 120 nm) show comparable values of the oil content in the skin. In practical applications, both emulsions disintegrate quickly, but at different speeds, due to water evaporation. (The velocity of the water evaporation, also measurable by using the NMR-Mouse® [41], was not determined.) After applying a macro-emulsion, the rapid evaporation leads to a reduced diffusion and a lower the oil content (however, only one formulation tested and, therefore, the result is not meaningful). The macro-emulsion seems to produce significantly increased oil values only in the stratum corneum, but not in the dermis (Figure 10.17), as observed after applying mini-emulsions. Measurements on W/O emulsions are pending – owing to the thickening of the oil phase, slow mass transport is expected.

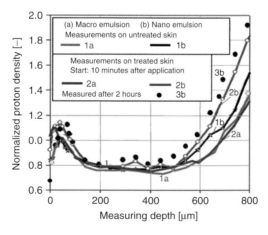

Figure 10.17 Treatment of the skin with a macro- (curves 1a, 2a) and a mini-emulsion (1b, 2b, 3b); the measurements were made 10 min after applying the cream on the left and right forearm; measuring time per curve about 25 min.

10.7
Targeted Product Design in the Course of Development

The three cases examined differ dramatically in practical application: Pure oil is unpleasant on the skin, and penetration through the skin too fast for healing actions. Better effects are produced with the creams. According to the measurements, macro-emulsions preferably reside in the upper layers of the skin, while natural oils of a mini-emulsion penetrate into deeper areas (dermis). The application of creams is easy; both water- and fat-soluble substances diffuse in a controlled manner into the skin without a greasy feeling. The lipophilic phase should contain key ingredients for the skin, such as di- and triglycerides with high levels of double or triple unsaturated acids (omega-6 fatty acids, C_{18}), possibly also the acids themselves, and squalene (triterpenes). Added oil-soluble vitamins, particularly a radical suppresser, support the reduction of the visible effects of skin aging. (The penetration behavior of major water-soluble substances cannot be assessed due to lack of appropriate *in vivo* investigation methods.)

The task of developing specific effects for skin care products can be solved chemically because there exist a variety of different ingredients. Desired mixtures of components both in type and quantity are possible. According to Figure 10.18, pre-optimization of the technology starts with a basic formula in combination with proven emulsifiers and viscosity regulators in the first cycle. During further optimization it is advisable to examine from time to time the emulsification by measuring the droplet size distribution, by testing the stability with a centrifuge, and checking the preservation by microbiological tests.

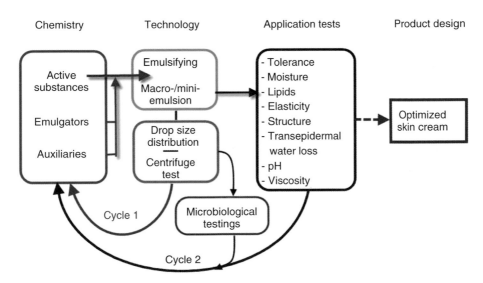

Figure 10.18 Cycles for optimizing product designs.

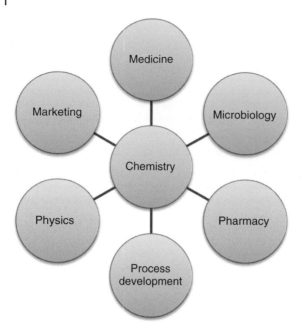

Figure 10.19 Cooperation with various departments in developing new formulations for skin care products.

Physical measurements on the skin and simultaneous application tests allow feedback concerning the use of active ingredients in terms of type and quantity (cycle 2, Figure 10.18). Furthermore there is the possibility of setting the properties of the developed skin care product. Using this optimization method in combination with known active ingredients, several skin properties are ultimately better than the market standard in terms of, for example, skin moisture, elasticity, barrier function, and surface structure. Targeted product designs presuppose laboratory equipment, measuring apparatus, and internal and external resources to test the skin properties of probands. In addition, such development is not possible without the knowledge and skills of different specialists (Figure 10.19) working synergistically in this process. In the end, the product goes to market, and for the next cream the same optimization process is started.

10.8
Production of Skin Care Products

Modern production facilities are designed according to the rules of good manufacturing practice (GMP). The GMP guidelines [42] ensure quality and hygiene in the pharmaceutical, food, and cosmetic industries. They refer in particular to raw materials, plant equipment, rooms, air/water, packaging, and finished products as well as to the staff. Laboratory controls and documentation secure the traceability of each batch. In the cosmetics industry, where the Cosmetic-GMP is valid, production takes place in largely sanitized facilities (pharmaceutical and biotechnological industries: sterile). Personnel entry into the production area is only possible after passing locks, in compliance with the GMP requirements of cleanliness and protective clothing. Owing to structural measures for the premises and for the production plant, cleaning and disinfecting inside and outside are generally easy. For disinfection of the inside surfaces against germs, bacteria, fungi, spores, and viruses, several liquids are suitable that contain chemicals such as peracetic acid, hydrogen peroxide, sodium hypochlorite and alcohols, quaternary ammonium compounds, and glyoxal, in acidic or alkaline solution, or mixtures thereof.

Owing to the high hygiene standards, the plant must consist of stainless steel with highly polished walls inside, dead space-free valves, and be equipped with spray nozzles for disinfecting and cleaning (CIP: cleaning in place) [43]. The system should be emptied easily and completely. Directly before each batch, a disinfection cycle with 2-propanol is run through the entire production equipment. In modern plants and filling lines, filters remove germs from the air, at the intake of outside air, and/or in the room by air circulation. Sensitive recipes require a low excess pressure of nitrogen during manufacturing and filling. The filling takes place preferably in "closed" systems, under purging with nitrogen, or in special cases with germ-free air. A thorough cleaning and disinfection of the entire system including the filling line follows after completion of each batch. The batch-wise manufacturing of creams and lotions is standard in the cosmetic industry.

The whole chain of production, from raw materials to the finished cream in a dispenser, expires in compliance with the cosmetic-GMP guidelines [42, 44] under careful documentation of all sub-steps. The heart of the production plant is a temperature-controlled vessel for emulsifying, which is equipped with agitators and with a rotor–stator machine. One stirrer works close to the wall with a scraper. Most often used is the anchor and helical ribbon stirrer or special structures, partially heated/cooled. If the vessel contains two stirrers, the second one runs counterclockwise, is smaller, and located centrally or in the field of half the radius. Common vessel sizes are 50–4000 l, and in exceptional cases up to $10 \, m^3$. All wetted parts consist of stainless polished steel (material numbers 1.4301, 1.4404, 1.4435, or 1.4462), to reduce surface roughness.

In the manufacturing, O/W emulsions arise by an energy-intensive division of the oil phase in an aqueous phase, in the simplest case by direct energy dispersion in a stirred tank. Here, driven by a metering pump, the lipid phase flows directly into the rotor–stator homogenizer, which works deep inside the vessel within the two-phase mixture. Figure 10.20a shows a simple facility, which is suitable for the production of an O/W-cream or lotion. The vessel employed for emulsifying also serves to regulate the product temperature and in some cases also as a reservoir for the filling plant. For a rapid emulsification, addition of the oil to the water phase directly within a separate multi-stage rotor–stator machine is more effective, because this action reduces the residence (manufacturing) time by increasing the local power density. In these plants, the different designed emulsifiers [45] work either under the vessel in the outlet flange area (Figure 10.20b) or are externally mounted in a cycle (Figure 10.20c). According to the level in the vessel, the emulsion flows, selectably, through the small and/or large circle. To decouple the energy input from the product flow, a metering pump in front of the rotor–stator machine is advisable.

The lower the throughput and the higher the viscosity of the continuous phase, the more energy dissipates by the gear rims of the homogenizer, lowering the danger of coalescence. The stirring arrangements for high viscosities in conjunction with commercially available emulsifiers normally meet the requirements for the manufacture of an O/W-skin cream or lotion, with viscosities of about 5 ± 3 Pa s, measured at room temperature. In practice (Figure 10.21) most creams are produced in facilities that are analogous to Figure 10.20b. To clean the apparatus, the lid opens hydraulically and lifts the stirrer (Figure 10.21a).

The production time depends on the batch size, recipe, and technology. A batch, starting with weighing and dissolving the substances up to discharging into the stirred reservoir of the filling line, lasts about 5–10 h, including the heating and cooling. The actual homogenization in the stirred tank takes about 5–25% of the

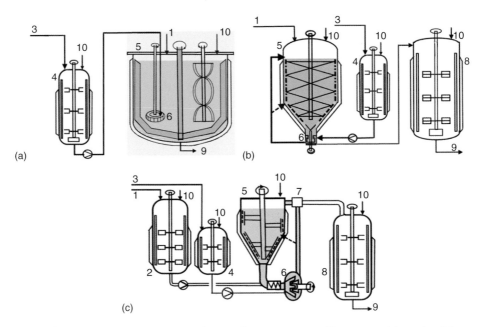

(a)

(b)

(c)

Figure 10.20 Discontinuous production of cosmetic emulsions (creams/lotions): (a) "all in one" emulsifying vessel; (b) flanged emulsifier with recirculation of the product; and (c) rotor–stator machine with dosing pump arranged in the external cycle. (1) Raw material supply of water and hydrophilic substances, (2) stirred tank for dissolving the hydrophilic substances and controlling the temperature, (3) raw material supply of the lipophilic substances, (4) stirred tank for dissolving lipophilic substances and controlling the temperature, (5) stirred emulsifying vessel, (6) rotor–stator machine, (7) three-way valve, (8) temperature controlled stirred tank as a reservoir for filling, (9) to the filling line, and (10) nitrogen supply; to remove air, emulsifying runs mostly in vacuum.

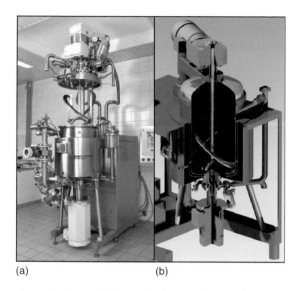

(a)   (b)

Figure 10.21 Small plants for the manufacture of creams/lotions: (a) the top-driven agitator is visible; (b) the bottom-driven rotor–stator dispergator, and the cycle. UNIMIX SRC 100, courtesy of Ekato Systems.

total completion time. To realize a mini-(nano)emulsion [46], the macro-emulsion should preferably run three times (in some cases five) through a high-pressure homogenizer [47]. A product cooler in front of the homogenizer controls the temperature to about 45 °C (Figure 10.22). To produce a nano-emulsion, the three runs through the homogenizer require, depending on plant design, and additional 2–4 h, and a total schedule of about 6–14 h.

Two alternately used stirred tanks ensure the processing of the batch. The first and the third run through the high-pressure homogenizer start from the emulsifying vessel, while in the second pass through the homogenizer the emulsion flows back from the reservoir of the filling line to the emulsifying tank. After the third stage, the cream runs from the stirred reservoir to the filling line, where bottling happens.

Figure 10.23 displays the drop size distributions after the first and second pass through a 600 bar homogenizer. Owing to the high-pressure differences, the microorganisms burst [48]; the product is completely sterile in every case. After the first run, numerous nano-droplets arise, as well as drops in the size range of a macro-emulsion. Already in the second run, both the extremely fine droplets below 50 nm and most droplets outside the micro-range disappear. The third pass leads to a close, stable distribution with very few or no micrometer-drops. On measuring the volume of the drops, few big drops show a large impact on the distribution. No significant progress (Figure 10.24) arises if, first, the pressure increases from 600 to 800 bar and, secondly, the number of passes at 600 bar increases from three to four, judged by the droplet size distributions.

There are several technical possibilities for producing an emulsion [17]. In the cosmetics industry, emulsification takes place preferably in rotor–stator machines.

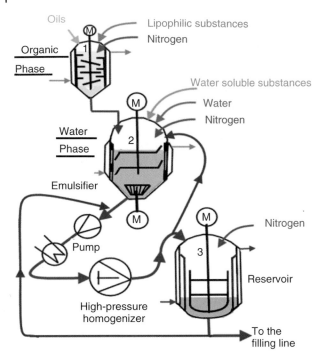

Figure 10.22 Batch plant with a high-pressure homogenizer to produce O/W mini-emulsions: (1) stirred vessel for preparation of the organic phase, (2) aqueous dissolving and emulsifying tank, which is also the reservoir for the first and third run through the homogenizer, and (3) reservoir for the second run and for the filling line.

The alternative emulsification by ultrasound with the same recipe leads to droplet size distributions that lie somewhere between the two possibilities, "Ultra-Turrax and high pressure" (d_{50} about 350 nm). The emulsification techniques, many more or less only applied in pilot plants, can be summarized by the terms "rotor–stator machines, colloid mills, pressure valves, jet disperser, membranes, ultrasound, and high-pressure machines/equipment." Owing to the rotor–stator principle, different mills also emulsify well. The specific energy input is approximately 10^4–10^{10} J m^{-3}. Lower values correspond to macro-emulsions, and higher values to mini-emulsions.

10.9
Bottles for Cosmetic Creams

In the cosmetic creams market, the following container types are common (Figure 10.25): for small quantities (10–50 ml) the tube or the crucible, for medium size (25–200 ml) the pump bottle (pump dispenser), and for large size (100 ml to about 500 ml) the bottles with a round opening, also available as an upside

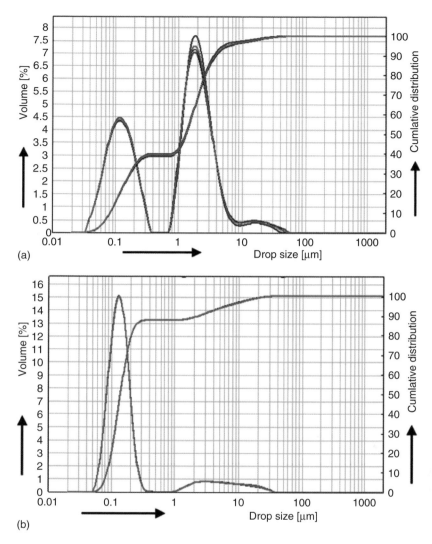

Figure 10.23 Droplet size distributions of a skin care product after (a) one and (b) two runs through a high-pressure homogenizer (the same medium with triple pass; see Figure 10.6); measured three times after dilution by laser diffraction/equipment: Malvern.

standing bottle. The large bottles are sealed with a flip cap or a screw cap. A dosage is possible by shaking or squeezing the bottle. In the airless dispensers, a new generation of hygienic bottles, a piston or a bag delivers the cream to the outlet. A vacuum forces the movement of the product.

Precious formulations contain high levels of polyunsaturated oils. These oils need multiple protection against chemical reactions. UV radiation initiates not only oxidation in the cream, which take place even in the presence of only small amounts

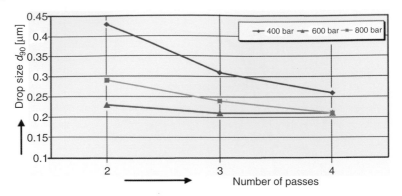

Figure 10.24 Diameter d_{90}-drop size as a measure of the width of the droplet size distributions after several runs through a high-pressure homogenizer (series of measurements on different days with freshly prepared macro-emulsion).

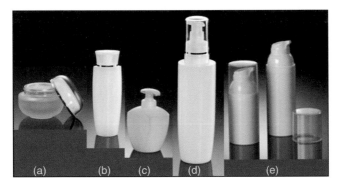

Figure 10.25 Bottles for cosmetic creams: (a) crucible made of glass with plastic lid; (b) bottle with screw cap; (c) bulbous pump bottle; (d) slim pump bottle; and (e) airless dispenser. Courtesy of Pohli.

of oxygen, but other chemical reactions (ester cleavage, condensation) also occur. Therefore, the use of UV light impermeable bottles is necessary. Depending on the type of container and an overpack, for cosmeceuticals transparent or translucent materials such as glass, poly(ethylene terephthalate) (PET), or polyethylene (PE)/polypropylene (PP) are suitable only in exceptional cases, because the sensitive ingredients must survive a storage and consumption time of two years and more.

Coloration with plastic dissolvable paints or with incorporated pigments protects the containers against UV rays. Pigments are suitable fine solids, such as titanium dioxide. One of the best materials for the bottles is the white, TiO_2-containing, opaque PP, because it is chemically neutral, physiologically harmless, and shows a smooth, haptically, and visually attractive surface. The gas permeability of the plastics used plays an important role. Depending on the temperature, high-density polyethylene (HD-PE) allows the diffusion of only 20–33% of the oxygen, compared with an equally strong wall of low-density polyethylene (LD-PE) [49]. PP exhibits comparable values to HD-PE. In both plastics, HD-PE and PP, the water vapor permeability is low. Owing to the exact dimensions of the piston, PP-airless dispensers need thicker walls, and so the gas permeability falls accordingly.

After careful filling of the almost germ-free cream under nitrogen into the containers, microorganisms do not multiply until use, even in weakly preserved products. After opening the containers, different hazards can infect the cream with germs. High-priced creams, filled in attractively shaped crucibles, are exposed to the ambient air over a large surface. In addition, a major disadvantage is removal of the cream with a finger. Microorganisms not only reach the cream over the air but, especially, through contact with the fingertips. Therefore, creams in jars must contain higher contents of preservatives. Similar conditions, but with reduced contaminations, occur in the big bottles. In the smaller pump bottles, only the ambient air contacts the cream during ordinary use. Therefore, from a hygienic point of view, pump dispensers are better than crucibles and big bottles.

The best choice for hygiene and convenience is provided by the airless dispensers (Figure 10.26). Available in sizes from 15 to 150 ml in round or oval shape they offer the greatest protection against contamination during both storage and application. Airless dispensers are filled free from air at room temperature, in most cases from above; the head with the valves is then pressed on the filled container. The alternative bottom filling requires a separate sealing station. During cream removal, the piston moves slowly upwards. The removal is stopped not only by closed valves

(a)   (b)   (c)

Figure 10.26 Airless dispensers: (a) function shown in a cross section: (1) cap (PP), (2, 3) head (PP), (4) upper valve (EVA = ethylene vinyl acetate), (5) bellows (LD-PE), (6) lower valve (POM = polyoxymethylene), (7) container for the cream (PP), (8) iston (HDPE), and (9) bottom (PP). (b) and (c) 100 and 50 ml market product after filling from the top: cosmeceutical skin care cream in an airless dispenser (ATS License GmbH [21]). Courtesy of Pohli.

but lips at the outlet also close. Contact of the cream inside the dispenser with the air is virtually impossible. Such dispensers are optimally suited for preservative-free formulations.

10.10
Design of all Elements

For cosmetics, product design includes the cream formulation and the dispenser. The essential elements are the product performance, convenient handling, and aesthetics. The optimal supply of appropriate active substances through the skin is given the highest priority in development. Additionally, the brand plays a major role (Table 10.9). The other points listed for an optimal design, including the need for essential marketing elements such as brand name and logo, colors, and lettering, are carefully considered and executed. They contribute to a consistent product appearance, with importance not only for simple cosmetic products but also for cosmeceuticals.

An unpleasant odor, a disagreeable color, or a greasy consistency may be tolerated in therapeutic agents, but are surely not accepted by the users of cosmeceuticals. However, the usage of alternative materials or changed processes overcomes such product failings. Optimal aesthetics and dosage with an airless dispenser increase the frequency of application and differentiate products in the market.

After applying an active skin cream (Figure 10.27) twice a day for some days, most people show a visible improvement in the appearance of their skin, and also of diseased skin. The results of multiple measurements (before application, after

Table 10.9 Design elements of active cosmetics.

Elements of design	Cream	Dispenser
Performance (responsible: the composition)	Effect tolerance stability	Hygiene
Convenience (responsible: the auxiliaries and the container)	Consistency productivity (range)	Removal dosage form
Aesthetics (responsible: the composition and the container)	Color gloss odor stickiness	Color form surface easy pumping
Brand/brand family (responsible: the composition, the container, and the label)	Typical ingredients; exclusion of substances (without perfume, dyes, etc.)	Name, lettering; logo, label, text, layout; colors

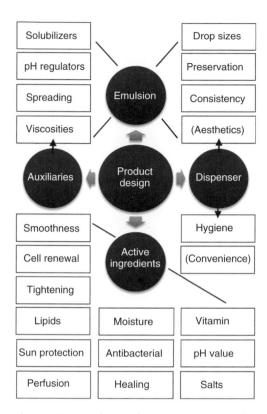

Figure 10.27 Product performance parameters of active cosmetics for the skin.

an hour, hours, days, weeks, and months) prove the effectiveness of a skin cream. Typical measurement parameters are the moisture and lipids content, skin texture, thickness, elasticity, and barrier function of the skin surface [50]. The chemist is usually responsible for the development. He or she tailors the recipes with suitable ingredients to solve a specific skin problem, taking into account the medical effects of the active substances. Thus, the gap between cosmetic creams, on the one hand, and ointments for therapeutic benefit, on the other hand, is closing.

References

1. Rähse, W. and Hoffmann, S. (2003) Product design – interaction between chemistry, technology and marketing to meet customer needs. *Chem. Eng. Technol.*, **26** (9), 1–10.
2. Rähse, W. (2007) *Produktdesign in der Chemischen Industrie*, Springer, Berlin.
3. Rähse, W. and Dicoi, O. (2009) Produkt-design disperser Stoffe: Emulsionen für die kosmetische Industrie. *Chem. Ing. Tech.*, **81** (9), 1369–1383.
4. Rähse, W. (2011) Produktdesign von Cosmeceuticals am Beispiel der Hautcreme. *Chem. Ing. Tech.*, **83** (10), 1651–1662.
5. Ellsässer, S. (2008) *Körperpflegekunde und Kosmetik*, 2nd edn, Springer-Verlag, Berlin.
6. Bouchemal, K., Briancon, S., Perrier, E., and Fessi, H. (2004) Nano-emulsion formulation using spontaneous emul-sification: solvent, oil and surfactant

optimization. *Int. J. Pharm.*, **280**, 241–251.

7. Gabard, B. (2000) *Dermatopharmacology of Topical Preparations: A Product Development-Oriented Approach*, Springer, Berlin.

8. European Commission (2009) EU Cosmetic Regulation – Consumer Affairs – Text of the Cosmetics Directive, *http://ec.europa.eu/consumers/sectors/cosmetics/documents/directive/* (accessed 7 March 2013).

9. Draelos, Z.D. (2009) *Cosmeceuticals*, 2nd Edition, Procedures in Cosmetic Dermatology, Elsevier Inc.

10. Walters, K.A. and Roberts, M.S. (eds) (2007) *Dermatologic, Cosmeceutic, and Cosmetic Development: Therapeutic and Novel Approaches*, Taylor & Francis, New York.

11. Draelos, Z.D. (2007) The latest cosmeceutical approaches for anti-aging. *J. Cosmet. Determatol.*, **6** (s1), 2–6.

12. Zuzarte, M. *et al.* (2011) Chemical composition and antifungal activity of the essential oils of Lavandula viridis L'Hér. *J. Med. Microbiol.*, **60**, 612–618.

13. Talakoub, L., Neuhaus, I.M., and Yu, S.S. (2009) in *Cosmetic Dermatology* (eds M. Alam, H.B. Gladstone, and R.C. Tung), Saunders, Philadelphia, p. 7.

14. Lupo, M.P. and Cole, A.L. (2007) Cosmeceutical peptides. *Dermatol. Ther.*, **20** (5), 343–349.

15. AMG (1976/2005) Gesetz über den Verkehr mit Arzneimitteln, *http://www.gesetze-im-internet.de/amg_1976/index.html* (accessed 7 March 2013).

16. Lüllmann, H., Mohr, K., and Hein, L. (2006) *Pharmakologie und Toxikologie*, 16 edn, Georg Thieme Verlag, Stuttgart ch. 2.6, p. 39 ch. 22, pp. 258–262.

17. Schuchmann, H.P. (2007) in *Product Design and Engineering*, Vol. 1 (eds U. Bröckel, W. Meier, and G. Wagner), Wiley-VCH Verlag GmbH, Weinheim, pp. 63–93.

18. Mollet, H. and Grubenmann, A. (1999) *Formulierungstechnik*, Wiley-VCH Verlag GmbH, Weinheim.

19. Griffin, W.C. (1949) Classification of surface active agents by HLB. *J. Soc. Cosmet. Chem.*, **1**, 311–326.

20. Mollet, H. and Grubenmann, A. (2008) *Formulation Technology: Emulsions, Suspensions, Solid Forms*, Wiley-VCH Verlag GmbH, Weinheim.

21. ATS License GmbH (2012) Cosmeceuticals: Product Portfolio, *http://www.ats-license.com/*.

22. Neerken, S., Lucassen, G.W., Bisschop, M.A., Lenderink, E., and Nuijs, T. (2004) Characterization of age-related effects in human skin: a comparative study that applies confocal laser scanning microscopy and optical coherence tomography. *J. Biomed. Opt.*, **9**, 274–281.

23. Fluhr, J.W. (ed.) (2011) *Practical Aspects of Cosmetic Testing*, 1st edn, Springer, Berlin.

24. Ostwald, W. (1900) *Z. Phys. Chem.*, **34**, 495.

25. von Rybinski, W. (2005) in *Emulgiertechnik* (ed. H. Schubert), Behrs Verlag, Hamburg, pp. 469–485.

26. Umbach, W. (ed.) (2004) *Kosmetik und Hygiene*, 3rd edn, Wiley-VCH Verlag GmbH, Weinheim.

27. Andersen, K.E., White, I.R., and Goossens, A. (2011) in *Contact Dermatitis* (eds J.D. Johansen, P.J. Frosch, and J.-P. Lepoittevin), Springer-Verlag, Berlin, ch 31, p. 560.

28. Lück, E. and Jager, M. (1995) *Chemische Konservierungsstoffe*, 3rd edn, Springer-Verlag, Berlin, ch. 19, p. 158 ff.

29. Sigg, J. (2005), in *Emulgiertechnik* (ed. H. Schubert), Behrs Verlag, Hamburg, ch. 21, p. 595 ff.

30. Elsner, P. and Maibach, H.I. (2005) *Cosmeceuticals and Active Cosmetics: Drugs Versus Cosmetics*, Cosmetic Science and Technology Series, 2nd edn, Vol. 27, Taylor & Francis Group, Boca Raton.

31. Schnuch, A., Uter, W., Lessmann, H., Arnold, R., and Geier, J. (2008) *Allergo J.*, **17**, 611–624.

32. Schnuch, A., Uter, W., Geier, J., Brasch, J., and Frosch, P.J. (2005) *Allergo J.*, **14**, 618–629.

33. Oakley, A., (2008) DermaNet NZ: Standard series of patch test allergens. *http://dermnetnz.org/dermatitis/standard-patch.html* (accessed 7 March 2013).

34. Rogiers, V. and Pauwels, M. (2008) Safety assessment of cosmetics in Europe, in *Current Problems in Dermatology*, Vol. 36 (ed. P. Itin), Karger, Basel.

35. Ebner, F., Heller, A., Rippke, F., and Tausch, I. (2002) Topical use of dexpanthenol in skin disorders. *Am. J. Clin. Dermatol.*, **3** (6), 427–433.

36. Heymann, E. (2003) *Haut, Haar und Kosmetik*, 2nd edn, Hans Huber Verlag, Bern.

37. Blümich, B. (2005) *Essential NMR for Scientists and Engineers*, Springer, Berlin.

38. Blümich, B. and Blümler, P. (1999) Verfahren zur Erzeugung von Messsignalen in Magnetfeldern mit einem NMR-Mouse-Gerät. DE 199 39 626.4, Aug. 20 1999.

39. Agache, P. and Humbert, P. (eds) (2004) *Measuring the Skin*, Springer, Berlin.

40. Kemenade, P.M. (1998) Water and ion transport through intact and damaged skin, PhD thesis, CIP-Data Library Technische Universiteit Eindhoven.

41. Dicoi, O., Walzel, P., Blümich, B., and Rähse, W. (2004) Untersuchung des Trocknungsverhaltens von Feststoffen mit kernmagnetischer Resonanz. *Chem. Ing. Tech.*, **76** (1–2), 94–99.

42. HAS (Health Sciences Authority) (2008) GMP Guidelines for the Manufacturers of Cosmetic Products, December 2008.

43. Gea Process Engineering Inc. (2011) CIP Cleaning-In-Place/SIP Sterilization-In-Place, Internet 2011, *http://www.niroinc.com/gea_liquid_processing/cleaning_in_place_sip.asp* (accessed 7 March 2013).

44. EFfCI (The European Federation for Cosmetic Ingredients) (2005) GMP Guide for Cosmetic Ingredients, 2005 Revision 2010. *http://www.effci.org/assets/snippets/filedownload/GMP/Archive%20-%20GMP%20Guidelines/EFfCI_GMP_2010.pdf* (accessed 7 March 2013).

45. Jafari, S.M., Assadpoor, E., He, Y., and Bhandari, B. (2008) Re-coalescence of emulsion droplets during high-energy emulsification. *Food Hydrocolloids*, **22** (7), 1191–1202.

46. Jafari, S.M., He, Y., and Bhandari, B. (2007) Optimization of nano-emulsions production by microfluidization. *Eur. Food Res. Technol.*, **225** (5–6), 733–741.

47. Gea Niro Soavi North America (2012) High-Pressure Homogenization Technology. *http://www.nirosoavi.com/high-pressure-homogenization-technology.asp* (accessed 7 March 2013).

48. Christi, Y. and Moo-Young, M. (1986) Disruption of microbial cells for intracellular products. *Enzyme Microb. Technol.*, **8**, 194–204.

49. Nentwig, J. (2006) *Kunststoff-Folien: Herstellung- Eigenschaften- Anwendung*, Carl Hanser, Munich, p. 100.

50. Förster, T. (2004) in *Kosmetik und Hygiene*, 3rd edn (ed. W. Umbach), Wiley-VCH Verlag GmbH, Weinheim, pp. 85–92.

11
Emulsion Gels in Foods

Arjen Bot, Eckhard Flöter, Heike P. Karbstein-Schuchmann, and Henelyta Santos Ribeiro

11.1
Introduction

Many foods contain oil or fat as an important energy source or carrier of aromas, vitamins, colorants, or active compounds. Palatability and bioavailability is improved when fatty compounds are dispersed into small droplets, typically in the range 0.5–100 µm. Thus, many food products are based on an emulsion structure. Such a finely dispersed phase is obtained by mixing two immiscible phases, usually – but not always – consisting of a lipid and an aqueous phase. Although the mixture has a homogeneous appearance, both constituents remain separated on a microscopic scale. The apparent homogeneity exists only by virtue of the fact that separation of both phases takes time, with the time scale involved being indicative of the stability of the emulsion droplets against coalescence.

The traditional classification of emulsions is in terms of the composition of dispersed and continuous phase. The most common examples are oil-in-water (O/W) and water-in-oil (W/O) emulsions, although biopolymer mixtures exist that can form water-in-water (W/W) emulsions. This simple subdivision tends to ignore the fact that large differences exist even within one class of emulsions. One of the additional criteria by which emulsions can be distinguished is whether droplets interact. Interactions, which are especially common in O/W food emulsions, may cause aggregation and even (partial) fusion of the emulsion droplets. Another very distinguishing feature is whether or not the continuous phase is gelled. Such phenomena may change the macroscopically observed rheological behavior of the emulsions completely, and turn a low-viscosity liquid emulsion into a thick or even solid material [1].

In fact, phenomena like droplet aggregation and gelling of the continuous phase are very common in food emulsions and are among the root causes for the qualitative differences between them. For example, regular margarine and liquid margarine both contain about 16% dispersed aqueous phase, but show completely different behavior. On the other hand, butter, margarine (16–20% aqueous phase), and low fat spreads (60% aqueous phase) contain very different amounts of aqueous phase but share a relatively similar rheology.

Product Design and Engineering: Formulation of Gels and Pastes, First Edition.
Edited by Ulrich Bröckel, Willi Meier, and Gerhard Wagner.
© 2013 Wiley-VCH Verlag GmbH & Co. KGaA. Published 2013 by Wiley-VCH Verlag GmbH & Co. KGaA.

These vast differences between emulsions suggest that it might be more fruitful to view food emulsions as composite materials, rather than as idealized emulsions consisting of liquid continuous and dispersed phases with non-interacting droplets. The dispersed phase would be considered to be the filler, and continuous phase would take the role of the matrix. The relative properties and amounts of both phases will determine the behavior of the combination. A soft filler will not change the properties of a stiff matrix, but a firm aggregated filler may have an dramatic impact on the rheology of a liquid continuous phase.

As mentioned above, emulsions are not stable over time. However, the time scale over which the instability develops may differ vastly between emulsions. One major factor determining the stability of the emulsion is the composition of the interface of the droplet. Generally, the interface of an emulsion is stabilized by the inclusion of surface active molecules or emulsifiers/surfactants. Depending on their molecular structure they may change the electrical charge of the droplets and thus influence electrostatic droplet interaction potentials. In addition large-molecular-weight molecules may either stabilize the emulsion by acting as a spacer between the droplets or promote inter-droplet interaction and, thereby, the tendency to flocculate. Thus, emulsifiers also strongly influence emulsion firmness – along side their important role in stabilizing the emulsion microstructure over the period of the shelf life.

In addition to the well-known O/W and W/O type emulsions, more complex structures exist. Multiple emulsion structures are found, for example, in pharma-ceutical or functional food applications. Water-in-oil-in-water ($W/O/W$) emulsion type formulations allow for the encapsulation of hydrophilic compounds, such as water or anthocyanins, with the former being used for fat-reduced products [2]. Oil-in-water-in-oil ($O/W/O$) structures offer the possibility of encapsulating lipophilic compounds such as vitamins, carotenoids, or long-chain poly-unsaturated fatty acids (*LCPUFA*s), and the corresponding oils.

This chapter will discuss the properties of dispersed and continuous phases, how processing can be used to create an emulsion, and how the resulting structure affects the food products that consist of these emulsions.

11.2
Food Emulsions

11.2.1
Dispersed Phase

The dispersed phase in a food emulsion may serve several purposes. It can be used to store ingredients that are not soluble in the continuous phase of the emulsion, or it can be used to modify the properties of the emulsion as a whole.

Food products are often intended to deliver mutually incompatible ingredients in a single product. An example is fat in milk, or water-soluble tastants like salt in margarine. More specialized examples are the delivery of specific components with a health benefit. For these applications it is of utmost importance that the ingredient remains nicely dispersed in the emulsion during the shelf life of the food product, because extensive coalescence of the dispersed phase may result in macroscopic separation of the product and hence in incomplete delivery of the desirable component. Factors that promote coalescence in emulsions will be discussed below.

With regards to the rheology-modification role of the dispersed phase, it is important to realize that emulsion droplets behave as more or less hard spheres as long as their radius is small enough to ensure that the surface energy contribution dictates a spherical shape. Thus, emulsion droplets can be considered to behave more or less as solid particles.

The presence of these droplets modifies the viscosity of the emulsion relative to that of a pure liquid continuous phase. These effects are quite small at low dispersed phase volume fraction, but become very significant at higher dispersed volume fraction, as can be appreciated by comparing milk (low dispersed volume fraction) with mayonnaise (high dispersed volume fraction). The increase in viscosity is not linear with dispersed volume fraction but becomes more pronounced the closer the emulsions gets to the random closed packing volume of the droplets φ_{rcp} (about 64% for monodisperse non-interacting rigid spheres, but other values can be found for polydisperse, deformable, or interacting systems). The behavior is described by the following equation [3]:

$$\eta/\eta_{continuous} = (1 - \varphi/\varphi_{rcp})^{-2.5\varphi_{rcp}} \tag{11.1}$$

This, the so-called Krieger–Dougherty equation, describes the divergence of the viscosity η when approaching close packing of the droplets. Above this close packing fraction the dispersion will behave as a (soft-)solid or a spoonable material, and other equations should be applied to describe such materials.

Furthermore, notably, the volume fraction in the Krieger–Dougherty equation is actually an effective volume fraction and the aggregation of droplets will lead to incorporation of some of the continuous phase in the droplet aggregate. This will increase the effective volume fraction of the dispersed phase, which can be accounted for by either using a higher (effective) volume fraction or an appropriate closed packing fraction for the system. An example of a system that shows aggregation of the dispersed phase is a sour cream, in which the protein stabilized droplets aggregate under the acidic conditions in the product, leading to a spoonable consistency. In such systems, reduction of the droplet size is usually a rather effective method for achieving a more tenuous aggregate and, thus, a higher effective volume fraction at a given amount of dispersed phase [4].

The nature of the dispersed phase also plays a role in such aggregated systems. It can be shown that the presence of crystallizing fats may result in an increase in the consistency of the emulsion gel compared to emulsions based on a liquid dispersed phase, especially if the dispersed phase has a tendency to promote partial coalescence [4].

11.2.2
Continuous Phase

The properties of the dispersed phase are especially important in emulsions for which the continuous phase contributes relatively little to the consistency of the emulsion. The situation is different, however, in systems that contain structuring agents.

11.2.2.1 Aqueous Continuous Phase

In aqueous continuous phases a wide range of potential structurants is available, from viscosifiers to gelling agents. The viscosifiers are usually soluble hydrocolloids or biopolymers with relatively little mutual interaction. Examples are guar gum or extracellular polysaccharides that are formed, for example, by some yogurt fermentation cultures. Gelling agents, however, are of most interest. These usually consist of partly soluble molecules, in which the less soluble part of the molecule engages in mutual interactions for at least some conditions (e.g., low temperature, high salt, after denaturation).

The iconic example of such a system is gelatin at sub-body temperatures (see, for example, References [5–8]), which lent its name to gelling phenomena in general. Gelatin is denatured collagen, and can be dissolved completely at high temperatures. Glycine/proline/hydroxyproline-rich hydrophobic patches form so-called triple helices at lower temperatures and aggregate to form a network.

Gelatin is quite unique as a structurant because of its steep melting behavior around body temperature, which makes the material very useful for food applications. One of the downsides of gelatin gels, however, is that they are quite deformable – too deformable for many food applications.

Pectins from sources like citrus fruits or apples are polysaccharide-based biopolymers (backbone is based on α-1,4-galacturonic acid), being able to gel upon cooling or addition of calcium ions. The possibility of modifying pectin molecules by amidification or esterification allows for the addition of extra interfacial active groups, which is of high interest in emulsion applications. Gelling and rheological characteristics are also influenced by chemical modification of these molecules [9].

Protein (particle) gels are usually less deformable and break at smaller deformations. The most common structurant of this type is milk protein. Milk contains mainly two types of proteins: caseins and whey proteins.

Casein occurs in small aggregates called *casein micelles*, typically 50–300 nm in diameter. The casein micelles can form gels through acidification or renneting. During renneting a specific subclass of caseins (*GMP*, glycomacropeptide) present on the surface of the micelle is split enzymatically, removing the relatively strongly charged part of these molecules and thereby their capacity to stabilize the micelle as a whole in solution. Renneting is the basis of hard cheese formation and will not be discussed further here. Acidification to pH values below approximately 4.6 (isoelectric point) neutralizes the charges in the same subclass of caseins, reduces electrostatic repulsive interaction forces, and renders them ineffective in stabilizing the micelle as a whole. The pH at which aggregation occurs depends on the history of the sample (heat treatment, temperature–pH route to destabilization) [10]. The properties of casein micelle networks are determined mostly by the interaction between the micelles and their spatial arrangement, and to a limited extent by the mechanical or structural properties of the micelles themselves. Important properties are firmness and resistance to syneresis (i.e., water exudation), both of which are promoted by having a finer network – for example, by applying a higher rate of acidification. As an alternative to casein gelling by means of destabilization of the surface of the micelle (rennet or acid), enzymatic crosslinking by means of transglutaminase can be used – although this approach has not found widespread application in the area of foods. It has been mainly applied in meat products.

Native whey proteins (mainly β-lactoglobulin) are acid stable, but do gel after denaturation of the protein, which involves a conformational change in which a hidden but reactive thiol moiety becomes available for a polymerization reaction amongst whey proteins [11]. This phenomenon is called *cold gelation*. In practice, milk proteins occur as approximately 80 : 20 mixtures of casein and whey proteins. The gel structure is therefore dominated by the casein network structure. The whey proteins affect the properties of the casein micelles, however, as heat-treated denatured whey protein that adsorbs at the micellar surface, modifying the aggregation properties of the micelles [12]. The denatured whey protein may strengthen the links between the casein micelles, making the gel firmer, and increase the pH at which the micelles start to aggregate. The stronger interaction between the micelles will also reduce the rate of rearrangement in the gel, resulting in a lower syneresis than in a pure casein gel. It is consequently recommended to add whey protein to the yogurt formulation to prevent syneresis after the gelation process.

A rather different way to structure aqueous phases is by means of monoglycerides, which form lamellar liquid crystalline phases upon dissolution in water. Upon cooling, these lamellar phases transform into the α-gel phase, a crystalline bilayer structure of α-crystals that is similar to α-crystals occurring during rapid crystallization of triglycerides. The firmness of this network is low, as it is in triglyceride networks. The α-phase transforms slowly into a so-called coagel phase of β-crystals, which is much firmer. These coagel phases show an uncanny similarity to fat crystal networks in oil [13–15].

11.2.2.2 Lipid Continuous Phase

In lipid continuous phases, the choice of structurants is far more limited. Crystallizing triglycerides (or triacylglycerols, *TAGs*) are the most common means to achieving a firm lipid phase in a food product. Fats can be obtained from a wide range of raw material sources (animal such as dairy and meat, as well as seed oils, tropical fruits, algae), and differ in their fatty acid composition in chain length distribution and degree of saturation. Since the physical properties of fats are determined by the triglycerides present, the organization of these fatty acid chains in the TAG molecules is of prime importance. In general the melting points of TAGs are higher with more saturated fatty acids (*SAFAs*) present in a TAG molecule. Additionally, they increase quite systematically with the chain length of the fatty acids present. It is fair to assume that in fat crystallization one practically always deals with multiple mixed solid phases, which makes adequate detailed description very difficult. Generally, one can simplify such a description by considering structuring of edible oils to originate especially from either TAGs containing three saturated fatty acids or TAGs with saturated fatty acids and one unsaturated fatty acid. If the crystals are sufficiently small, they can effectively structure a liquid oil phase by aggregating in a tenuous but firm network. The combination of actual size, shape, and number of the crystals determines to a large extent the mechanical properties of the network. Fast and deep cooling stimulates the formation of large numbers of very small crystals, which interact with strong "primary" sintered connections and weak "secondary" van der Waals bonds between the crystals. The mechanical network properties, and thus the final product properties, can be tuned by processing via the ratio between primary and secondary bonds. Additionally, proper choice of crystallizing triglycerides may result in a crystal network in which the crystallites melt at mouth temperature, giving rise to desirable sensorial properties for a food emulsion.

The downside of the crystallizing triglycerides, however, is that the SAFAs are considered to be unhealthy because they contribute to the risk of cardiovascular diseases when present in the diet [16]. This has raised the question of whether it is possible to achieve structuring of edible oils without the use of crystallizing triglycerides [17, 18]. In this quest it has been realized that an edible oil can be viewed as an organic solvent, and that the problem is very similar to finding organogelling compounds. Potential candidate systems to achieve non-triglyceride structuring systems for food-grade vegetable oils include fatty acids, fatty alcohols, monoglycerides, diglycerides, phospholipids, ceramides, waxes, wax esters, or mixtures thereof [19–22] as well as polymers [23].

An even more interesting system is the self-assembling plant sterol + γ-oryzanol mixture in edible oil [24]. The sterol and oryzanol molecules can stack their sterane cores, but the presence of a hydrogen bond prevents perfectly parallel stacking of the molecules. As a consequence, the aggregate does not form a perfectly straight one-dimensional crystallite but, instead, the assembly curves somewhat, much like a staircase. On a supramolecular length scale this leads to the formation of a helical ribbon in triglyceride oil [25]. The resulting tubules aggregate into a firm "organogel" network [26]. Sitosterol, ergosterol, stigmasterol, cholesterol, or cholestanol are all capable of taking the role of the plant sterol. The typical diameter of a double-walled tubule is almost 10 nm and the typical wall thickness of the inner wall is 1.5 nm and of the outer wall just under 1 nm, and the tubules are much longer than they are wide [27, 28]. The tubules are thin enough to ensure that the gel is still transparent.

11.2.2.3 Emulsion Stabilization by Emulsifiers or Particles
Emulsions are stabilized by surface-active molecules, emulsifiers, adsorbed at the interface. Chemically, emulsifiers form a diverse group that includes low-molecular-weight compounds like monoglycerides and phospholipids and high-molecular-weight components like proteins. The molecules share the tendency to

adsorb at the interface, thus reducing the interfacial energy. According to the so-called Bancroft rule [29] an emulsifier that predominantly dissolves in the aqueous phase tends to promote an O/W emulsion, whereas an emulsifier that dissolves primarily in the oil phase promotes a W/O emulsion – though the rule may not be valid for extreme volume ratios of dispersed to continuous phase.

An extreme form of stabilization of the interface is by means of surface active particles [30]. The energy of detachment in such systems is so high that adsorption of the particles is effectively irreversible. For O/W emulsions, particle stabilization has been observed with starch granules that have been hydrophobized by means of the octenyl succinic anhydride (*OSA*) [31, 32]. For W/O emulsions, stabilization can be achieved by means of fat crystals [33, 34]. The latter can be observed during processing of high-fat W/O emulsions like margarine, for which it can be shown that is possible to stabilize emulsion products without emulsifiers.

11.2.2.4 Interaction between Continuous and Dispersed Phase: Hydration of Structurant in Organogel-Based Emulsions

One of the first aspects to consider when creating an emulsion is whether the water phase and lipid phase ingredients interact mutually. This is a serious consideration, because gelling of phases is usually based on limited solubility of the structuring agent in one of the phases. Fortunately, aqueous and lipid phases are so dissimilar that ingredients that are partly soluble in one phase (e.g., at high temperatures) are usually insoluble in the other phase. However, some water can dissolve in a relatively polar lipid phase like triglyceride oil, and structurants with a tendency to form hydrogen bonds may be sensitive to this water.

An example of such a system would be the plant sterol–oryzanol structuring system, mentioned above. If such emulsions are prepared with pure water as an aqueous phase, the water interferes with the hydrogen bond formation in the supramolecular aggregates. Under these conditions, the formation of sterol hydrates is energetically more favorable than the formation of the mixed sterol–oryzanol complexes. As a result the firmness of these emulsion gels drops dramatically. The structuring capability can be preserved by either reducing the water activity of the aqueous phase considerably (water activity $a_w < 0.9$) or by replacing the triglyceride oil with more apolar oils with a lower water solubility (dielectric constant $\varepsilon < 2.5$) [35]. Neither route is extremely attractive for broad application of this structuring technique in food products.

Another example of a structuring system that is severely compromised in an emulsion product is the application of monoglycerides in an aqueous phase. Monoglycerides preferably organize at the oil–water interface and dissolve primarily in the lipid phase of an emulsion. As a consequence, exposure of an aqueous monoglyceride gel to an oil phase leads to (partial) dissolution of the structurant in the oil. The firmness and emulsion stability of such emulsion gels is low, and therefore this approach not suitable for product structuring unless very high amounts of structurant are added to the emulsion.

11.3
Creating a Food Emulsion

11.3.1
Basic Principles

Emulsification in general combines the creation of fine droplets and their stabilization (Figure 11.1) [36].

Standard processing starts with a premix process. Coarse emulsion droplets in the size range above 10 μm are created by pre-mixing the hydrophilic and lipophilic phases. These are subsequently finely dispersed to droplets in the micrometer-range or even smaller by deforming and disrupting them at high specific energy using several types of emulsifying machines (Figure 11.2).

In each of these machines, energy is used to induce laminar and/or turbulent flow, which in turn is responsible for local shear and elongational stresses that are transferred from the continuous phase to the droplets. As a consequence, droplets deform and finally break-up once a critical deformation is exceeded for longer than a critical time [37, 38]. In addition, droplets of micrometer-size or bigger may also be formed directly by pressing the inner phase of the emulsion through micropores into the continuous phase [39].

Whether directly formed or disrupted, the small droplets have to be stabilized against coalescence by adsorbing emulsifier molecules (Figure 11.3). This adsorption takes some time [36, 40], especially when emulsifiers of high molecular weight and/or a complex molecular structure are applied, as is often the case in foods [41]. As previously described, proteins – typical food emulsifiers – denature, (re-)orientate, and crosslink at interfaces. Thus, they build a stable viscoelastic interfacial layer that is responsible for food emulsion stability [42]. Adsorption, orientation, and reorientation, as well as crosslinking may take up to several minutes (Figure 11.3).

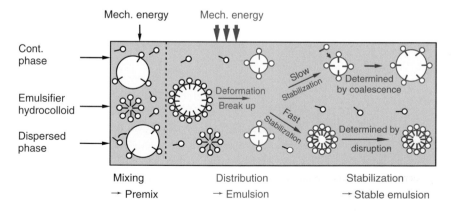

Figure 11.1 Creating an emulsion: droplets have to be disrupted and stabilized [2].

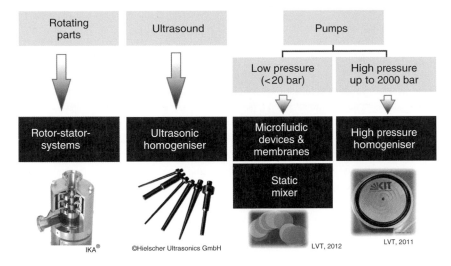

Figure 11.2 Typical emulsifying machines and energy source.

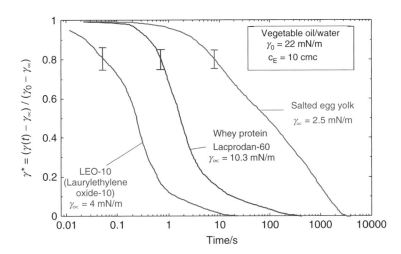

Figure 11.3 Kinetics of the adsorption, orientation, and arrangement of some emulsifiers at an oil–water interface: dimensionless interfacial tension versus time [2].

Even when stabilization is rather fast, coalescence of newly created droplets is found in technical emulsification processes [43]. Owing to coalescence, droplet diameters in emulsions are always bigger than predicted [40, 44] by disruption theory, as published by e.g. Walstra [38]. The coalescence rate may be decreased by decreasing the dispersed phase content to extremely small values (<1 vol.%, see [42]), by increasing droplet interaction forces, by solidifying droplets (at low vol.%) or at least interfacial layers [2], or by increasing the continuous phase viscosity [45]. The latter principles are applied in food emulsions.

11.3.2
Emulsification Processes for Gel-Like Food Emulsions

11.3.2.1 Emulsification Machines

Rotor–Stator Type Machines In the production of gel-like food emulsions, rotor–stator systems have been used for emulsification for decades and are still broadly applied today. Rotor–stator machines consist of a rotating and a fixed machine part. Rotor and stator geometry differ depending on supplier and application. The simplest rotor–stator machine is a vessel with a stirrer, which is used to produce emulsions batchwise or quasi-continuously. A stirred vessel has several advantages in emulsification: Different process steps (such as, for example, heating for pasteurization, cooling before packaging, or later adding of shear sensible ingredients such as herbs) can be combined with the emulsification step. It is relatively easy to produce small quantities of several recipes within one machine, enhancing process line flexibility [2]. Therefore, stirred vessels are widely used in the food industry, although there are some drawbacks. The power density is low and broadly distributed. Therefore, small mean droplet diameters (e.g., <1 μm) can rarely be produced. In addition, a long residence time and, thus, emulsification time of several minutes is required, often resulting in broad droplet size distributions or even the production of undesired by-products (especially at higher temperatures).

To solve these problems, the disruption zone has to be small and well defined, and the power density should be increased. This is realized in continuously working rotor–stator machines, such as colloid mills, or gear rim or toothed discs dispersing machines. In a colloid mill, the rotating part is a truncated cone, most often toothed. It rotates within a stator, also of toothed surface. In gear rim or toothed discs dispersing machines, rotor and stator consist of one or several discs having pins or teeth of different design (Figure 11.4).

Rotor–stator machines are relatively easy to handle and of low cost. They can be operated at throughputs from about $50\,l\,h^{-1}$ to several tons per hour. Droplet disruption is mainly due to turbulent flow within the dispersing zone between rotor and stator [40]. Typical rotor–stator emulsification machines also act as a pump due to the centrifugal forces created by the rotating part, with the throughput depending on the rotor tip speed. Thus, shear rate and volume throughput cannot be varied independently. This disadvantage is overcome by a rotor–rotor machine type, which is being developed for toothed discs dispersing machines. Here, not only the rotor itself but also the former stator rotates either in the same or opposite direction. Throughput and shear rate, that is, specific energy, can be adapted independently. Both can thus be significantly increased compared to a standard rotor–stator-system. However, machine investment is roughly doubled.

High-Pressure Homogenization High-pressure homogenizers have been traditionally used in the dairy industry. These machines are operated continuously at throughputs up to several thousand liters per hour, but are limited in emulsion

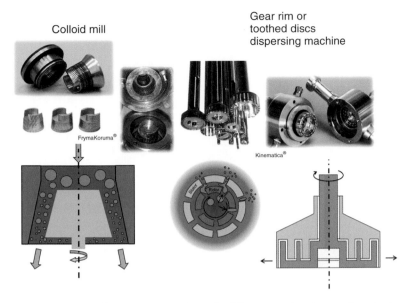

Figure 11.4 Typical rotor–stator machines for laboratory, pilot plant, and production-scale emulsification at increased viscosity.

viscosity. They consist essentially of a high-pressure pump and a homogenizing nozzle. The pump creates the pressure, which is then transferred within the nozzle into kinetic energy that is responsible for droplet disintegration. The design of the homogenizing nozzle influences the flow pattern of the emulsion in the nozzle and hence droplet disruption. Recent developments in high-pressure homogenizing have, thus, concentrated on nozzle design. Examples of new homogenizing nozzles are opposing jets (e.g., Microfluidizer®), jet dispersers, and simple orifice valves as well as other developments based on these. Droplet disruption in high-pressure homogenizers is predominantly due to inertial forces in turbulent flow, shear forces in laminar elongational flow, as well as cavitation. More information on high-pressure homogenization can be found in Schuchmann and Köhler [2] and Schuchmann *et al.* [46]. As the above-mentioned type of emulsification machines are difficult to apply for gel-like emulsions, this principle is not discussed here.

Cavity Transfer and Controlled Deformation Dynamic Mixers   Cavity transfer mixers (*CTMs*) and controlled deformation dynamic mixers (*CDDMs*) are classes of liquids mixing apparatuses characterized by closely-spaced surfaces that move relative to each other during operation, and by arrays of cavities machined into each of those surfaces, which, in conjunction with the relative positions of those arrays, define the flow pathways for the liquids being mixed. These pathways cause the liquids to pass between the surfaces during mixing, but while there is little variation in the cross-sectional area for flow as liquids pass between those surfaces in the case of CTMs there is a significant variation in that area in the case of CDDMs. Thus, while both CTMs and CDDMs affect distributive mixing via the rotational shear flow induced by the relative movement of those surfaces, CDDMs further affect dispersive mixing via the extensional shear flow induced by the change in cross-sectional area for flow between the surfaces.

While the earliest designs of CTMs appeared in the 1950s, it was not until circa 1980 that a commercially deployable design became available with the invention by Gale (1980) of the so-called "RAPRA CTM" for the compounding of rubber and plastics compositions. That invention was followed by the discovery by Akay *et al.* [47] of the CDDM technology for the manufacture of structured liquids. In addition, Bongers *et al.* [48] discovered that the CDDM technology provides a novel and energy efficient, compared with conventional rotor–stator systems and high-pressure homogenizers, means for the production of several classes of emulsions, including the classes of W/O compositions with slowly stabilizing emulsifiers and O/W compositions consisting of colloidal dispersions of narrow size distribution. Regarding the latter, the CDDM technology may provide an excellent opportunity for the industrial-scale manufacture of colloidal dispersions with/without active molecules via melt emulsification.

11.3.2.2 Emulsification in Theory: Dimensionless Numbers and Process Functions

For shear flow, droplet breakup can be described by the capillary number (Ca) – see Equation 11.2 (viscosity of continuous phase η_c, shear rate $\dot{\gamma}$, interfacial tension γ, and drop diameter x). The critical capillary number (Ca_{crit}) is the value that has to be reached for droplet breakup given that the critical deformation time depending on stress, viscosity, and capillary pressure is reached as well. Grace [49] demonstrated that the viscosity ratio λ (ratio of the viscosities of disperse η_d and continuous phase η_c – see Equation 11.3) is a crucial factor influencing the critical capillary number. Grace also stated for laminar shear flow of Newtonian fluids that the critical capillary number reaches a minimum for a viscosity ratio between 0.5 and 1 and no droplet breakup is possible above a viscosity ratio of 4, as shown in Figure 11.5 with $\alpha = 0$. The parameter α describes the flow type acting on the droplet surface. A value of 0 represents pure shear flow while a value of 1 represents pure elongational flow [50]:

$$Ca = \frac{\eta_c \dot{\gamma} x}{2\gamma} \tag{11.2}$$

$$\lambda = \eta_d / \eta_c \tag{11.3}$$

Bentley and Leal [50] investigated the impact of elongational flow at varied fractions of shear and elongational flow on the droplet breakup in Newtonian fluids. The capillary number has to be calculated in this case with a velocity gradient instead of the shear rate to account for both shear and elongational stress. With increasing elongational stress – that is, increasing values of α – an overall decrease in Ca_{crit} was observed. They reported that with increasing elongational stress the droplet breakup occurred at increasing viscosity ratios above 4. The data are presented in Figure 11.5 with $\alpha > 0$.

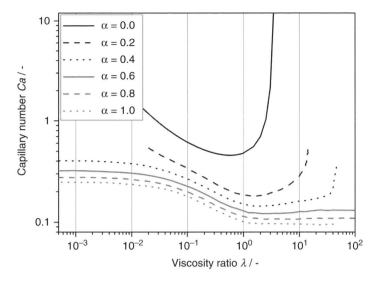

Figure 11.5 Values of critical capillary number Ca as a function of viscosity ratio λ for different rates of shear and elongational stress (α = 0 only shear, α = 1 only elongation). Adapted from References [49, 50].

With increasing ratio between actual Ca number and the critical one for break-up Ca/Ca_{crit}, droplets are deformed more intensely into long filaments that consequently break into finer droplets (filamental break-up) (Figure 11.6).

Armbruster [51] and later Jansen *et al.* [52] stated that in the case of a highly concentrated system the viscosity of the emulsion η_e has to be used instead of the viscosity of the continuous phase for calculation of λ:

$$\lambda = \eta_d/\eta_e \qquad (11.4)$$

Most gel-like emulsions show non-Newtonian, shear thinning flow behavior. Here, knowledge of the occurring shear rates in the process is essential for correct consideration of rheological properties. Several researchers investigated the effect of viscoelastic properties on droplet breakup, as reviewed amongst others by Guido [53]. He also reported that even though theoretical investigations showed

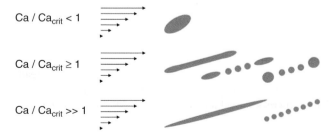

Figure 11.6 Droplet break-up depending on the Ca/Ca_{crit} ratio according to Reference [49].

comparably lower deformation of Newtonian droplets in viscoelastic matrices some experimental investigations showed the opposite trend.

In technical emulsification processes, coalescence competes with droplet disruption, being responsible for an increase in droplet sizes compared to values calculated using Ca_{crit}. Figure 11.7 gives an example using a high viscous food emulsion, which is processed in a colloid mill. Therefore, technical emulsification is often described using a simple process function [36, 40], such as:

$$\bar{x} \propto E_V^{-b} \eta_d^{0...0.75} \tag{11.5}$$

where $\bar{x}$ is the mean droplet diameter, E_V is the specific energy (energy input per volume), and η_d is the droplet phase viscosity. The value of b is determined by the mechanism of droplet disruption and the rate of coalescence found in the process [2]. For droplet disruption in pure laminar flow, b has a value of 1 while it is in the range of 0.25–0.4 for fully developed turbulence. When coalescence governs emulsion, b values may even be positive. This is referred to by practitioners as *"over-processing."*

11.3.3
High Internal Phase Emulsions (HIPEs)

High internal phase emulsions (*HIPEs*) have several applications in the food industry. Water dispersible HIPEs are desirable as they provide better microbiological stability due to their very low water activity. In the 1970s Princen [54] introduced an HIPE produced via centrifugal force, and has proven that after deformation the dispersed phase normally assumes a polyhedral shape with rounded corners and edges. Sonneville-Aubrun *et al.* [55] have shown HIPE O/W emulsions concentrated

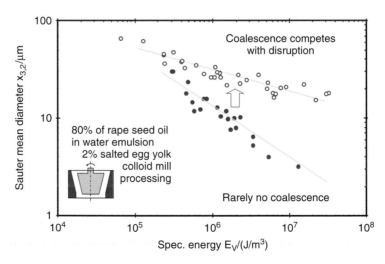

Figure 11.7 Technical emulsification as a result of droplet disruption and coalescence for the processing of an 80% oil-in-water emulsion in a colloid mill [40].

under ultrahigh centrifugal force. A rigid gel-like, elastic, and transparent structure was formed after centrifugation. Microscopic images showed the morphology to be as described by Princen [54]. The higher the dispersed phase fraction, the more pronounced is the polyhedral shape of the droplets.

HIPE carotenoid-loaded O/W emulsions containing different oil phase fractions stabilized by Tween® 20 and xanthan gum were studied by Ribeiro *et al.* [56]. Firstly, pre-emulsions were manufactured by a rotor–stator system, and then the droplet sizes were reduced below 1 µm using high-pressure homogenization. Final HIPE emulsions were produced using the method described by Sonneville-Aubrun *et al.* [55]. The addition of xanthan was very important to keep the same droplet size distribution profile after centrifugation. The study reported a dispersed phase fraction up to 94%; however, subsequent work carried out with these systems has shown that the oil droplets could be concentrated up to 99%. The highly concentrated O/W emulsions containing carotenoids present a viscoelastic flow behavior.

Another approach is shown by Romoscanu and Mezzenga [57], where a concentrated emulsion is dried and oil is turned into an elastic solid without modifying the oil itself. The process is an alternative to oil hydrogenation, avoiding the production of *trans* fatty acids. It is a non-energy input process where monodisperse self-assembled O/W emulsions are stabilized by a crosslinked protein monolayer adsorbed at the oil–water interface.

11.3.4
Production of Emulsion Gel Foods

11.3.4.1 Margarine

The continuous fabrication of margarine essentially consists of two unit operations. These are the initialization of crystallization in scraped surface heat exchangers, so-called A-units, and the continuation of crystallization under shear in pin-stirrers, so-called C-units. The processing set up is typically referred to as a *Votator unit*. Prior to detailing the processing steps it has to be mentioned that the crystallization of fats is characterized by monotropic polymorphism. This means that TAGs can assume essentially three different crystal structures. The monotropic nature of the polymorphism implies that only one of these crystal structures is stable. The least stable form is characterized by a hexagonal crystal structure and is usually called α. The more stable orthorhombic β' and triclinic β forms are characterized by a denser packing and their relative stability depends on the composition of the material to be crystallized.

Fat crystallization typically obeys Ostwald's rule stages [58]. This rule states in short that "it is easier to convert to an energetically similar state than to the energetically most favorable". This polymorphic behavior of fats dictates the choice of the manufacturing process for spreads. In the A-unit the coarse water and oil emulsion is rapidly cooled under vigorous agitation. At this stage α crystals are formed and the emulsion droplet size is reduced to between 3 and 7 µm. This droplet size is stabilized if sufficient solid material, in the form of α crystals, is present to cover the oil–water interface. In general it is assumed that the α state cannot be undercooled. The residence time in the A-unit is typically of the order of a few seconds and the desired processing temperatures can be reached by utilization of a series of A-units. Owing to the polymorphic nature of fats the α crystalline state is only transient and a polymorphic transition to the more stable β′ or, less likely in spread products, the β state proceeds. These polymorphic transitions, which are solution mediated and generate an increase in actual solids level, take place at composition and processing parameters over specific time scales. Usually, values between one and several minutes are observed. The residence times in C-units are of the same order of magnitude. While the droplet size is practically fixed in A-units, the C-unit is necessary to allow the polymorphic transformation to proceed while the emulsion is still under the mixing process. This is desirable because the crystallization and recrystallization of large amounts of solid material after packaging would result in uncontrolled and possibly brittle product structures. The undesired excessive change of structure after packaging is referred to as *post-hardening* and needs to be managed carefully during process design. Even though crystallization is reasonably far progressed during processing, the consistency of the product at packaging can be considered as liquid to allow filling but the crystal network quickly settles after packaging.

The combination of process and composition design aims to generate a large number of small crystals in order to make best use of the solid fat for structuring and emulsion stabilization purposes. The minimum levels necessary to fulfill these functions are as low as 1% to create a space-filling crystal network in a bulk lipid phase and 1.5% to stabilize a 20% W/O emulsion. When manufacturing fat continuous products with significantly reduced fat levels (<50%) it is likely that during manufacturing one finds initially a water continuous emulsion. Even though these products are processed in a similar fashion as the high fat products in an inversion unit, a high speed pin-stirrer is introduced to invert the emulsion after initial changes of the viscosity of the phases have taken place due to cooling.

11.3.4.2 Mayonnaise

Mayonnaise has been produced for decades via a colloid mill, where a coarse emulsion is pumped through a narrowing gap between a stator and a rotor, reaching high shear rates between 10^4 and 10^5 s^{-1}, and where the breakup of droplets takes place. Egg yolk (containing lecithin) is normally applied as the emulsifier, which reduces the interfacial tension at the oil–water interface. Commercially, droplet sizes lie between 1 and 5 μm. Further reduction in droplet size is possible by, for example, the careful application of high-pressure homogenization.

11.3.4.3 Cream Cheese

Cream cheese is manufactured at the factory scale by using high-pressure homogenization as the emulsification process. The equipment contains an orifice or a valve where the fluid if forced to pass through the narrow gap, causing high shear forces. For example, at 100 bar, droplet sizes below 1 μm can be reached.

Proteins are used as the emulsifier in the premix of milk and cream. When the premix is mainly stabilized by casein micelles, homogenization clustering may occur due to the low rate of adsorption at the oil–water interface, creating a cluster of fat droplets coated by micelles. This allows an increase in viscosity. This phenomenon happens only if large proteins (e.g., casein micelles) are used and if droplets collide in the low pressure zone beyond the homogenization valve.

11.3.4.4 Dairy Cream

A batch process is used at the factory scale to produce dairy creams. A coarse premix at approximately 60 °C is prepared first. The elevated temperature allows the fat to melt and prevents the growth of microorganisms.

The premix is subsequently pasteurized or sterilized in-line, depending on the pH of the premix. The final droplet size of approximately 1 μm is achieved through high pressure (100–200 bar) or, if clusters are formed, after a second homogenization step (10–20 bar).

Dairy cream emulsions at neutral pH are cooled in-line and filled in sterile packs. For acidified dairy creams, the above premix needs to be either chemically acidified (e.g., lactic and/or citric acid) or fermented (fermentation culture). Chemical acidification is a short-time process, where low temperatures can be applied, leading to finer protein networks. In the case of fermented dairy cream, firstly the premix is pasteurized and then a fermentation culture is inoculated. Thermophilic (or yogurt) cultures give a simpler and acidic profile, while mesophilic cultures provide a more complex and milder profile. Thermophilic cultures multiply optimally at about 40 °C and mesophilic cultures at approximately 25 °C.

Gums can be added in low-fat acidified creams, resulting in a product that is more stable against syneresis. Any type of dairy creams needs to be stored at chilled temperatures.

11.4
Applications of Gel-Like Type Emulsions

Emulsions show a few characteristics that determine their usefulness in food applications. First of all, the rheology of the emulsion is important. Rheological properties of the emulsion include the deformation behavior at small deformations, which can be expressed in terms of the elastic and loss moduli as measured in dynamic rheological assessment. More importantly, however, they also include the large deformation characteristics like yield or breaking stress and strain. These properties are more important for a food material than the small deformation rheology because a food product is designed to fail during use [59]. The mode of failure may depend on the application or even the stage of use: a spread should be spreadable when removed from the tub, it should melt away and release certain taste components in the mouth. Obviously, these properties are affected by the ingredients in the product, by the structure in which these ingredients have been assembled, and the stability of this assembly as a function of temperature or shear.

The use of different types of food applications depends somewhat on their microstructure. Although taste and texture are important for all products, it can be seen that fat-continuous emulsions are more commonly used if lubrication is important, whereas water-continuous products are more common in applications where an immediate taste impact is desired. The relation of the microstructure to the rate of taste release is obvious if one considers that the continuous phase of a water-continuous product immediately mixes with the saliva in the mouth during consumption, whereas the aqueous phase in a fat-continuous product first has to be released from the product before it can be detected by the taste buds in the mouth.

11.4.1
Water Continuous Food Products

Water-continuous emulsion gels rely far less often on the structuring of the continuous phase. Although hard cheeses like Gouda or Cheddar might be considered as counter-examples of such a system, this introduces a relatively deformable texture that is not always considered desirable. Quite often, therefore, the structure relies more on structuring by means of the dispersed phase. A typical example of structuring via the dispersed phase is mayonnaise, an emulsion containing typically 80% liquid vegetable triglyceride oil, for example, sunflower oil. The size of the droplets is usually between 1 and 5 µm. In mayonnaise, the contact between the droplets in the crowded high-fat emulsion prevents the emulsion from behaving like a viscous liquid ($\varphi > \varphi_{\mathrm{rcp}}$). The polydispersity of the droplets determines the amount of oil that can be emulsified: if a wide range of droplet sizes is present in the emulsion (as is usually the case for home-made mayonnaise), somewhat more oil can be dispersed when a narrow range of droplet sizes is produced. Again, the risk in such a high phase emulsion is coalescence. This can be brought about relatively easily by introducing solids in the oil phase. For example, the presence of large amounts of waxes in the oil phase may promote coalescence.

It is therefore important to control the emulsification process. This is relatively difficult, as anyone who has failed to make mayonnaise at home is likely to agree. On an industrial scale, however, the process can be controlled adequately by using specialized equipment. In such a process, homogenization occurs usually in two steps. In the first step, oil is dispersed coarsely to 20–100 µm droplets in a slow mixing process in a stirred tank. In the second step, this pre-emulsion is dispersed more finely in the vigorous mixing process, for example, in a colloid mill, resulting in droplet sizes of approximately 5 µm.

The droplet size of the mayonnaise controls the texture of a product, because the resistance of the close-packed emulsion droplets against deformation is determined by the Laplace pressure of the droplets. This effect becomes less in lower-fat salad dressings (Figure 11.8), because the packing of the droplets becomes less important for the firmness of the product. However, in such systems it is possible to maintain most of the characteristics of the mayonnaise texture by replacing fat with swollen starch particles. The starch fills the volume that was previously occupied by the

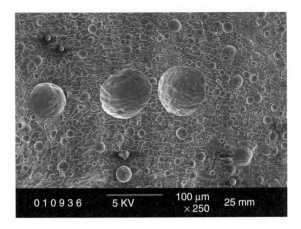

Figure 11.8 SEM image of a salad dressing.

fat droplets, retaining essentially a locally close-packed droplet structure in the remainder of the structure.

Low-fat water-continuous emulsions may actually benefit somewhat from the presence of destabilizing components like crystalline fat in the oil phase, because the amount of dispersed phase in itself is too low to affect the rheology of the continuous phase considerably. Here, aggregation without coalescence may help to inflate the actual amount of dispersed phase to a higher apparent amount of dispersed phase because the droplet aggregates of the dispersed phase include a certain amount of continuous phase. In extreme cases, this may even lead to a sample-spanning droplet network. An example is the whey-protein stabilized O/W model emulsion gels studied by Kiokias *et al.* [4], Kiokias and Bot [60, 61], which represent a model system for certain fresh cheese-type products. Their firmness is partly based on partial coalescence, and the phenomenon can be aggravated by temperature cycling. This behavior is of interest because destabilization occurs at relatively high protein content, under conditions at which neutral emulsions based on native whey protein are usually very stable (in the absence of any low-molecular weight emulsifiers). The destabilization is caused by denaturation of the whey protein in the emulsion, but the reduced functionality of the denatured whey protein is hardly apparent in the non-cycled emulsion. This is explained by the competition between W/O tending fat crystals and O/W tending proteins, which seem to dominate unless they are highly aggregated. The sensitivity to destabilization increases with reduced average distance between the droplets [61].

Dairy creams are whitish fine O/W emulsions with fat droplet sizes of approximately 1 μm. Neutral creams are sold as a viscous liquid, acidified creams as a spoonable or spreadable soft solid. As well as the difference in pH, the difference in dry matter (mainly fat, protein, lactose) in the product also matters: firmer products tend to have higher amounts of dry matter, especially fat.

In these products three aspects are of key importance: (i) emulsion droplet stability, (ii) droplet aggregation/(partial) coalescence, and (iii) protein aggregation.

11.4.1.1 Emulsion Droplet Stability

The emulsion droplets are primarily stabilized by dairy protein (casein micelles, whey protein). However, depending on the formulation, low molecular weight emulsifiers may play a role, too. These may be of dairy origin, like the milk phospholipids, or added separately. Bulky emulsifiers, especially like the casein micelles, tend to stabilize the emulsion very well during prolonged storage. They are not very efficient in quickly stabilizing newly formed interfaces during creation of the emulsion; however, this role is fulfilled much more efficiently by low molecular weight emulsifiers, like phospholipids and monoglycerides. Unfortunately, these are not very good at stabilizing the emulsion during prolonged storage or under the influence of shear.

11.4.1.2 Droplet Aggregation

Destabilization of emulsions under shear is exploited in the case of whipping cream. The presence of a substantial amount of low molecular weight emulsifiers (in the case of non-dairy whipping cream unsaturated monoglycerides) [62] at the droplet interface allows droplet aggregation and partial coalescence to occur during whipping. The resulting fat droplet aggregates assemble at the air–water interface, and can stabilize air bubbles very well.

Under acid conditions, protein stabilized emulsions aggregate as well, but usually this does not lead to (partial) coalescence of the droplets. It is likely that the occurrence of droplet aggregation is an important factor in the formation of texture in acidified creams – otherwise one would be pressed to explain why the composition of the fat phase in the emulsions tends to have so much influence on the firmness of the emulsion.

11.4.1.3 Protein Aggregation

Protein aggregation may occur as a result of heat treatment (affecting mainly the whey proteins) and as a result of acidification (affecting the casein micelles and the denatured whey protein). In products with high amounts of protein, like fresh cheese, the effect on texture is very important. In acidified creams, the effect of protein concentration is generally not the leading contribution to firmness, except if one considers soft fat blends. The protein network is very important in controlling the exudation of water from the product (i.e., syneresis); however, in products with low dry matter (usually low-fat products), gums are often added. These introduce the thermodynamic incompatibility of proteins and gums. Phase separation occurs and the emulsion phase is concentrated in separate domains.

The stability of (neutral) low-fat cooking cream is mainly determined by the stability against creaming. It has been observed that creaming is more rapid at ambient storage temperature, but overall the stability of cooking creams is seldom an issue.

The lack of stability of (neutral) whipping cream is a serious and recurring issue, because the system is by nature a product that is easily destabilized: if the product is too stable it does not whip (which results in consumer complaints), if it is not stable enough it destabilizes in the pack (thick in pack). Often it is difficult to point out the exact cause of the problems, but several rules of thumb exist to rectify the problem. The instability of whipping creams is sometimes used to generate additional consistency by a process known as *tempering*. The product is subjected to a specific temperature cycle in the warehouse. Because the heat transfer in the pallet is not particular efficient, a typical tempering cycle can take up to a week.

In some cases, the dairy fat in creams is replaced by vegetable fat. This allows for an additional degree of freedom in formulation by changing the triglyceride composition of the blend. If consistency is the most important aspect of the product, higher SAFA blends are often used. If nutritional considerations dominate the formulation choices, lower SAFA blends are chosen. The stability of high-SAFA acidified creams is mainly determined by their stability against post-firming (i.e., in the packaging). Often, blends rich in palm (kernel) oil or coconut fat are preferred due to their high stability against oxidation. However, such blends may destabilize the emulsion due to re-crystallization, especially under the influence of temperature fluctuations during storage or consumer use. As a consequence, partial coalescence of the emulsion droplets may occur, resulting in very firm products. The microscopic changes associated with partial coalescence also induce damage to the protein network and the resulting coarse pores in the protein network may result in extensive syneresis in the product.

The taste stability of low-SAFA acidified creams is usually determined by the rate of oxidation of the fat. Notably, light oxidation plays a role due to the presence of riboflavin in the dairy ingredient. This is especially important in high *PUFA* (polyunsaturated fatty acid)-oil based products, and light-tight packs are desirable in such cases.

11.4.2
Fat Continuous FoodProducts

The behavior of high-fat spreads like margarine are mostly determined by the properties of their continuous phase, and the composition of the dispersed phase only starts to play a role when the fat crystals in the continuous phase have melted away at mouth temperature. The product properties are determined mainly by the melting behavior of the fat crystals, which depends on the TAG composition of the lipid phase and to a lesser extent on the manufacturing process. The desired TAG composition can be manufactured through application of oil modification techniques such as hydrogenation, fractionation, and chemical or enzymatic interesterification.

Low-fat W/O spreads are much more complex in this aspect. Though still fat-continuous, the amount of crystalline fat in such emulsions is much lower, such that one cannot (and would not like to) ignore the contribution of the dispersed phase to the overall rheology. These emulsions face two risks. First, the low amount of structured continuous phase may render them too soft, and in such cases one might want to gel the dispersed phase, too – for example, by means of adding gelatin, starch, or a combination of thickener compounds. Second, there is a risk that the emulsion droplets coalesce because of the high amount of dispersed phase and their shear proximity. The addition of a gelling agent to the dispersed phase is an appropriate mitigation.

The lipid phase of margarine and low-fat spreads is typically composed of a mixture of TAGs (or triglycerides) referred to as *fats* and *oils*. Oils do not show any crystallized TAGs at ambient temperature, whereas fats do. Fats and oils are usually classified in terms of the source from which they are retrieved (e.g., soy bean oil, sunflower oil, butter fat), but the fundamental differences between lipids can be best explained in terms of the constituent TAGs. For example, the amount of crystalline fat ("solid fat content" or *SFC*) is related to the level of SAFAs in the specific TAG mixture. The temperature-dependent SFC is often referred to as the *N-line* of the fat. The TAG composition of the lipid phase can be modified through simple blending of fats and/or the use of oil modification techniques that include hydrogenation, dry or wet fractionation, and chemical or enzymatic catalyzed interesterification. Fat crystallization is special in two aspects: (i) polymorphism of the TAG crystals (i.e., the formation of α, β', and β crystal structures) and (ii) the presence of a wide range of chemically different TAGs in any fat, reflecting the complex fatty acid composition of the raw materials.

The consistency of a lipid phase is based on the structuring effect of fat crystals in the lipid phase, which in turn depends on the amount of crystals, their size, morphology (shape), and polymorph, and on the strength of the interactions between the crystals. The strength of the interaction between the single crystals is of key importance for the strength of the network and its plasticity or brittleness. For a plastic structure, crystals should predominantly be linked through secondary bonds based on van der Waals forces. When crystals are more or less sintered together through solid bridges, also referred to as *primary bonds*, the resulting network is usually harder and brittle, and thus less plastic.

In fat continuous food products, emulsifiers are used as processing aids and to confer properties to the final product. For W/O emulsions, typically, a low molecular weight emulsifier of either a mono- or diglyceride of a fatty acid is used, or a lecithin (phospholipid) is employed. These hydrophobic fat-soluble emulsifiers quickly adsorb at the interface and thus lower the interfacial tension. Monoglycerides also help to stabilize the newly created water droplets by forming a shell around them, which also facilitates contact with the stabilizing TAG network.

The composition of the water phase depends on the type of product. In high-fat spreads the water phase is often empty, but in low-fat spreads structurants are added. Historically, proteins were used, but nowadays starches are the ingredient of choice. Proteins are also added to promote destabilization of the emulsion in the mouth.

Figure 11.9 shows a cryo-SEM (scanning electron microscope) image of a low-fat spread from which the oil and water have been removed and the fat crystals remain as shells. The water is normally found within the crystalline shells and the oil is the continuous phase between the stabilized water droplets.

Margarine leaving the Votator process is still considerably supercooled, a significant portion of the fat crystallizing quiescently (in the tub, i.e., under "unworked"

Figure 11.9 Crystalline fat at the water–oil interface in a fat low fat spread. Picture scale = 5 μm width.

conditions). This may lead to the development of brittleness. Other defects may develop as a result of recrystallization processes, which change the fat crystal network owing to polymorphic transitions and/or de-mixing of the mixed crystals. These processes typically take place slowly and can be stimulated by exposure to shear or fluctuating temperatures. They result typically in reduced product hardness, coarser crystals, or the development of very distinct crystal agglomerates of sizes sometimes as large as millimeters.

During storage, transport, and use of spreads, several undesirable textural changes may occur. These changes are partly due to the fact that spreads are not in equilibrium when leaving the production line and partly due to temperature fluctuations during storage and transport. These changes may result in product defects, caused by modifications in the fat crystal network, and emulsion instability.

Coarsening of the crystals in the fat crystal network may occur spontaneously or because of temperature cycling, for example, during transport or during use, as a result of the lower surface free energy of bigger crystals. The phenomenon is known as *Ostwald ripening*. The process is greatly accelerated by temperature cycling, because the redissolution of crystallized fat in oil at higher temperatures followed by subsequent recrystallization at lower temperatures helps the transport of fat molecules from the smaller to the bigger crystals.

The product stability is largely governed by the properties of the continuous phase. A W/O emulsion will not suffer from re-coalescence unless the margarine is severely abused during storage, for example, by extreme temperature cycling. The emulsion can undergo some initial destabilization by Ostwald ripening since water is slightly soluble in oil and the difference in Laplace pressure between large and small droplets is enough to drive the process. However, such changes are minimal if the process control (droplet size distribution) is narrow and a short tempering step is included in the storage period.

11.4.3
Chemical Properties

One of the critical aspects of food emulsions is that the large oil–water interfacial area promotes oxidation of monounsaturated fatty acids (*MUFAs*) and particularly polyunsaturated fatty acids (PUFAs) in the triglyceride oil phase, because oxidation progresses particularly effectively near an interface as it involves both polar and apolar compounds. In addition, the ions in the aqueous phase act as a catalyst for lipid oxidation [63].

The complex chemistry of lipid oxidation involves initiation, propagation, and termination reactions. Initiation refers to a reaction step that transforms peroxides that were generated under the influence of light, heat, or metal-ion catalyzed reactions into fatty acid radicals. During propagation the fatty acid radicals transform into lipohydroperoxides. Finally, termination requires reactions between two radicals, which generate compounds like dimers, polymers, alcohols, and ketones. The typical off-flavor of oxidized edible oil is not caused by the initial reaction products but by volatiles such as alkanes, aldehydes, and ketones that form later in this cascade of reactions [64, 65]. The typical oxidation off-taste is usually described as rancid or cardboard-like, although a wide range of flavors can develop depending on storage conditions and the type of oil.

The most obvious factor determining the sensitivity of the oil against oxidation is the degree of unsaturation: the sensitivity increases from TAGs containing SAFAs, to MUFAs, to PUFAs. For example, the autoxidation rates of stearic acid (C18:0) and linolenic acid (C18:3) have been reported to differ by a factor of 10^5–10^6.

For a given type of oil composition, there is a limited number of means that can be used to limit oxidation. As with all chemical reactions, reducing the temperature of the emulsions will reduce the oxidation rate. More specifically, the initiation reaction can be slowed down by removing access of light to the emulsion and/or by removing ingredients that contribute metal ions to the formulation (e.g., proteins) and/or by adding sequestrants (like EDTA, ethylenediametetraacetic acid) to render metal ions inactive. The propagation reaction can be slowed down by removing oxygen as much as possible from the emulsion. However, notably, in many cases the amount of oxygen dissolved in the emulsion is greater than the amount present in the headspace. Finally, the termination reaction can be slowed by means of addition of "chain-breaking" antioxidants in the aqueous phase such as butylated hydroxyanisole (BHA) and butylhydroquinone (BHQ), and/or fat soluble antioxidants (e.g., tocopherol, carotenoids). Obviously, choosing high-quality ingredients is the most effective measure to prevent off-taste formation, but anything that can be done to reduce the initiation reaction rate tends to be rather effective, too.

There are some indications that the microstructure of the emulsion can be used to confine oxidation, for example, by removing oxidation promoting ingredients from the interface of the emulsion droplets [66].

Many of these routes are indeed applied for food emulsions, but not all at the same time. Experience has taught the product developer which aspect of the oxidation reaction is most critical for each application. In some mayonnaises nitrogen blanketing of the oil is applied. In some low-fat spreads EDTA or citric acid are added as sequestrant. Glass bottles will filter out the most energetic part of the light spectrum. Some spreads with sensitive oils are always distributed chilled. On the other hand, some aspects are deemed less critical in application. Some margarines are sold in rather permeable plastic tubs (when compared to, for example, glass). Mayonnaise is distributed at ambient temperature. Chain-breaking antioxidants are usually not added to the formulation of food emulsions.

11.4.4
Microbiological Properties

The major risk for a food product is contamination by microorganisms, especially by pathogenic microorganisms. There are two ways to counteract this threat – either through a chemical or through a physical route.

The chemical route involves the formulation. Many organic acids are detrimental for the survival of microorganisms. The best known example is the use of acetic acid in mayonnaise, or the use of sorbic acid to prevent mold growth. In fact, many organic acids have some activity in this respect. Reduction of the water activity by adding salt is also a much-used route. The alternative application of specific proteins, like nisin, is a less-common route.

The non-chemical routes usually involve a suitable heating step during processing of the product. Typical examples are pasteurization or sterilization [e.g., through ultra-high temperature processing (*UHT*) or autoclaving], although all kinds of intermediate processes exist too [e.g., extended shelf life (*ESL*), which is a heat treatment that is milder than UHT but more intense than pasteurization]. An intriguing and less-trodden path is compartmentalization of the aqueous phase, as happens in some margarines and low-fat spreads. If the aqueous phase is dispersed in small enough droplets in a continuous oil phase, any potentially dangerous microorganisms cannot reproduce sufficiently to become a danger to the consumer. To achieve this, the water droplets need to be smaller than approximately 7 µm [67, 68], and the droplets should not coalesce extensively during storage or temperature cycling.

11.5
Final Considerations

Over recent decades, there has been increasing research into understanding the microstructure of food, which is basically driven by the physical chemistry of food-grade materials. There are still potentially challenging areas of research – such as alternative structuring ingredients, the mechanism of network formation, emulsion stability, both physical and chemical, smarter ways to cover surfaces, and others compared to current practice – that have not been fully explored yet and deserve further investigation. Novel emulsification processes have been patented since the 1990s for the purpose of improving functionality of ingredients. They also allow the production of more homogeneous systems and higher monodispersity of emulsions' dispersed phase than conventional emulsification processes. Material science and process engineering complement each other and should work together more closely to explore the innovative potential for new food structuring systems.

References

1. Dickinson, E. (1992) *An Introduction to Food Hydrocolloids*, Oxford University Press, Oxford.
2. Schuchmann, H.P. and Köhler, K. (2012) *Emulgiertechnik*, 3 Aufl edn, Behrs, Hamburg.
3. Krieger, I.M. and Dougherty, T.J. (1959) A mechanism of non-Newtonian flow in suspensions of rigid spheres. *Trans. Soc. Rheol.*, **3**, 137–152.
4. Kiokias, S., Reiffers-Magnani, C.K., and Bot, A. (2004) Stability of whey protein stabilized oil in water emulsions during chilled storage and temperature cycling. *J. Agric. Food Chem.*, **52**, 3823–3830.
5. Clark, A.H. and Ross-Murphy, S.B. (1987) Structural and mechanical properties of biopolymer gels. *Adv. Polym. Sci.*, **83**, 57–192.
6. Djabourov, M., Bonnet, N., Kaplan, H., Favard, N., Favard, P., Lechaire, J.P., and Maillard, M. (1993) 3D analysis of gelatin gel networks from transmission electron microscopy imaging. *J. Phys. II*, **3**, 611–624.
7. Bot, A., van Amerongen, I.A., Groot, R.D., Hoekstra, L.L., and Agterof, W.G.M. (1996) Large deformation rheology of gelatin gels. *Polym. Gels Networks*, **4**, 189–227.
8. Bot, A., van Amerongen, I.A., Groot, R.D., Hoekstra, L.L., and Agterof, W.G.M. (1996) Effect of deformation rate on the stress-strain curves of gelatin gels. *J. Chim. Phys.*, **93**, 837–849.
9. Phillips, G.O. and Williams, P.A. (2000) *Handbook of Hydrocolloids*, Woodhead Publishing Ltd., Abington, Cambridge.
10. Vasbinder, A.J., Rollema, H.S., Bot, A., and De Kruif, C.G. (2003) Gelation mechanism of milk as influenced by temperature and pH; studied by the use of transglutaminase cross-linked casein micelles. *J. Dairy Sci.*, **86**, 1556–1563.
11. Roefs, S.P.F.M. and de Kruif, C.G. (1994) A model for the denaturation and aggregation of beta-lactoglobulin. *Eur. J. Biochem.*, **226**, 883–889.
12. Vasbinder, A.J. and De Kruif, C.G. (2003) *Int. Dairy J.*, **13**, 669–677.
13. Heertje, I., Roijers, E.C., and Hendrickx, H.A.C.M. (1998) Liquid crystalline phases in the structuring of food products. *Lebensm. Wiss. Technol.*, **31**, 387–396.
14. Sein, A., Verheij, J.A., and Agterof, W.G.M. (2002) Rheological characterization, crystallization and gelation behavior of monoglyceride gels. *J. Colloid Interface Sci.*, **249**, 412–422.
15. van Duynhoven, J.P.M., Broekmann, I., Sein, A., van Kempen, G.M.P., Goudappel, G.J.W., and Veeman, W.S. (2005) Microstructural investigation of monoglyceride–water coagel systems by NMR and Cryo-SEM. *J. Colloid Interface Sci.*, **285**, 703–710.
16. Mensink, R.P., Zock, P.L., Kester, A.D.M., and Katan, M.B. (2003) Effects of dietary fatty acids and carbohydrates on the ratio of serum total to HDL cholesterol and on serum lipids and apoproteins: a meta analysis of 60 controlled trials. *Am. J. Clin. Nutr.*, **77**, 1146–1155.
17. Pernetti, M., van Malssen, K.F., Flöter, E., and Bot, A. (2007) Structuring of edible oils by alternatives to crystalline fat. *Curr. Opin. Colloid Interface Sci.*, **12**, 221–231.
18. Marangoni, A.G. and Garti, N. (eds) (2011) *Edible Oleogels: Structure and Health Implications*, AOCS Press, Urbana, IL.
19. Daniel, J. and Rajasekharan, R. (2003) Organogelation of plant oils and hydrocarbons by long-chain saturated FA,

fatty alcohols, wax esters, and dicarboxylic acids. *J. Am. Oil Chem. Soc.*, **80**, 417–421.

20. Dassanayake, L.S.K., Kodali, D.R., Ueno, S., and Sato, K. (2009) Physical properties of rice bran wax in bulk and organogels. *J. Am. Oil Chem. Soc.*, **86**, 1163–1173.

21. Rogers, M.A., Wright, A.J., and Marangoni, A.G. (2009) Oil organogels: the fat of the future? *Soft Matter*, **5**, 1594–1596.

22. Toro-Vazquez, J.F., Morales-Rueda, J.A., Dibildox-Alvarado, E., Charó-Alonso, M., González-Chávez, M., and Alonzo-Macias, M.M. (2007) Thermal and textural properties of organogels developed by candelilla wax in safflower oil. *J. Am. Oil Chem. Soc.*, **84**, 989–1000.

23. Laredo, T., Barbut, S., and Marangoni, A.G. (2011) Molecular interactions of polymer oleogelation. *Soft Matter*, **7**, 2734–2743.

24. Bot, A. and Flöter, E. (2011) in *Edible oleogels: Structure and Health Implications* (eds A.G. Marangoni and N. Garti), AOCS Press, Urbana, IL, pp. 49–79.

25. Bot, A., den Adel, R., and Roijers, E.C. (2008) Fibrils of γ-oryzanol + β-sitosterol in edible oil organogels. *J. Am. Oil Chem. Soc.*, **85**, 1127–1134.

26. Bot, A. and Agterof, W.G.M. (2006) Structuring of edible oils by mixtures of γ-oryzanol with β-sitosterol or related phytosterols. *J. Am. Oil Chem. Soc.*, **83**, 513–521.

27. Bot, A., den Adel, R., Roijers, E.C., and Regkos, C. (2009) Effect of sterol type on structure of tubules in sterol + γ-oryzanol-based organogels. *Food Biophys.*, **4**, 266–272.

28. Bot, A., Gilbert, E.P., Bouwman, W.G., Sawalha, H., den Adel, R., Garamus, V.M., Venema, P., van der Linden, E., and Flöter, E. (2012) Elucidation of density profile of self-assembled sitosterol + oryzanol tubules with small-angle neutron scattering. *Faraday Discuss.*, **158**, 223–238.

29. Bancroft, W.D. (1913) Theory of emulsification. *J. Phys. Chem.*, **17**, 501–519.

30. Pickering, S.U. (1907) Emulsions. *J. Chem. Soc. Trans.*, **91**, 2001–2021.

31. Garrec, D.A., Frasch-Melnik, S., Henry, J.V.L., Spyropoulos, F., and Norton, I.T. (2012) Designing colloid structures for micro and macro nutrient content and release in foods. *Faraday Discuss.*, **158**, 37–49.

32. Rayner, M., Sjöö, M., Timgren, A., and Dejmek, P. (2012) Quinoa starch granules as stabilizing particles for production of Pickering emulsions. *Faraday Discuss.*, **158**, 139–155.

33. De Bruijne, D.W. and Bot, A. (1999) in *Food Texture: Measurement and Perception* (ed. A.J. Rosenthal), Aspen, Gaithersburg, MD, pp. 185–227.

34. Ghosh, S. and Rousseau, D. (2011) Fat crystals and water-in-oil emulsion stability. *Curr. Opin. Colloid Interface Sci.*, **16**, 421–431.

35. Sawalha, H., den Adel, R., Venema, P., Bot, A., Flöter, E., and van der Linden, E. (2012) Organogel-emulsions with mixtures of β-sitosterol and γ-oryzanol: influence of water activity and type of oil phase on gelling capability. *J. Agric. Food Chem.*, **60**, 3462–3470.

36. Schuchmann, H.P. (2007) in *Product Design and Engineering – Best Practises, Basics and Technologies*, Vol. 2 (eds G. Wagner, W. Meier, and U. Bröckel), Wiley-VCH Verlag GmbH, Weinheim, pp. 63–93.

37. Walstra, P. (1983) in *Encyclopedia of Emulsion Technology*, Vol. 1 (ed. P. Becher), Marcel Dekker, New York, pp. 57–128.

38. Walstra, P. (1993) Principles of emulsion formation. *Chem. Eng. Sci.*, **48** (2), 333–349.

39. Vladisavljevic, G.T., Lambrich, U., Nakajima, M., and Schubert, H. (2004) Production of O/W emulsions using SPG membranes, ceramic α-aluminium oxide membranes, microfluidizer and a silicon microchannel plate – a comparative study. *Colloids Surf., A*, **232** (2–3), 199–207.

40. Karbstein, H. (1994) Untersuchungen zum herstellen und stabilisieren von öl-in-wasser-emulsionen. Dissertation. Universität Karlsruhe (TH).

41. McClements, D.J. (1999) *Food Emulsions*, CRC Press, Boca Raton, FL.

42. Dickinson, E. (1999) Adsorbed protein layers at fluid interfaces: interactions, structure and surface rheology. *Colloids Surf., B*, **15** (2), 161–176.

43. Danner, T. (2001) Tropfenkoaleszenz in Emulsionen. Dissertation. Universität Karlsruhe (TH), GCA Verlag.

44. Danov, K.D., Kralchevsky, P.A., and Ivanov, I.B. (2001) *Encyclopedic Handbook of Emulsion Technology*, Marcel Dekker, New York, pp. 621–659.

45. Tesch, S. and Schubert, H. (2002) Influence of increasing viscosity of the aqueous phase on the short-term stability of protein stabilized emulsions. *J. Food Eng.*, **52** (3), 305–312.

46. Schuchmann, H.P., Hecht, L.L., Gedrat-Winkelmann, M., and Köhler, K. (2012) High pressure homogenization for the production of emulsions, in *Industrial High-Pressure Applications: Processes, Equipment and Safety* (ed. M. Eggers), Wiley-VCH Verlag GmbH, Weinheim, pp. 97–122.

47. Akay, G., Irving, N.G., Kowalski, A.J., and Machin, D. (1997) Process for the production of liquid compositions. WO 96/20270, filed Dec. 27, 1995 and issued Oct. 8, 1997.

48. Bongers, P.M.M., Ribeiro, H.S., Irving, N.G., and Egan, M.J. (2012) Method for production of an emulsion. WO 2012/089474, filed Dec. 28, 2010 and issued July 5, 2012.

49. Grace, H.P. (1982) Dispersion phenomena in high-viscosity immiscible fluid systems and application of static mixers as dispersion devices in such systems. *Chem. Eng. Commun.*, **14** (3–6), 225–277.

50. Bentley, B.J. and Leal, L.G. (1986) An experimental investigation of drop deformation and breakup in steady two-dimensional linear flows. *J. Fluid Mech.*, **167**, 241–283.

51. Armbruster, H. (1990) Untersuchungen zum kontinuierlichen emulgierprozeß in kolloidmühlen unter berücksichtigung spezifischer emulgatoreigenschaften und der strömungsverhältnisse im Dispergierspalt. Dissertation. Universität Karlsruhe (TH), Karlsruhe.

52. Jansen, K.M.B., Agterof, W.G.M., and Mellema, J. (2001) Droplet breakup in concentrated emulsions. *J. Rheol.*, **45** (1), 227–236.

53. Guido, S. (2011) Shear-induced droplet deformation: effects of confined geometry and viscoelasticity. *Curr. Opin. Colloid Interface Sci.*, **16** (1), 61–70.

54. Princen, H.M. (1979) Highly concentrated emulsions. *J. Colloid Interface Sci.*, **71**, 55–66.

55. Sonneville-Aubrun, O., Bergeron, V., Gulik-Krzywicki, T., Jönsson, B., Wennerström, H., Lindner, P., and Cabane, B. (2000) Surfactant films in biliquid foams. *Langmuir*, **16**, 1566–1579.

56. Ribeiro, H.S., Cruz, R.C.D., and Schubert, H. (2005) Biliquid foams containing carotenoids. *Eng. Life Sci.*, **51** (1), 84–88.

57. Romoscanu, A.I. and Mezzenga, R. (2006) Emulsion-templated fully reversible protein-in-oil gels. *Langmuir*, **22**, 7812–7818.

58. Ostwald, W. (1907) Zur Systematikl der kolloide. *Kolloid Z.*, **1**, 291–341.

59. Lillford, P.J. (2000) The material science of eating and food breakdown. *Mater. Res. Soc. Bull.*, **25** (12), 38–43.

60. Kiokias, S. and Bot, A. (2005) Effect of denaturation on temperature cycling stability of heated acidified protein-stabilised o/w emulsion gels. *Food Hydrocolloids*, **19**, 493–501.

61. Kiokias, S. and Bot, A. (2006) Temperature cycling stability of pre-heated acidified whey protein-stabilised o/w emulsion gels in relation to the internal surface area of the emulsion. *Food Hydrocolloids*, **20**, 245–252.

62. Pelan, B.M.C., Watts, K.M., Campbell, I.J., and Lips, A. (1997) The stability of aerated milk protein emulsions in the presence of small molecule surfactants. *J. Dairy Sci.*, **80**, 2631–2638.

63. Ribeiro, H.S., Ax, K., and Schubert, H. (2003) Stability of lycopene emulsions in food systems. *J. Food Sci.*, **68** (9), 2730–2734.

64. Allen, J.C. and Hamilton, R.J. (eds) (1994) *Rancidity in Foods*, 3rd edn, Blackie Academic & Professional, Glasgow.

65. Chan, H.W.-S. (1987) *Autoxidation of Unsaturated Lipids*, Academic Press, London, pp. 1–16.

66. McClements, D.J. and Decker, E.A. (2000) Lipid oxidation in oil-in-water emulsions: impact of molecular environment on chemical reactions in heterogeneous food systems. *J. Food Sci.*, **65**, 1270–1282.

67. Verrips, C.T. and Zaalberg, J. (1980) The intrinsic stability of water-in-oil emulsions. 1. Theory. *Eur. J. Appl. Microbiol. Biotechnol.*, **10**, 187–196.

68. Verrips, C.T., Smid, D., and Kerkhof, A. (1980) The intrinsic stability of water-in-oil emulsions. 2. Experimental. *Eur. J. Appl. Microbiol. Biotechnol.*, **10**, 73–85.

Index

Product Design and Engineering: Formulation of Gels and Pastes, First Edition.
Edited by Ulrich Bröckel, Willi Meier, and Gerhard Wagner.
© 2013 Wiley-VCH Verlag GmbH & Co. KGaA. Published 2013 by Wiley-VCH Verlag GmbH & Co. KGaA.